Embedded Systems Security

For Hannah and Aaron

Always Reach

—DK

For My Grandchildren

—MK

Embedded Systems Security

Practical Methods for Safe and Secure Software and Systems Development

David Kleidermacher

Mike Kleidermacher

ELSEVIER

AMSTERDAM • BOSTON • HEIDELBERG • LONDON • NEW YORK • OXFORD
PARIS • SAN DIEGO • SAN FRANCISCO • SINGAPORE • SYDNEY • TOKYO
Newnes is an imprint of Elsevier

Newnes

Newnes is an imprint of Elsevier
The Boulevard, Langford Lane, Kidlington, Oxford, OX5 1GB
225 Wyman Street, Waltham, MA 02451, USA

First published 2012

British Library Cataloguing in Publication Data
A catalogue record for this book is available from the British Library

Library of Congress Number: 2012931463

ISBN: 978-0-12-386886-2

For information on all Newnes publications visit our website at www.elsevierdirect.com

Printed and bound in the United States

12 13 14 15   10 9 8 7 6 5 4 3 2 1

Contents

Foreword

At last! Finally, a book about building secure embedded systems.

Read the newspaper and you get an almost daily litany of stories about leaks and exploits. Credit card numbers seem to leap out of databases. Personal information spews from corporate servers.

But the elephant in the room, the real threat vector, has been largely ignored. For every PC in the world, there are hundreds of embedded systems, interconnected via an ever-growing array of communications channels. WiFi, Bluetooth, Ethernet, RFID, FireWire—the list is endless. A smartphone has at least four antennas and reflashable memory, and, if compromised, could be an electronic Typhoid Mary.

Anything with a connection can be a threat vector. A USB toothbrush (yes, at least one is on the market) might carry both biological and computational infectious elements. As can a smart mousepad (which uses the connection to display custom pictures). That WiFi electronic picture frame showing cute photos of your loved ones may have a wicked heart as it poisons the neighbor's network.

Consider that some automobiles currently have brake-by-wire. Tractors have had steer-by-wire for some time; the steering wheel merely spins an encoder, and that technology will certainly make it into cars. But consumers want Internet connections at work, at play, and on the road. How hard will it be for a bad guy halfway around the world to bring the I-5 to a disastrous crunch by issuing TCP/IP packets?

Not hard at all, apparently. Researchers have already taken control of cars using a variety of vectors.

Engineers at the DHS destroyed a large diesel generator by infiltrating it remotely. Stuxnet showed that a boring old SCADA control system could be a state-sponsored weapon. SCADA systems control power plants, factories, and, well, just about any industrial process. A sewage plant in Queensland was successfully attacked, causing it to leak noxious goop over a wide area. The Feds admit that foreign states have infiltrated the USA's aging power grid, perhaps leaving behind rogue code.

What happens when the grid becomes smart?

At least one company sells bearings that use a magnetic field to suspend a shaft; a DSP performs some 15,000 calculations per second to keep things running smoothly. And the bearing controllers have Ethernet connections! A coordinated attack on bearings—bearings!—could cripple manufacturing.

Internet-connected toasters are on the market, as are networked coffee makers. When 100 million bleary-eyed, and possibly hung over, Americans wake one Monday morning and find their caffeine fix unavailable, business will come to a standstill. But the truth is they'll still be blissfully sleeping since a teenager in Nigeria reset all the alarm clocks.

There's another security issue: non-malware bugs. As embedded systems become increasingly part of the fabric of civilization, any bug that compromises the reliability of a system can become a mission-critical security threat. When the automated jail control doors fail to close, the prison becomes a hotel. And not Hotel California, because the guests are sure to check out. A task that errantly consumes too many resources (like memory) or CPU cycles can prevent other activities from running. The traffic light fails to turn red, the railroad signal remains open, or the ATM's bill counter fails to stop spewing money (one can hope).

Most embedded systems developers have little training in security and are largely unaware of both the threats and the techniques and technologies needed to make their creations robust. For the first time, there's a book on the subject. Dave and Mike Kleidermacher clearly describe the issues and offer a five-point "Principles of High Assurance Software Engineering" guide. Need a clear but complete description of encryption techniques? Or details about protecting data? There's a lavishly illustrated chapter on each.

Many developers are resigned to building insecure systems, figuring the enemy will always be one-step ahead. And embedded systems aren't upgraded every 15 minutes, like PCs are, so improved defenses always lag new attacks. But Dave and Mike show that it is indeed possible to build embedded software that is inherently secure.

The authors don't claim that making secure embedded systems is easy. In some cases even the tools themselves must be evaluated and qualified. But secure firmware will be both a national priority and a competitive advantage. This book offers answers. I hope it serves as a wake-up call as well.

Jack Ganssle

Preface

About this Book

The goal of this book is to help embedded systems developers improve the security and reliability of their products. While several books target embedded systems security, the subject matter of these works has been rather narrow, focusing almost exclusively on hardware-related issues and/or network security protocols and their underlying cryptography. In contrast, *Embedded Systems Security* aims for a comprehensive, *systems* view of security: hardware, platform software (such as operating systems and hypervisors), software development process, data protection protocols (both networking and storage), and cryptography. While no title can realistically cover every embedded systems security topic, this book does attempt to address the major security-relevant building blocks of contemporary embedded systems.

The reader will gain a solid understanding of the critical systems software and hardware issues that must be considered when designing secure embedded systems. Most embedded systems developers do not write their own operating systems and network protocols, nor do they design their own microprocessors. Therefore, a proficient knowledge in the security-relevant aspects of these platform components is critical to making good embedded design choices, particularly with respect to meeting security objectives given a particular operating environment.

The reader will learn an efficient methodology for developing secure embedded software. In addition to properly utilizing platform components, embedded developers must design their own software and integrate the overall system in such a manner as to ensure the highest possible levels of security. Thus, an important goal of this book is to discuss proven and practical techniques for developing secure software. We also seek to debunk the myth that it is not possible to dramatically increase software assurance without a commensurate growth in development time and cost. The methodology presented is derived from a combination of industry standards and practical experience. It is our belief that embedded systems developers are sorely lacking in high-quality literary guidance in this area. We aim to fill that gap.

Audience

Embedded Systems Security is written primarily for engineering professionals involved in the development of embedded systems. Hardware, software, and systems engineers and architects all have a hand in embedded systems security.

One of the most important tenets of computer security is that it is difficult, unwise, and often financially and/or technically infeasible to retrofit security capability to a system that was not originally designed for it. Therefore, the only hope for improving security across the world of embedded systems is to educate the developers, who must learn to think about security issues as much as they already think about functionality, memory footprint, and debugging.

This book also provides an important reference for professionals involved in the testing and quality assurance of embedded systems. These engineers must learn to test for security strength, a challenge that is inherently more difficult than testing against a functionality specification because the scope of potential security threats is often difficult, if not impossible, to fully enumerate. Security testing requires tremendous creativity and determination. Yet there are many tools that the quality control engineer can bring to bear on the problem, and a key focus of this book is to provide practical instruction regarding this toolbox.

Concern about building secure embedded systems must permeate the organization. Developer training must include books like this, attendance at technical conferences with security instruction, training from key suppliers of hardware and software to a project, and regular exposure to pertinent current events. Thus, this book is intended for the management teams responsible for making sure that developers design for security. Management must understand embedded security issues and be sold into the need for this training. In the example of the automobile, this book is relevant to managers responsible for individual components, such as an infotainment system and integration of those components (e.g., major tier 1 suppliers and the automobile manufacturers themselves), as well as the overall success of the product. Indeed, the VPs, GMs, and officers of businesses building cars, aircraft, trains, industrial control systems, or any other kind of electronic product with significant embedded systems content will benefit from reading this book and keeping it on the shelf for reference.

In addition to security-concerned professionals, this book provides guidance to developers of reliability-critical embedded systems, including life-critical medical devices, avionics and other transportation systems, telecommunications systems, and even high-volume, sophisticated consumer devices such as smartphones. There is tremendous commonality across requirements for building secure, safe, and reliable embedded systems.

Finally, many of the security concepts in this book, especially software development practices, are relevant to the engineering professional beyond the embedded systems field. For example, this book teaches the web application developer to be concerned not only with scripting and database vulnerabilities, but also with a comprehensive understanding of how the web server could impact security of the operating system, other applications, and the underlying computer hardware (and vice versa).

The relative seniority of engineers and architects reading this book is irrelevant. This book is equally useful for the entry-level programmer as well as the most seasoned developer who may have relatively light experience with security issues. Even engineers with significant security background and knowledge are likely to find the guidance in this book a useful addition to their arsenal.

While not written in textbook style, *Embedded Systems Security* will benefit academic instructors and students who are teaching or learning about embedded systems within a computer science or engineering regimen. While embedded systems development is a sorely lacking discipline in the world's technical colleges, embedded security instruction is practically non-existent.

Organization

We recommend all readers process all chapters sequentially from beginning to end. For those pressed for time (who isn't?), we provide the following guidance to help prioritize material relative to job function.

Chapter 1, "Introduction to Embedded Systems Security," discusses the trends driving the need for improved security in embedded systems and then provides foundational definitions for threats and the corresponding protective policies to counter them. In essence, we take the distinct notions of security and embedded systems and summarize their intersection within the modern embedded system. Chapter 1 also provides sample applications of the security concepts and guidance in some of the most exciting, growing, and important new embedded systems technology areas, including smartphones and the smart grid, as a foretelling of how the rest of the book may apply to current and future projects. This chapter is strongly recommended for all audiences.

Chapter 2, "Systems Software Considerations," provides a comprehensive discussion of security concerns and recommended best practices related to platform software, including operating systems, hypervisors, and the Multiple Independent Levels of Security (MILS) architecture. This chapter also discusses key security concerns and recommended best practices related to overall system security architectures and impacts of available hardware capabilities, such as MMUs, IOMMUs, and virtualization acceleration. This chapter is most important for engineers, architects, and technical managers.

Chapter 3, "Secure Embedded Software Development," is dedicated to maximizing assurance of embedded software. Key principles and guidance are provided throughout and hence the chapter is most critical for software developers and their technical management.

Chapter 4, "Embedded Cryptography," provides an overview of the most important cryptographic algorithms, key management, and related U.S. government guidance facing embedded systems. In addition to this general coverage, we discuss cryptographic concepts in the context of embedded systems and their unique constraints and requirements. Any reader lacking a firm grounding in cryptography should read this chapter. Even if current projects do not make use of cryptographic functionality, there is a strong probability that future projects will. Encryption and authentication form the backbone of all data confidentiality and network access protections.

Chapter 5, "Data Protection Protocols for Embedded Systems," covers the most important, widespread network security protocols, such as IPsec and Transport Layer Security (TLS), as well as storage encryption approaches, and emphasizes implementation concerns for resource-constrained embedded systems. This chapter is suggested for professionals looking for an introduction to or a refresher on the latest security protocol standards and revisions. One of the core goals of this chapter is to help developers make intelligent decisions regarding the systems layer (network and storage) at which to apply security. This chapter also includes advanced topics, such as network timing channels, that will be of interest to readers looking to protect the confidentiality of high-value resources.

Chapter 6, "Emerging Applications," provides extended case studies that build upon and apply the concepts of the preceding chapters. The discussion covers the security concerns and sample solution architectures for a handful of emerging applications and environments. Chapter 6 is recommended for all audiences.

Approach

The authors use concrete metrics and historical events to introduce the reader to the growing need for improved security and the challenges likely to be faced during next generation embedded systems projects. Throughout the course of discussion, we present critical issues followed by practical and simple advice on how to approach them. Nuggets of wisdom (***Key Points***) are set apart from the rest of the prose with special typeface and summarized at the ends of chapters. The chapter summaries act as a quick reference guide that will give readers an effective way to use the book throughout their careers. The book is not intended as a one-time-read.

We make generous use of real case studies, both from the authors' personal experience as well as from the authors' extensive knowledge of the embedded security world.

Since so many systems security concerns are architectural in nature, the book contains numerous technical diagrams to help explain concepts. For example, a case study of a radio design includes a set of diagrams to explain the cryptography and key management approaches used to achieve the desired security result.

Source code examples are provided to demonstrate the secure embedded software development techniques covered in the book (especially Chapter 3).

While the book covers much ground in the area of embedded systems security, it does this by focusing on the issues most likely to affect the modern embedded systems designer and avoiding too broad a discussion of introductory topics. For example, the network security discussion in Chapter 5 give focuses more on the most recent versions of protocols and algorithms and less on the basics of network security. Nor does Chapter 5 attempt to provide a comprehensive list of protocols and algorithms. Rather, the reader is referred to sources for more basic overview material. In this sense, the book should generally be thought of as an intermediate-level approach, not overly introductory or exceedingly advanced. We feel that this approach is best suited for the general embedded systems readership.

Acknowledgements

The authors would like to thank Guy Broadfoot for his contribution of the Model Driven Development section in Chapter 3.

The authors would like to thank Jack Ganssle for writing the book's Foreword and for his review and feedback.

The authors would like to thank Michael Barr for his review, feedback, and contribution of the Netrino *Embedded C Coding Standard* discussed as a case study in Chapter 3.

The authors would like to thank Thomas Cantrell, Jack Greenbaum, Dan Hettena, and Philippa Hopcroft for their thoughtful review and feedback.

The authors would like to thank Elsevier editors Tim Pitts and Charlotte Kent for their support throughout the project.

The authors would like to thank Tamara Kleidermacher, the book's illustrator, who painstakingly created the diagrams throughout this work. The ability to add visualization to aid a technical discussion is critical, and Tamara's flair for visual style and consistency has improved this book immeasurably.

Mike would like to thank his son Dave for the invitation to co-author this work.

Mike would like to thank Ellwood (Chip) McGrogan who "taught me all I know about cryptography."

David would like to thank his family—Tamara, Hannah, and Aaron—for their constant support and extraordinary patience during the writing of this book over a great many nights and weekends.

David would like to thank his brother Paul for being a lifelong role model and for his steadfast encouragement, support, and honest counsel.

David would like to thank Dan O'Dowd and Mike Kleidermacher for generously sharing and imparting their knowledge and passion about security over decades.

David would like to thank the employees, both past and present, of Green Hills Software, for their support over the years. It has been a privilege and honor to work with such a talented group to make the world a better, safer, more secure place.

CHAPTER 1

Introduction to Embedded Systems Security

Chapter Outline

1.1 What is Security?

Any book about security must start with some definition for it. If ten security professionals are asked to define the term, ten different results will be forthcoming. To attain validity for the innumerable variety of embedded systems and their functions, our brush uses a broad stroke:

> **Key Point**
>
> Security is the ability of an entity to protect resources for which it bears protection responsibility.

Embedded Systems Security. DOI: 10.1016/B978-0-12-386886-2.00001-1

In an embedded system, this protection responsibility may apply to resources within or resources of the overall system to which the embedded system is connected or in which it is subsumed. As we discuss later in this chapter, the protective properties of a component or system are embodied in its *security policy.*

1.2 What is an Embedded System?

Attempts to define "embedded system" are also often fraught with controversy. For the purposes of this book, we define embedded system as follows:

> **Key Point**
>
> An embedded system is an electronic product that contains a microprocessor (one or more) and software to perform some constituent function within a larger entity.

Any definition of embedded system must be augmented with examples. We do not claim an aircraft is an embedded system, but its flight control system; traffic collision avoidance system (TCAS); communication, navigation, and surveillance system (CNS); electronic flight bag system (EFB); and even in-flight entertainment system are all examples of embedded systems within the aircraft (see Figure 1.1).

We do not claim the automobile is an embedded system. But its infotainment "head-unit," anti-lock breaking system, powertrain engine control unit, digital instrument cluster, and a plethora of other electronic subsystems—dozens in the typical modern car—are all examples of embedded systems (see Figure 1.2).

Embedded systems are often characterized by what they are not: the antithesis of the embedded system is the desktop personal computer whose main Intel Architecture (IA)-based microprocessor powers the human interface and application environment that serves as the entity's sole purpose. Similarly, a rack-mounted server's main microprocessor performs a dedicated service, such as hosting a website.

A gray area causes the aforementioned controversy. Some argue whether a smartphone is an embedded system or just a miniature desktop computer. Nevertheless, there is little debate that individual components within the phone, such as the radio with its own baseband microprocessor and software, are embedded systems. Similarly, some servers contain auxiliary daughter cards that perform health monitoring and remote management to improve overall availability. Each card contains a microprocessor and software and hence meets our definition of embedded system.

The scope of this book liberally includes smartphones whose overall security is highly dependent upon embedded hardware and software.

Figure 1.1:
Embedded systems within modern commercial aircraft.

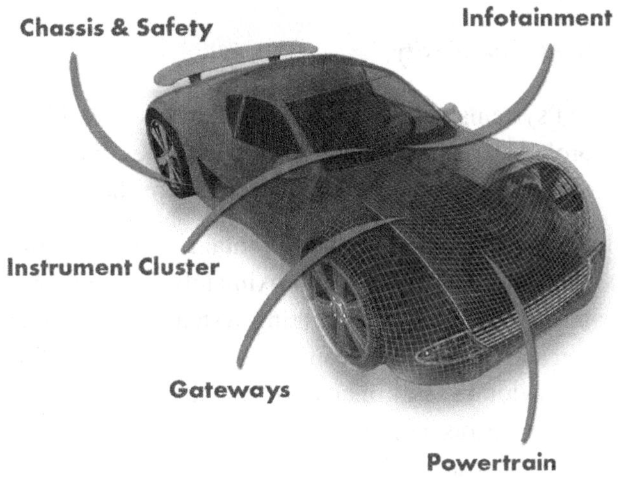

Figure 1.2:
Some embedded systems within a typical automobile.

Of course, this book is concerned about embedded systems that are involved in some security-critical function, and some embedded systems lack security requirements altogether. This book generally does not concern itself with a standalone, battery-powered thermostat run by an 8-bit microcontroller and a few kilobytes of software programmed in assembly code. The largest security challenge in embedded systems lies in network-connected, sophisticated electronic products that are managed by an embedded operating system running significant software applications written in high-level programming languages such as C, C++, Ada, and Java.

1.3 Embedded Security Trends

The MP944, what many consider to be the world's first microprocessor, ran the flight control system aboard the U.S. Navy's F-14 Tomcat fighter jet and began what has been more than 40 years of advancement in embedded systems technology. Depending on the particular analyst asked, embedded computers account for 94% to 98% of the world's computers. Practically every major multinational corporation—firms such as Lockheed Martin, Exxon, General Motors, Hewlett Packard, and Johnson & Johnson—builds and depends on embedded systems within its most important products. And, of course, the average consumer depends on the embedded applications within aircraft, automobiles, games, medical equipment, and so on, constantly.

At the same time, software and hardware complexity, network connectivity, and malicious attack threat continue to grow in embedded systems, which are increasingly relied upon for consumer safety and security. The smart grid—with its smart appliances and sensors, smart meters, and network gateways (all embedded systems)—is a good example, but only one of many. The complex set of embedded systems and networks in a smart grid is shown in Figure 1.3.

1.3.1 Embedded Systems Complexity

One of the first embedded systems within an automobile was the 1978 Cadillac Seville's trip computer, run by a Motorola 6802 microprocessor with 128 bytes of RAM and two kilobytes of ROM. The printed source code could not have occupied more than a handful of pages.

In contrast, even the lowest-end automobile today contains at least a dozen microprocessors; the highest-end cars are estimated to contain approximately 100 microprocessors. With infotainment systems running sophisticated operating systems such as Microsoft Windows and Linux, the total embedded software content can easily exceed 100 million lines of code. The F-35 Joint Strike Fighter's avionics is estimated to host approximately 6 million lines of code, driven by fly-by-wire controls, complex situational-awareness capabilities, sensor processing, and high-resolution graphical displays for the pilot. Enterprise network switches and routers routinely contain millions of lines of code for network protocol processing, management and configuration, anti-virus rate limiting, and access controls.

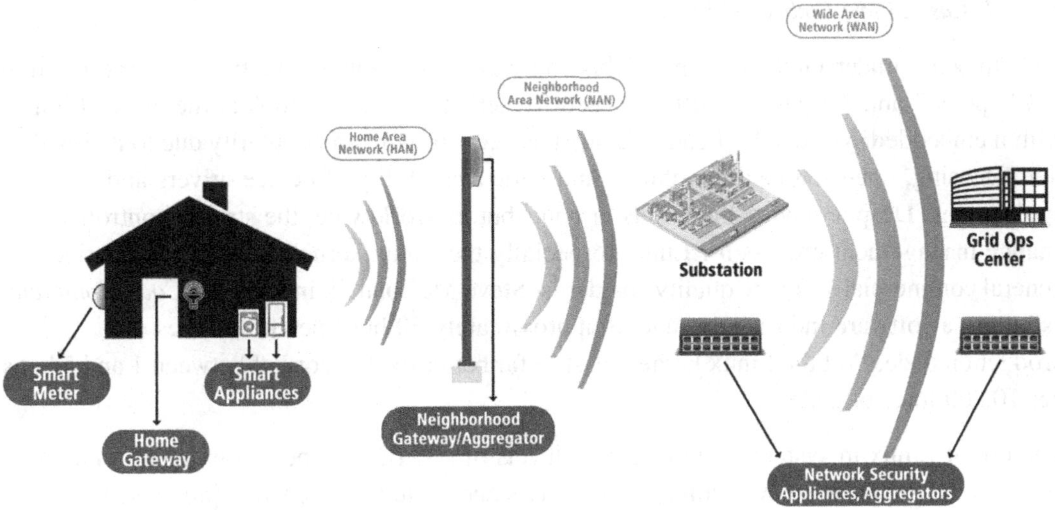

Figure 1.3:
Smart grid, embedded systems content, and sample network topology.

In short, complexity is driven by the inexorable demand for better capabilities, the digitization of manual and mechanical functions, and the interconnection of our world. While this growth in electronic content has been beneficial to society, that growth is also a key source of our security woes.

> **Key Point**
>
> Many of the problems relating to loss in quality, safety, and/or security in electronic products can be attributed to the growth of complexity that cannot be effectively managed.

It is well known that operational flaws, such as a buffer overflows (when software fails to validate the length of an input, permitting the input to overwrite beyond the end of an allocated memory area that is used to hold the input), are often the means by which attackers are able to circumvent system security policies. Complexity, of course, cannot be measured only by code size or transistor count.

> **Key Point**
>
> Linear growth in hardware/software content creates far more than linear growth in overall complexity due to an exponential increase in interactions between functions and components.

Complexity breeds flaws, and flaws can be exploited to breach system security. Controlling complexity from a security perspective is one of the foremost concerns of this book.

1.3.1.1 Case Study: Embedded Linux

To help better understand the scope of this complexity problem and motivate the information in Chapters 2 and 3 regarding software security, let's take a closer look at the use of Linux within embedded systems. Embedded Linux has been growing in popularity due to its royalty-free licensing, open source accessibility, and wide availability of device drivers and applications. Despite having thousands of contributors worldwide, the strictly controlled change management process for Linux (especially the Linux kernel) is excellent relative to general commercial software quality standards. Steve McConnell, in his book *Code Complete,* estimates a software industry average of approximately 30 bugs per 1,000 lines of production code.[1] Yet the Linux kernel boasts a far better track record of between 1 and 5 bugs per 10,000 lines of code.

The use of Linux in systems requiring high levels of security has been a frequent topic of controversy. Supporters have claimed that Linux's open source approach improves security due to exposure to a worldwide community of developers and users (sometimes called the "many eyes" theory). Detractors have maintained that the complexity and architecture of Linux make it unsuitable for high criticality applications.

Two recent events are shining a bright light on this debate. In August 2009, the Linux Foundation published a paper, *Linux Kernel Development*, detailing the massive, rapidly growing development and deployment of Linux in everything from mobile phones to television sets and video cameras.[2] About a year later, researchers published details of a severe kernel vulnerability, which had existed in Linux for the preceding eight years.

With a Linux lifetime of over 15 years, there are now plenty of public statistics with which to analyze the operating system's robustness. This case study looks at the aforementioned recent events as well as other public sources of information to conclude on the current state and outlook for Linux in high-security systems.

1.3.1.1.1 Linux in Government Systems

Some powerful organizations have been supporters of Linux in security-critical government computer systems. Linux is the trusted operating system in HP's NetTop, the Raytheon/TCS Trusted Thin Client, and the General Dynamics Trusted Virtual Environment (TVE)—a product of NSA's High-Assurance Platform (HAP) program. All these products are designed to consolidate computers used by government personnel to access classified and unclassified networks. The specialized computer provides multiple "virtual" desktops and is trusted to protect sensitive information. To prepare it better for the task of becoming the "touching point" between physically distinct networks, Linux was enhanced by the NSA's National Information Assurance Research Laboratory with additional security controls, known as Security-Enhanced Linux (SELinux). The SELinux extensions have been adopted by the enterprise Linux community and employed within the aforementioned computer systems.

Along with their investment in Linux, these military suppliers have made bold claims about the trustworthiness of these products. According to General Dynamics, the TVE provides "high robustness" and a "quantum leap in the way military and government security levels are accessed."[3]

It is interesting to note, however, that the NSA's developers were careful not to claim suitability for high criticality systems, stating that SELinux is "very unlikely by itself to meet any interesting definition of secure system." Furthermore, the SELinux effort has included "no work focused upon increasing the assurance of Linux itself."[4]

While many discussions about the security of Linux have been clouded by hyperbole and commercial agendas, a number of independent resources, many published by the Linux community, are painting a more complete, unbiased picture about Linux security.

Linux development follows general commercial practices, not compliant with any stringent safety or security standard. While Linux's open source exposure has enabled it to achieve a low defect rate relative to most commercial software, the size of the kernel assures a large and continuous dose of flaws. In 2004, an automated static analysis tool discovered almost 1,000 bugs in the Linux kernel.

The U.S. National Institute of Standards and Technology and the U.S. Department of Homeland Security's National Security Cyber Division publish a catalog, the National Vulnerability Database (NVD), of security defects in commercial software products.[5] As of August 16, 2009, a search on Linux yielded 1,288 entries, 457 of which are considered "High Severity." One hundred thirty-four high severity vulnerabilities are associated with the Linux kernel. The NVD reports 91, 77, 87, 111, and 115 Linux kernel vulnerabilities for each of the years 2006, 2007, 2008, 2009, and 2010, respectively. It is statistically assured that a similar number will be found in future years, implying that numerous vulnerabilities exist in today's shipping version. These numbers, of course, do not account for unreported defects.

On August 10, 2009, a memory leak in the SELinux security extensions was published in the NVD. A few days later, five more vulnerabilities were published. One of these, CVE-2009-2692, reports a severe kernel defect that can be trivially exploited by a user to take complete control of the system. This vulnerability was latent in the Linux kernel for eight years!

1.3.1.1.2 Linux Rate of Change

Some critical software components gain assurance over time. This occurs when the software is relatively simple, changes very little (except perhaps for bug fixes), and is deployed for a long period of time in a variety of environments. The Linux kernel, however, undergoes continuous modification, including in the field (e.g., over-the-air patching). The latest major version of Linux, 2.6, has changed more rapidly than previous versions and regularly undergoes major modifications to the "stable branch" of the kernel. As an example, Linux developer Greg Kroah-Hartman reported that the 2.6.24 kernel saw approximately 5,000 lines of code added per day during a three-month period, prompting his lament, "It's fricken scarily amazing that things are still working at all...."[6]

The rate of change has been accelerating. Kroah-Hartman reported that over 12,000 lines of code were added per day on average during the 2.6.30 development cycle. Since 2005, the Linux kernel has been modified by more than 5,000 different people at a rate that now exceeds six changes per hour. In the first release of Linux 2.6 (2.6.0), the Linux kernel consisted of more than 5 million lines of code. In 2.6.30, that number had grown to more than 11 million lines.

Another good example was provided by Jim Ready, founder of embedded Linux vendor MontaVista, who discussed NVD defect CVE-2006-1528, which was patched in Linux version 2.6.13. To get the bug fix in a supported release, a user running 2.6.10 would have been forced to take in 846,233 new lines of code, representing the changes between versions 2.6.10 and 2.6.13.[7]

1.3.1.1.3 CVE-2009-2692—An Illustration of a Total Loss of Security

The following is a detailed description of the aforementioned severe kernel vulnerability. The subtlety of the flaw, coupled with the contrastingly simple form of programs used to exploit it, provides readers with a flavor for the difficulty facing Linux security. Readers who want to skip the details should proceed to the next section.

Security researchers Tavis Ormandy and Julien Tinnes discovered a set of NULL function pointer dereferences in the Linux kernel. The function pointers come from Linux networking code. Linux uses a structure, called *proto_ops*, to hold a set of prototypical operations associated with network sockets. The implementations of each operation vary across socket families (e.g., AppleTalk, Bluetooth, Infrared, Stream Control Transmission Protocol for IPv6). The function pointers associated with these operations are not consistently initialized or validated at their kernel call sites. When a socket service call is executed by an application, the Linux kernel determines the proper *proto_ops* structure to use based on the socket type. The kernel executes the function pointer for the requested operation within this structure.

The following Linux kernel example shows a *proto_ops* pointer (the *splice_read* operation) that is NULL checked at the call site:

```
static ssize_t sock_splice_read(struct file *file, loff_t
   *ppos, struct pipe_inode_info *pipe, size_t len, unsigned int flags)
{
  struct socket *sock = file->private_data;
  if (unlikely(!sock->ops->splice_read))
```

```
    return -EINVAL;
  return sock->ops->splice_read(sock, ppos, pipe, len, flags);
}
```

And here is an example where the same check is missing:

```
static ssize_t sock_sendpage(struct file *file, struct page *page, int offset, size_t
    size, loff_t *ppos, int more)
{
    struct socket *sock;
    int flags;
    sock = file->private_data;
    flags = !(file->f_flags & O_NONBLOCK) ? 0 :
      MSG_DONTWAIT;
    if (more)
      flags | = MSG_MORE;
    return sock->ops->sendpage(sock, page, offset, size, flags);
}
```

Another method used in Linux to avoid *proto-ops* NULL dereferences is to pre-initialize the pointer to a stub function. If the *sendpage* pointer in the preceding example is initialized to the Linux stub *sock_no_sendpage*, then the unprotected call in the preceding code will execute the stub function and avoid the NULL deference. Many socket family *proto-ops* are initialized in this manner, for example, the one used for Infrared datagrams:

```
static const struct proto_ops
      SOCKOPS_WRAPPED(irda_dgram_ops) = {
    .family =         PF_IRDA,
    .owner =          THIS_MODULE,
    .release =        irda_release,
    .bind =           irda_bind,
    .connect =        irda_connect,
    .socketpair =     sock_no_socketpair,
    .accept =         irda_accept,
    .getname =        irda_getname,
    .poll =           datagram_poll,
    .ioctl =          irda_ioctl,
#ifdef CONFIG_COMPAT
    .compat_ioctl = irda_compat_ioctl,
#endif
    .listen =         irda_listen,
    .shutdown =       irda_shutdown,
    .setsockopt =     irda_setsockopt,
```

```
    .getsockopt =   irda_getsockopt,
    .sendmsg =      irda_sendmsg_dgram,
    .recvmsg =      irda_recvmsg_dgram,
    .mmap =         sock_no_mmap,
    .sendpage =     sock_no_sendpage,
};
```

However, other instances are not properly initialized; for example, the following Bluetooth Network Encapsulation Protocol (BNEP) *proto-ops*:

```
static const struct proto_ops bnep_sock_ops = {
    .family         = PF_BLUETOOTH,
    .owner          = THIS_MODULE,
    .release        = bnep_sock_release,
    .ioctl          = bnep_sock_ioctl,
#ifdef CONFIG_COMPAT
    .compat_ioctl   = bnep_sock_compat_ioctl,
#endif
    .bind           = sock_no_bind,
    .getname        = sock_no_getname,
    .sendmsg        = sock_no_sendmsg,
    .recvmsg        = sock_no_recvmsg,
    .poll           = sock_no_poll,
    .listen         = sock_no_listen,
    .shutdown       = sock_no_shutdown,
    .setsockopt     = sock_no_setsockopt,
    .getsockopt     = sock_no_getsockopt,
    .connect        = sock_no_connect,
    .socketpair     = sock_no_socketpair,
    .accept         = sock_no_accept,
    .mmap           = sock_no_mmap
→MISSING SENDPAGE STUB INITIALIZATION
};
```

Thus, an attacker simply needs to invoke a *sendpage* operation on an open BNEP socket to force a NULL dereference.

Kernel NULL execution can have a variety of effects. For example, execution of the 0 page could cause the system to completely crash (a simple but total denial-of-service attack). However, several unrelated flaws in the Linux kernel allow an attacker to inject arbitrary executable code into the kernel at address 0. In addition, as reported in NVD candidate CVE-2009-2695 and detailed by Red Hat, the SELinux extensions exacerbate the address 0 execution vulnerability.

The combination of NULL pointer and kernel code injection vulnerabilities enable an attacker to trivially take complete control of the system by forcing the kernel to execute the attacker's code.

Several exploits of the *proto-ops* flaw were immediately published. These exploits are simple programs that a Linux user could run to demonstrate total security failure. The general flow of the exploit is as follows:

```
/* Code the attacker wants to execute at kernel */
/* Privilege to do absolutely anything!         */
Attacker_code(void)
{
   kprint("*** You've Been Hacked!!\n");
   ...
}
...
/* Service call that causes address 0 to be mapped */
unsigned char *Kernel0 = mmap(0, 0x1000, flags, 0, 0);
/* Install instructions at 0 to run Attacker's code */
Kernel0[0] = '\xff'; /* JMP DWORD –Jump indirect */
Kernel0[1] = '\x25';
*(unsigned long *)(Kernel0+2) = 6;/* Jump target is @6 */
*(unsigned long *)(Kernel0+6) = (unsigned long)&Attacker_code;
/* Create socket of type containing flawed proto_ops */
fd = socket(PF_BLUETOOTH, SOCK_DGRAM, 0);
/* Use socket operation to cause NULL dereference */
/* This triggers the attacker's code to run       */
sendfile(fd, ...);
```

Linux has a "feature" that makes this exploit a bit easier: for improved system call performance, Linux keeps the user application's memory mappings enabled during system call servicing. This precludes the need to perform manual address translation on service call parameters located in application memory. Thus, the application's memory page at address 0 becomes the kernel's page at address 0, and *Attacker_code* is automatically mapped and executable by the kernel.

1.3.1.1.4 Case Study Wrap-Up

While Linux fulfills the vision of its creators as a high-performance general-purpose operating system, it was never designed for the level of robustness demanded for high criticality systems. This is no different from other general-purpose operating systems. Imagine using Windows to run the flight control system in a modern airliner—an act that could give new meaning to the term *blue screen of death.*

Solaris is another general-purpose operating system that has been deployed for network consolidation (it is used in the military's DoDIIS Trusted Workstation). CVE-2007-0882 (high severity) describes an embarrassing defect in Solaris 10 and 11, which enables any attacker to log in remotely, completely bypassing authentication.

So, to be clear, while the statistics in this case study reference Linux, the conclusions are common to large-scale, monolithic, general-purpose software products. In addition, the security problems in general-purpose operating systems are due not only to software complexity; kernel architecture is another important factor, as we discuss in Chapter 2.

Key Point

Embedded Linux is a good example of generally well-crafted software that, due to its immense code complexity and rapid rate of change, is riddled with vulnerabilities and hence is not suitable for the protection of high-value resources against sophisticated attack threats.

1.3.2 Network Connectivity

Another clear trend in embedded systems is the addition of network connectivity. Embedded systems have traditionally been thought of as standalone entities immune to the risks of the Internet. There are many reasons why embedded systems are being networked. Connectivity enables remote management of an embedded system. Management tasks might include upgrading software to address a flaw. Network connectivity adds new capability. For example, home ovens built by TMIO allow owners to connect from a cell phone or the Internet to remotely turn on and off heating. Suddenly, home appliances are security critical.

In 2010, General Motors introduced a feature to enable car owners to manipulate the locks and start the engine from anywhere on the planet using a smartphone. This remote connection piggybacks on GM's OnStar telematics system, which became standard in all GM models in 2007. Should consumers be worried about the security impact of this connectivity?

Just prior to GM's announcement of the smartphone feature, a team of university researchers published a study demonstrating how such a car's critical systems—brakes, engine throttling, and so on—could be maliciously tampered with by exploiting vulnerabilities in the car's embedded systems.[8]

The researchers commandeered the car's brakes, engine, and door locks via a diagnostics port. They learned how to bridge from the low-security network to the critical systems using fuzzing techniques. The researchers showed admirable determination; practically every major critical subsystem of the car was discovered, learned, and then totally subverted. Brakes and engine were disabled while the car was in motion, demonstrating that the attacks could indeed place passengers in extreme peril. The research paper is fantastic, a must-read for embedded security professionals and enthusiasts.

Many articles and blogs have been penned in response to the research, but the overall reaction has been muted, almost soporific. This may be caused by the authors' diligent attempt to preempt panic:

- "We're not interested in taking an alarmist tone."
- "We have no reason to believe this is an issue today."
- "Today everyone is focusing on Web security and botnets. We want to make sure that in 5 or 10 years we don't add cars to that list."

The absence of alarm is surprising and concerning in and of itself. Are the researchers advocating security by obscurity? They refuse to reveal the hacked car's make and model and are not releasing their "car shark" tool used to implement the subversions. OnStar has always provided a remote connection. Attaching to the cellular networks simply opens up more avenues of attack. Some may ask why anyone would want to attack the car network. That's like asking why would anyone want to attack the power grid? What better way to guarantee catastrophe than disabling the brakes on millions of cars simultaneously? The bad guys have really smart and dedicated researchers, too.

We need to take an alarmist tone. The research demonstrates that we have millions of vulnerable cars on the road. We now know attackers are sophisticated enough to disable a driver's brakes while he is barreling down the highway. The only question is whether attackers are sophisticated enough to find a way in remotely.

The researchers state: "In our car we identified no fewer than five kinds of digital radio interfaces accepting outside input, some over only a short range and others over indefinite distance.... Taken together, ubiquitous computer control, distributed internal connectivity, and telematics interfaces increasingly combine to provide an application software platform with external network access."

Ironically, emergency connectivity services, like GM's OnStar, may now provide the means for distributed, remote attacks. Passengers want the Internet inside and smartphone apps to control convenience functions, but they never expected these interfaces to be connected to the drivetrain. The researchers continue: "The CLS [Central Locking System] must also be interconnected with safety critical systems such as crash detection to ensure that car locks are disengaged after airbags are deployed to facilitate exit or rescue."

What the researchers do not talk about is what we can do about embedded automotive security today. Most likely small changes could be made to better isolate the network subsystems. Strong cryptographic authentication must be used for all network connections. Trusted platforms and remote attestation must be used to prevent rogue firmware installs from exposing the car network to attackers. Electronic Control Units (ECUs) with mixed criticality functionality must employ high-assurance partitioning and access control: the rear-view camera must not be affected by iTunes.

Car manufacturers and tier-1 OEMs may not have been thinking a lot about security when they designed the cars hitting roads today, but clearly that must change. Manufacturers must work closely with embedded security specialists early in the design and architecture of in-car electronics and networks. Many of these topics are explored in detail throughout the book.

1.3.3 Reliance on Embedded Systems for Critical Infrastructure

Earlier, we mentioned the smart grid as an important, emerging source of embedded systems with security requirements. One of the predominant concerns in this case is financial: attackers could manipulate metering information and subvert control commands to redirect consumer power rebates to false accounts.

Smart grids imply the addition of remote connectivity, from millions of homes, to the back-end systems that control power generation and distribution. These back-end systems are protected by the same security technologies (firewalls, network access authentication, intrusion detection, and protection systems) that today defend banks and governments against Internet-borne attacks. Successful intrusions into these systems are a daily occurrence. The ability to impact power distribution has obvious safety ramifications, and the potential to impact a large population increases the attractiveness of the target. The smart grid, if not architected properly for security, may provide hostile nation states and cyber terrorists with an attack path from the comfort of their living rooms. Every embedded system on this path—from the smart appliance to the smart meter to the network concentrators—must be secure.

Key Point

Utilities and their suppliers are still early in the development of security strategy and network architectures for smart grids; a golden opportunity now exists to build in security from the start.

Transportation systems—planes, trains, elevators, and industrial vehicles—provide perhaps the most obvious example of safety-critical embedded technology. In 2008, the U.S. Federal Aviation Administration (FAA) expressed concern about accessibility of the flight control system in Boeing's newest commercial airliner, the 787 Dreamliner, from the passenger in-flight entertainment system by way of the aircraft's internal network. The FAA's notification to Boeing to address the potential for malicious access to the critical embedded avionics from passengers is likely the first of its kind in commercial aircraft history.

Financial infrastructure, including banks and credit card processors, has been a target for hackers since the Internet became popular, and embedded systems are taking a larger role in high-value financial transactions. For example, sophisticated point-of-sale kiosks in retail establishments provide an attractive target: the ability to subvert the embedded system may provide access to locally stored credit card numbers as well as a pathway to the back-end payment networks where hundreds of thousands to millions of sensitive financial records are stored.

Some medical devices have security requirements. Life-critical embedded systems with significant software content include patient monitoring systems, imaging equipment, and implantable wireless devices. Digital radiography systems contain million-line source code bases and network connectivity for remote management and data archival. Security researchers have recently demonstrated the ability to commandeer pacemakers over a wireless network.[9] Electronic medical records (EMRs) introduce a threat to sensitive medical records; physicians use mobile devices to access patient information over the Internet.

Industrial process control is another important example of security-critical embedded systems. We discuss this further in the following section.

1.3.4 Sophisticated Attackers

The increasing reliance on embedded systems in commerce, critical infrastructure, and life-critical functions makes them attractive to the well-funded and determined attacker. Embedded industrial control systems managing nuclear reactors, oil refineries, and other critical infrastructure present opportunity for widespread damage.

Stuxnet infiltrated Siemens process control systems at nuclear plants by first subverting the Microsoft Windows workstations operators use to configure and monitor the embedded control electronics (see Figure 1.4).

Key Point

The Stuxnet worm is likely the first malware to directly target embedded process control systems and demonstrates the incredible sophistication potential in modern embedded security attacks.

Much of the security community discussion about Stuxnet has been speculation about the attacker's identity and motive as well as the unprecedented level of attack sophistication, which includes clever rootkit construction and the employment of no fewer than four zero-day Windows vulnerabilities. These vulnerabilities enabled Stuxnet to gain access to and download malware to the Siemens controller itself, implying that the attackers had intimate knowledge of its embedded software and hardware.

Stuxnet demonstrates the need for improved security skills within the embedded development community, but it also elucidates the requirement for a higher level of assurance in critical infrastructure than with standard commercial IT practices. In Chapter 3, we discuss the levels of security assurance appropriate to the level of threat and how the international Common Criteria standard approaches this mapping.

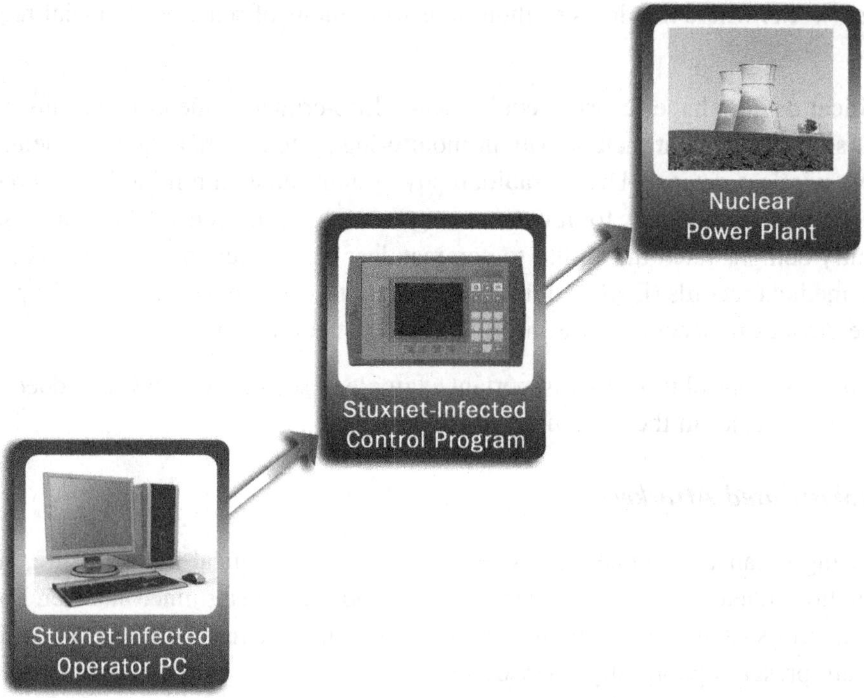

Figure 1.4:
Stuxnet infiltration of critical process control system via operator PC.

Stuxnet also demonstrates that embedded systems and IT systems are often interdependent. SCADA networks are controlled by common PCs. As a response to Stuxnet, the U.S. Department of Defense Chief of Cyber Command and NSA Director, General Keith B. Alexander, recommended in September 2010 the creation of an isolated network for critical infrastructure. This may sound like a heavy-handed approach, but it is precisely how many governments protect their most sensitive, compartmentalized classified networks. Physical isolation introduces some inefficiency that can be ameliorated with the application of high-assurance access solutions that enables a client computer to securely access multiple isolated virtual desktops and back-end networks. These access control systems use the latest and greatest Windows or Linux human-machine interfaces but do not depend on Windows or Linux for their security. Rather, they rely on a technique called MILS Virtualization. MILS Virtualization is discussed in a case study involving software-defined radio (SDR) in Chapter 6.

1.3.5 Processor Consolidation

Another security-impacting trend is processor consolidation within embedded systems. The automobile provides yet another good example. As the automobile continues its inexorable transformation into an electronic system of systems, electronic component counts and associated wiring content within the car have skyrocketed. This electronics growth poses a significant production cost, physical footprint, and time-to-market challenge for automotive manufacturers. The response is to reverse the growth trend and instead merge disparate functions into a lesser number of electronic components. Consolidation requires the proper systems architecture to ensure that these components do not interact in unforeseen ways, posing a reliability risk to critical systems. Chapter 2 discusses the core of this architecture: a partitioning operating environment capable of providing strict resource guarantees, deterministic execution, strict access controls, application hardening, and high assurance. Examples of emerging consolidated applications are found later in this chapter and especially in Chapter 6.

Processor consolidation is closely aligned with the trend toward *mixed criticality systems* in which safety, security, or real-time critical components must coexist with less-critical components.

Key Point

Based on an informal survey of embedded systems developers attending recent Embedded Systems Conferences, we predict that mixed criticality systems will account for at least 60% of all embedded projects by 2020.

For example, consolidating the infotainment head-unit with the rear-view camera component results in a mixed criticality embedded system (see Figure 1.5).

Because it can share the center stack computer's audio and video capability, the rear-view camera module is a natural candidate for consolidation. While the rear camera is still optional on many car models, it is still considered a safety-critical function due to its ability to alert the driver to hazards (think tricycle and small child) that may otherwise go unnoticed with human eyesight alone.

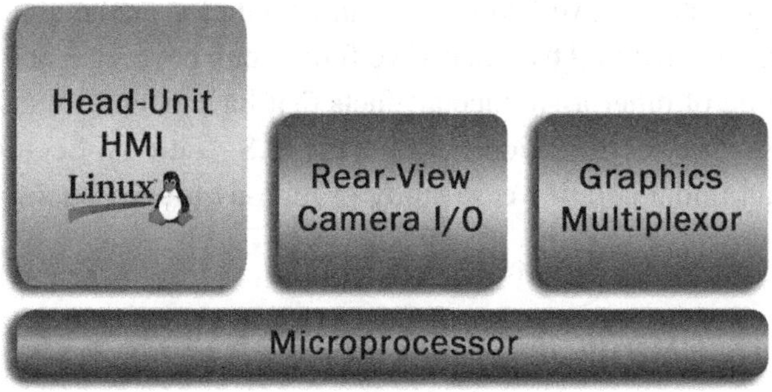

Figure 1.5:
Mixed criticality automotive embedded system.

1.4 Security Policies

Up until now, we've talked about security in a somewhat general sense, without formalizing the specific security properties that we need our embedded systems to enforce. The properties and requirements of a security-enforcing component or system are encompassed in what is called the component's *security policy*. Security policies can have widely varying granularity, depending on the level of detail required. For example, the purchaser of a firewall product may be interested in a high-level security policy that states, "Ingress packets exhibiting a format found on a configurable blacklist of disallowed traffic shall not be forwarded through the firewall egress port." The firewall's administrator is interested in the implementation-specific security policy configured in the blacklist. This policy might consist of hundreds of individual rules that block specific ports, protocols, origination and destination addresses, and permutations of such parameters.

Security policies are necessary to make any practical evaluation of product security. In a security evaluation, evaluators judge the assurance that product users can have with respect to claimed security policies. Security policy is created to counter threats.

1.4.1 Perfect Security

Readers have probably heard this statement before: "There's no such thing as perfect security," or its corollary, "The only thing that is perfectly secure is a brick." This sentiment originates from users of technology, developers of technology, and even security pundits worldwide.

Unfortunately, this attitude is harmful. By starting with a broad negative assertion, developers are provided a built-in excuse not to strive for the best achievable security. Furthermore, the assertion is simply false. In fact, it is possible to achieve perfect security. But it is important to realize that perfection can apply only to specific security policies, as we have just explained. While it is not feasible to prove the security policies of a billion-transistor multicore microprocessor, a small number of important pieces of software, such as an embedded microkernel operating system, have had their security policies mathematically proven. And these formal proofs are augmented by exhaustive functional, coverage, and hardware platform testing and a large suite of other assurance artifacts that have been evaluated and affirmed by foremost experts in security certification (such as the U.S. National Security Agency). In Chapter 3, readers can find a detailed case study of high-assurance operating system security policies.

1.4.2 Confidentiality, Integrity, and Availability

Most security policies can be mapped to one or more confidentiality, integrity, and availability (CIA) protection elements. Confidentiality is the prevention of unauthorized disclosure of information. Integrity is the prevention of modification or corruption of a resource. Availability is the prevention of attacks that would keep a resource from being accessed or used according to its intended function.

Imagine a security policy that requires strong public key authentication to guard remote network access to an automobile infotainment system. This policy could be in place to prevent exposure of digital rights managed assets (confidentiality) as well as to prevent remote attackers from bridging across the infotainment network to disable the car's critical systems (availability).

Key Point

When determining the appropriate security policy to implement within an embedded system, developers must start by thinking of the highest-level confidentiality, integrity, and availability (CIA) protections that must be upheld.

In Chapter 2, we discuss systems software security in detail, but to help better describe the concept and application of security policy, here we summarize some of the relevant security policies for the hardware and software components, such as operating system and memory management unit, in managing the physical resources of an embedded platform (memory, CPU time, I/O access).

1.4.3 Isolation

On embedded microprocessors that provide memory protection hardware (e.g., a memory management unit), an operating system or hypervisor uses these hardware facilities and its own software techniques to enforce an isolation policy between software components.

Key Point

Component isolation is a foundational embedded systems security policy required to be able to realize higher-level system security policies.

For example, the operating system kernel must be isolated from its hosted applications so that the operating system can enforce access control policies for I/O resources under its control. Without this isolation, a faulty or malicious application could corrupt the kernel and prevent it from providing any other security services. Isolation, in this context, refers to temporal as well as spatial isolation; a runaway application must not deny the execution of other applications or the kernel itself.

Another good example of the compositional power of isolation policy can be seen in the effective deployment of cryptography services within a system. Applications such as IPsec or an encrypting file system, which employ cryptography to realize their security policies, depend on the isolation of the cryptographic subsystem (especially private keys) from other applications or subsystems. Without this isolation, private keys can be exposed, rendering impotent the confidentiality policy intended by data-in-motion (DIM) or data-at-rest (DAR) protection. Data-in-motion, also sometimes referred to as data-in-transit (DIT), security is enabled by network security protocols, and data-at-rest by storage media encryption. Both are discussed in Chapter 5.

Isolation can also limit damage resulting from a security failure. For example, if a file system is corrupted and fails to provide its file services, an isolated network stack can continue to provide its communications services. Damage limitation can be considered a security policy in and of itself. Chapter 2 discusses the relative security advantages in operating system architectures that promote (or discourage) component isolation between typical operating system services such as file systems and networking.

1.4.4 Information Flow Control

Operating system services, such as requests to spawn threads or manipulate I/O devices, are examples of a strictly controlled information flow between applications and the operating system kernel. To assure the integrity of its own services and higher-level security policies, the kernel must configure applications and validate service call parameters according to the system's information flow control policy. For example, the information flow policy may permit components Alice and Bob to communicate, but not Alice and Charlie. An operating system could enforce this policy by permitting runtime requests by Alice to send messages to Bob but denying requests by Alice or Bob to send messages to Charlie (and vice versa), as in Figure 1.6. Alternatively, the system designer may statically configure Alice and Charlie such that they lack any communications descriptor upon which to make any such requests to the operating system. Embedded operating system access control and other key security policies and mechanisms are discussed further in Chapter 2.

Information Flow Policy		Receiver		
		Alice	Bob	Charlie
Sender	Alice	Yes	Yes	No
	Bob	Yes	Yes	No
	Charlie	No	No	Yes

Figure 1.6:
Sample information flow security policy.

1.4.5 Physical Security Policies

In contrast to digital software-based attacks, some embedded systems must be protected against physical attacks. For example, a smart grid aggregator, attached to a neighborhood power pole, guards access (via an authenticated, encrypted network connection) to the smart grid from a large number of neighborhood smart meters. An attacker wishing to gain unauthorized entry into the back-end infrastructure can use physical attacks to tamper with the utility's aggregator, attempting to steal private keys that will yield the desired network access. Radios used in a battlefield environment for classified communications must be protected against physical attack that would be possible if the enemy captured the radio.

Physical attacks come in two forms: invasive and non-invasive. Invasive attack on the smart grid aggregator might include breaking open the aggregator's enclosure to procure its flash memory and steal its stored private keys and software.

A non-invasive attack might include placing an energy monitor near the aggregator to perform differential power analysis (DPA) of the system; power consumption patterns can be used to derive information about software functionality such as the execution of a well-known cryptographic algorithm like the Advanced Encryption Standard (AES). DPA and other techniques have been applied successfully to recover secret keys. TEMPEST is a codename, coined in the 1960s, referring to non-invasive eavesdropping on electromagnetic emanations.

Another class of physical attack involves the use of malicious components within an embedded system. The malicious component may be installed at any weak point in the supply chain or manufacturing process of an embedded product or any of its subsystems. For example, in 2008, the U.S. government seized large quantities of Chinese-manufactured counterfeit Cisco routers, leading to speculation that the routers were intended to weaken internal security and steal sensitive information. Embedded designers must be concerned about the provenance of all materials used to build a product, including microprocessors, cryptographic keys, and software. Furthermore, supply chain security must encompass the transportation and delivery of components and products between supplier and consumer.

1.4.6 Application-Specific Policies

A practically limitless set of security policies may be required to protect application-specific resources against the universe of potential threats. System designers must define application-specific policies that make sense for a particular embedded system implementation, component, and/or process. One example of application-specific policy is a role-based access control (RBAC) policy that defines (and therefore limits) the precise resource accesses permitted by software processes and/or human users based on their job responsibility and need to know. For example, a smart grid aggregator may require remote network access to upgrade software in the field. The aggregator contains a management process whose role permits an ability to reprogram the internal flash memory for upgrade purposes. No other process within the system has a role that permits this write access. The management role also requires access to the network from which to receive management requests. The management application can then enforce its own application-specific policy that permits certain authenticated remote users to make upgrade requests.

1.5 Security Threats

As discussed earlier, security policies exist to counter security threats.

> **Key Point**
>
> System designers must first determine what threats are feasible and then what security policies make economic sense relative to the value of resources exposed to a particular threat.

An embedded system within Fort Knox and its army of security guards may not require protection against physical attacks. An embedded system without a network connection does not require protection against network-borne attacks like packet storms, protocol replay attacks, and port probing.

An individual's MP3 player is unlikely to be targeted by nation states or other sophisticated attackers and hence does not justify expensive firewalls and NSA-certified operating systems and cryptography. On the other hand, critical infrastructure systems must consider sophisticated threats such as side channel software attacks and memory/CPU denial-of-service attacks.

An embedded system may be insecure if the system designer errs in judging threats or how the system may be vulnerable to those threats. Such lapses occur for many reasons, including sheer ignorance about the capabilities of the underlying hardware, embedded operating system, or third-party middleware. Does the system designer know what kind of cryptography (algorithm, key length, storage protection for private keys) the Transport Layer Security (TLS) protocol implementation uses? Does the designer know what ports the embedded operating system might be leaving open? Does the designer know how microprocessor and firmware boot are locked down to guarantee the establishment of a secure initial state?

1.5.1 Case Study: VxWorks Debug Port Vulnerability

In 2010, a critical and widespread vulnerability was discovered in embedded systems running VxWorks, an embedded real-time operating system (RTOS). Examples of affected products include DSL concentrators, SCADA industrial automation systems, video conferencing systems, Fiber Channel switches, and Wi-Fi routers. Researcher H.D. Moore used a debug communications port, lacking strong authentication, to commandeer the systems running VxWorks. Using the debug interface, a remote attacker could read or write any physical memory location: admin passwords could be extracted and supervisor mode malware trivially installed. Moore revealed during a B-Sides security conference talk that a quarter million devices accessible directly from the Internet were found to be vulnerable. Most likely, the total number of affected devices is orders of magnitude higher.

Embedded systems, historically thought of as autonomous, closed systems, are increasingly constituents of the *Internet of Things*, with requirements for in-field upgrade, remote debugging and diagnostics, and other management functionality.

Embedded developers who fail to accurately assess threat, expose their systems, resources, missions, and attached networks to risk of attack. Embedded security is not just about choosing the right hardware and software technology. Equally important are security design and architecture, including configuration, isolation of security functions, platform vulnerability assessment, establishment of a secure initial state, fail-security, and so on. While the answer to this VxWorks hack may seem as simple as disabling the diagnostics connection or adding strong authentication capability to it, a more systemic approach is often needed. What other open ports, buffer overflow flaws, unprotected data, or other weaknesses may be present in a legacy software environment?

1.6 Wrap-up

Despite the growing roles of embedded systems, their security posture remains weak, and the vast majority of embedded systems professionals have only rudimentary knowledge of embedded security issues. Many embedded systems professionals have worked with relatively small code bases and simple hardware architectures and are struggling to deal with the trends toward more sophisticated designs, network connectivity, system consolidation, and determined and well-funded attackers. Furthermore, most embedded software professionals are not trained in the art and science of designing systems for security. Security is an afterthought and almost never can be feasibly retrofitted.

The result is that embedded systems are shipped with vulnerabilities that leave critical infrastructure exposed. Embedded systems professionals need to be educated on the state-of-the-art in embedded systems security so that they can integrate and apply these practices to improve security. It is the authors' fervent hope that this book takes the embedded development community a significant step forward in this goal.

1.7 Key Points

1. Security is the ability of an entity to protect resources for which it bears protection responsibility.
2. An embedded system is an electronic product that contains a microprocessor (one or more) and software to perform some constituent function within a larger entity.

3. Many of the problems relating to loss in quality, safety, and/or security in electronic products can be attributed to the growth of complexity that cannot be effectively managed.
4. Linear growth in hardware/software content creates far more than linear growth in overall complexity due to an exponential increase in interactions between functions and components.
5. Embedded Linux is a good example of generally well-crafted software that, due to its immense code complexity and rapid rate of change, is riddled with vulnerabilities and hence is not suitable for the protection of high-value resources against sophisticated attack threats.
6. Utilities and their suppliers are still early in the development of security strategy and network architectures for smart grids; a golden opportunity now exists to build in security from the start.
7. The Stuxnet worm is likely the first malware to directly target embedded process control systems and demonstrates the incredible sophistication potential in modern embedded security attacks.
8. Based on an informal survey of embedded systems developers attending recent Embedded Systems Conferences, we predict that mixed criticality systems will account for at least 60% of all embedded projects by 2020.
9. When determining the appropriate security policy to implement within an embedded system, developers must start by thinking of the highest-level confidentiality, integrity, and availability (CIA) protections that must be upheld.
10. Component isolation is a foundational embedded systems security policy required to be able to realize higher-level system security policies.
11. System designers must first determine what threats are feasible and then what security policies make economic sense relative to the value of resources exposed to a threat.

1.8 Bibliography and Notes

1. McConnell S. *Code Complete*. 2nd ed. Redmond, WA; Microsoft Press; 2004.
2. Kroah-Hartman G, Corbet J, McPherson A. *Linux Kernel Development: How Fast It Is Going, Who Is Doing It, What They Are Doing, and Who Is Sponsoring It: An August 2009 Update.* Linux Foundation, 2009.
3. General Dynamics. website: http://www.gdc4s.com/content/detail.cfm?item=570bc11a-1d6d-4acf-a68e-c256a615808d&page=2.
4. NSA SELinux Frequently Asked Questions: http://www.nsa.gov/research/selinux/faqs.shtml.
5. http://web.nvd.nist.gov/view/vuln/search.
6. http://kerneltrap.org/mailarchive/linux-kernel/2008/2/2/700024.
7. Corbet J. An Interview with Jim Ready (June 11, 2008), LWN.net.
8. Koscher K, Czeskis A, Roesner F, Patel S, Kohno T. *Experimental Security Analysis of a Modern Automobile*. Oakland, CA: 2010 IEEE Symposium on Security and Privacy; May 19, 2010.
9. Halperin D, Heydt-Benjamin TS, Ransford B, Clark SS. *Pacemakers and Implantable Cardiac Defibrillators: Software Radio Attacks and Zero-Power Defenses*. Oakland, CA: 2008 IEEE Symposium on Security and Privacy; March 12, 2008.

CHAPTER 2

Systems Software Considerations

Chapter Outline

Embedded Systems Security. DOI: 10.1016/B978-0-12-386886-2.00002-3

2.1 The Role of the Operating System

Because the operating system controls the resources (e.g., memory, CPU) of the embedded system, it has the power to prevent unauthorized use of these resources. Conversely, if the operating system fails to prevent or limit the damage resulting from unauthorized access, disaster can result. In an operating system context, unauthorized actors may refer to applications/processes, collections of applications, and/or human users that attempt to access computer resources not permitted by the system security policy.

Key Point

The operating system bears a tremendous burden in achieving safety and security.

Operating system security is not a new field of research. Yet practically all the deployed embedded operating systems are unable to meet meaningfully high levels of security certification. One of the reasons for the lack of secure operating systems is the historical approach taken to achieve security. In most cases, security is bolted on as an afterthought. However, even operating systems designed for security attempted to provide a kitchen sink of services—protection and partitioning, device access controls, secure file systems, and secure network services. As a result, these systems were simply too large and complicated to evaluate at high levels of security.

We would all agree that it is a bad idea to trust our critical embedded systems to insecure operating systems. Unfortunately, much of the world's computer systems used to monitor and control plants and equipment in industries such as water and waste control, energy, and oil refining are running such operating systems, the same as those running a run-of-the-mill desktop PC. As stated by Michael Vatis, executive director of the Markle Foundation's Task Force on National Security in the Information Age, "The vulnerabilities are endemic because we have whole networks and infrastructures built on software that's insecure. Once an outsider gains root access, he could do anything. Any given day, some new vulnerability pops up."[23]

2.2 Multiple Independent Levels of Security

Recently, a small number of embedded operating system technologies have taken a new approach that attempts to divide and conquer the problem of operating system security. These operating systems adopt the Multiple Independent Levels of Security (MILS) architecture that stipulates a layered approach to security.

Key Point

The foundation of a MILS-based embedded system is the separation kernel, a small microkernel that implements a limited set of critical functional security policies, including data isolation, information flow control, damage limitation, and periods processing.

2.2.1 Information Flow

Information cannot flow between partitioned applications unless explicitly permitted by the system security policy.

2.2.2 Data Isolation

Data within partitioned applications cannot be read or modified by other applications.

2.2.3 Damage Limitation

If a bug or attack damages a partitioned application, this damage cannot spread to other applications.

2.2.4 Periods Processing

Periods processing is a policy that ensures that information within one component is not leaked into another component through resources, such as kernel-managed memory buffers and CPU registers, which may be reused across execution periods. For example, if component A stores private information in a memory page and then releases that memory page back to the operating system kernel, the kernel must ensure that the page is cleared before it can be reused by another component B requesting a memory page allocation (see Figure 2.1). Similarly, if microprocessor registers are written with data during A's execution, the operating system must ensure that these register values are cleared when context switching to B on the same core. Without periods processing, the confidentiality of A's information would be violated by disclosure of information to B through these computer resources.

The MILS separation kernel realizes these MILS policies by using the microprocessor's memory management hardware to prevent unauthorized access between partitions and by implementing resource allocation mechanisms that prevent one partition's operation from affecting another (e.g., by exhausting a resource such as memory or CPU time). The information flow policy prevents unauthorized access to devices and other system resources by employing an efficient capability-based object model that supports both confinement and revocation of these capabilities when the system security policy deems it necessary. Capabilities and the associated revocation and confinement properties are discussed later in this chapter.

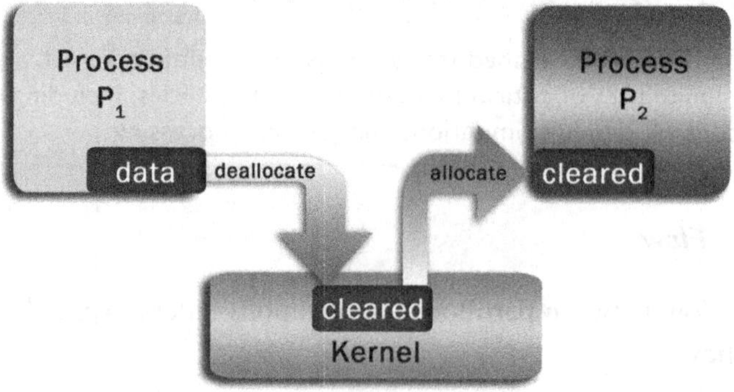

Figure 2.1:
Periods processing example: sanitization of memory prevents information leakage across security domains.

The MILS architecture also adopts the *reference monitor* or reference validation mechanism, first described in a 1970s U.S. government computer security report.[1]

> **Key Point**
>
> A separation kernel is considered a reference monitor when the kernel's MILS policy enforcement mechanisms are always invoked, tamper-proof, and evaluable.

These three properties are described further in the following sections.

2.2.5 Always Invoked

Applications requesting resources or services from the operating system must not be able to bypass the system security policies. For example, let's consider an operating system that enforces the following access control policy for processes A and B over files X, Y, and Z: process A is permitted to access files X, Y, and Z, but process B is not permitted to access any of X, Y, or Z. However, let's suppose that process B is able to obtain access to /dev/hdd, the physical hard disk device used by the file system. B can use this physical device to bypass the file system altogether (see Figure 2.2) and access any file on disk.

2.2.6 Tamper Proof

Applications must not be able to tamper with the security policy or its enforcement mechanisms. Using the preceding file system example, the access control policy for a file is encoded in the file's metadata or in a special policy file maintained by the file system. Once again, given direct access to the underlying hard drive, B can tamper with the access control policy data and give itself or its confederates access to files X, Y, and Z.

2.2.7 Evaluable

Security claims are a dime a dozen. Anyone can make claims about security. The question is: how confident are end users and other stakeholders that a product actually fulfills its claims? This confidence is referred to as *security assurance*.

> **Key Point**
>
> The MILS architecture requires the use of security-enforcing components whose functional requirements meet a high level of assurance.

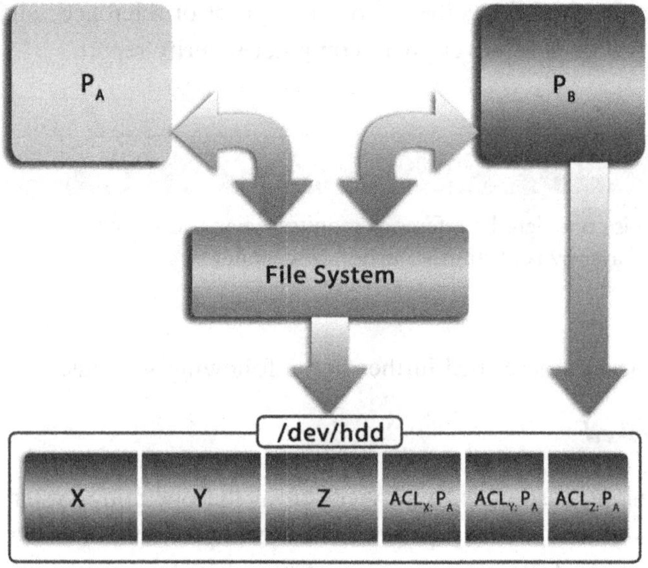

Figure 2.2:
Bypassing file system policy via direct media access.

Highly reputable firms make unsubstantiated claims about security that fall flat in the face of discovered vulnerabilities. For example, let's look at a recent security certification for the VMware enterprise hypervisor.

As VMware virtualization deployments in the data center have grown, security experts have voiced concerns about the implications of "VM sprawl" and the ability of virtualization technologies to ensure security. On June 2, 2008, VMware attempted to allay this concern with its announcement that its hypervisor products had achieved a Common Criteria Evaluated Assurance Level (EAL) 4 security certification. VMware's press release claimed that its virtualization products could now be used "for sensitive, government environments that demand the strictest security."

On June 5, just three days later, severe vulnerabilities in the certified VMware hypervisors were posted to the U.S. Computer Emergency Readiness Team (U.S. CERT) National Vulnerability Database. Among other pitfalls, the vulnerabilities "allow guest operating system users to execute arbitrary code." In Chapter 3, we discuss the Common Criteria security standard and the relative assurance levels; but suffice it to say that EAL 4 does not meet the high level of assurance sought by MILS. Chapter 3 discusses Common Criteria in more detail, focusing on the differences in evaluated assurance levels.

MILS stipulates that core kernel security policy enforcement be evaluable to the highest levels of assurance. This assurance is absolutely critical and is the reason why the separation kernel enforces a focused set of policies and does not provide higher-level security policies such as role-based access control for files or network security. Since a high-assurance security evaluation requires a formal model of the system, formal proof of correspondence between the model and actual implementation, and proof of security theorems against this model, a system of more than approximately 10,000 lines of code becomes too difficult and expensive to evaluate. The MILS security policies can be implemented with a microkernel that is small enough to be evaluated at the highest assurance level.

Under the MILS concept, higher-level secure software, such as a secure communications mechanism, web server, or file system, can be layered on top of the separation kernel. The MILS security policies are recursive: a MILS file system, using the fact that an underlying separation kernel enforces its partitioning security policies, can be used to ensure file system data isolation, information flow, and damage limitation properties. System designers can select, as needed, the MILS components that make up an actual system. If the system does not require a secure web server, then there is no need to go through the pain of evaluating one. MILS components can be independently evaluated at the highest assurance level and can come from multiple vendors. This modular approach to security is a key reason why MILS-based systems have been widely deployed with a lower overall life-cycle cost than early attempts to create verified operating systems.

Another major advantage of the separation kernel is that it allows software at varying levels of criticality to run on a single microprocessor. For example, an application containing classified data and algorithms can occupy one partition while another partition is connected to an unclassified network such as the Internet. The MILS security policies, if highly assured, make this possible. This can lead to enormous cost savings in product development because complicated multi-function applications can run on a single powerful microprocessor without requiring all these applications to be evaluated at the highest assurance level.

2.3 Microkernel versus Monolith

In computer security, the term *Trusted Computing Base* (TCB) is used to refer to those portions of a system (software and hardware) that are critical to security and therefore must be trustworthy.

Monolithic operating systems contain system software, such as networking stacks, file systems, and complex device drivers that share a single memory space and execute in privileged (supervisor) mode. This results in a large TCB, providing a plethora of opportunities for hackers to find and exploit vulnerabilities. Examples of monolithic operating systems include Windows, UNIX, Linux, and VxWorks.

Key Point

Microkernel operating systems provide a better architecture for security than monolithic operating systems.

A microkernel operating system runs only a minimal set of critical system services, such as thread management, exception handling, and inter-process communication, in supervisor mode and provides an architecture that enables complex systems software to run in user mode, where they are permitted access only to the resources deemed appropriate by the system designer. A failure in one component cannot cause damage to the hard drive because the infected component simply does not have access to that resource. Because the microkernel is simple, its security can be more easily verified and assured. If a network driver does not perform a security function, then it is not considered part of the TCB and need not be held to the same security standard as the microkernel or other critical components. In a monolithic system, the network driver is part of the kernel and therefore security critical by definition. Figure 2.3 depicts the stark difference between the microkernel and monolithic approaches in which typical operating system services are either isolated from each other (and the kernel), or they are not.

Most general-purpose operating systems are monolithic because they focus foremost on performance. User applications are able to access most services—TCP/IP networking, files, and I/O devices—with a simple and efficient system call into the monolithic kernel. In contrast, a microkernel implements these services in separate processes, requiring inter-process

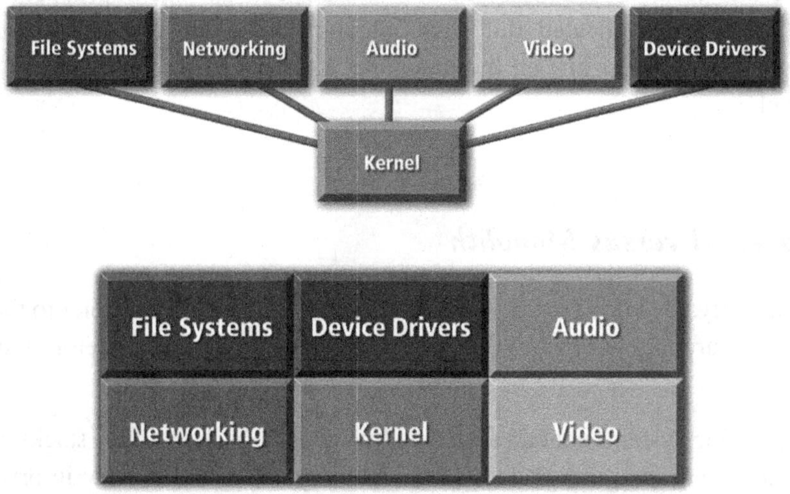

Figure 2.3:
Microkernel (top) versus monolith (bottom).

communications (IPC) between the requesting process and one or more of the relevant system service processes. For example, an application that wishes to access a remote, network-resident file system using Network File System (NFS) may require communications between the application, a TCP/IP process, a network interface device driver process, and an NFS process. This work is behind the scenes: the user application employs the same *read()* or *write()* applications programming interface (API) that a monolithic operating system provides, and the microkernel takes care of routing data between the appropriate system processes.

The chief historical complaint against microkernels is the loss in performance due to the extra context switches and message-passing overhead of this service process architecture. Two trends have dissolved this complaint. First, faster microprocessors have reduced the impact of IPC overhead. Second, commercial microkernel developers have become adept at optimizing IPC. For example, Jochen Liedtke, the original author of the open source L4 microkernel, published studies demonstrating that microkernel IPC could be improved dramatically by design over first-generation efforts from the 1980s.[2] The most compelling evidence, however, that microkernels have become practical is the widespread successful embedded deployments using microkernels since 2000. Microkernels run practically every type of embedded product, including smartphones, avionics, networking gear, process control systems, and medical devices. Examples of microkernel-based embedded operating systems include INTEGRITY, LynxSecure, Neutrino/QNX, OKL4, PikeOS, and Symbian.

Over the years, operating system designers have expressed differences of opinion, sometimes vociferously, regarding the relative merit of these two major approaches to operating system design. O'Reilly published portions of a now-infamous debate between Linus Torvalds, the founder of Linux, and Andrew Tanenbaum, a professor and researcher of operating systems, which started in 1992.[3] Unfortunately, much of this debate has failed to focus on the core architectural impact of operating system design on security. Rather, the debate has meandered across open source licensing issues, patching protocols, and, at times, personal insults. While Linux has had and will to continue to have tremendous adoption due to its open source licensing and development model, it should be obvious to all readers that a component-based microkernel is superior for security and robustness. Even Torvalds admitted such in one of his e-mails to Tanenbaum: "True, Linux is monolithic, and I agree that microkernels are nicer. From a theoretical (and aesthetical) standpoint Linux loses." Practically every notable university operating system researcher—Andrew Tanenbaum, Gernot Heiser, Jochen Liedtke, Jonathan Shapiro, Kang Shin, Ken Thompson, Rajkumar Buyya, and many others—have asserted the architectural superiority of the microkernel approach.

Furthermore, as we discuss later in this chapter, the decision to use a microkernel or monolithic operating system is not necessarily mutually exclusive; virtualization provides the ability to incorporate the best implementations of both architectures in a single system.

2.3.1 Case Study: The Duqu Virus

In November 2011, security researchers discovered a dangerous computer virus, dubbed *Duqu*. Similar to the Stuxnet attack described in Chapter 1, Duqu is believed to have been written by sophisticated attackers and takes advantage of a zero-day Windows vulnerability to gain access to embedded critical control infrastructure. As described in the Microsoft security advisory: "An attacker who successfully exploited this vulnerability could run arbitrary code in kernel mode. The attacker could then install programs; view, change, or delete data; or create new accounts with full user rights. We are aware of targeted attacks that try to use the reported vulnerability."[4] The National Vulnerability Database has assigned the vulnerability a severity rating of 9.3 (high),[5] primarily due to its remote exploitability and the impact implied by malicious root access.

According to online reports, and corroborated by the aforementioned National Vulnerability Database entry overview, common Windows operating systems (XP, Vista, Windows 7, Windows Server) all execute the font-parsing engine in the kernel. In fact, the font parsing occurs in *win32k.sys*, the massive Windows system device driver module that has been the cause of many "blue screens of death" (BSOD) over the years.

Duqu's authors crafted Word documents that would exploit the font-parsing vulnerability, enabling remote code execution at superuser privilege. As we have learned from many such vulnerabilities (Stuxnet being just one other case in point), once the attacker has an effective malware vehicle like this, it is a simple matter to socially engineer attack targets into opening the file. The attacker needs only one successful endpoint intrusion; the infected PC is then used as a launching point against electronics and computers accessible to the PC locally and across the network.

Duqu provides a wonderful example of the deleterious repercussions of monolithic operating system design. In some cases, *win32k.sys* consists of millions of bytes of device driver code executing in kernel/superuser mode.

2.4 Core Embedded Operating System Security Requirements

The use of microkernel architecture, of course, is only one important part of the security story for embedded operating systems. In the following sections, we discuss key technical features of embedded operating systems that impact security.

2.4.1 Memory Protection

Key Point

Memory protection is a fundamental requirement for robust embedded systems.

The operating system utilizes memory protection hardware of the microprocessor to isolate unrelated functions. The protected components are often called *processes*. A software application may be mapped to one or more operating system processes. With memory protection, malicious code is unable to crash an application or the operating system by corrupting its memory. Similarly, an errant pointer caused by a programming error in one application cannot affect another application executing in a separate memory-protected process.

2.4.2 Virtual Memory

Most modern operating systems support virtual memory in which the addresses of a process's code and data are logical rather than physical. The kernel uses memory management hardware to map the logical memory to physical memory. For example, let's consider three processes: A, B, and C, each of which requires four kilobytes of storage. In a virtual memory system, A, B, and C may all occupy the same logical address range, say 0x1000 to 0x2000, within separate virtual address spaces. The operating system maps these four-kilobyte blocks to physical locations, say 0x10000, 0x11000, and 0x12000 that are hidden from the processes themselves (see Figure 2.4).

Several embedded operating systems, such as OSE and VxWorks, employ memory protection without virtual memory. In the example of A, B, and C, a flat memory model operating system may locate these processes at physical addresses 0x10000, 0x11000, and 0x12000. These addresses are visible to the program; code and data objects are linked and loaded within these physical ranges.

> **Key Point**
>
> Virtual memory provides additional security features, including guard pages and location obfuscation, on top of basic memory protection.

2.4.2.1 Guard Pages

One big advantage of virtual memory is the ability to selectively map and unmap pages into a virtual address space. Of course, physical memory pages are mapped in to hold the application's code and data. In addition, pages are mapped in to hold the runtime stacks of application processes. The operating system should provide the ability to leave a page's (or more) worth of the virtual addresses below each stack area unmapped so that if a process overflows its stack, a hardware memory protection fault will be triggered. The kernel will suspend the process instead of allowing it to corrupt other important memory areas within the address space. In a flat memory model, each physical memory page represents actual RAM (or

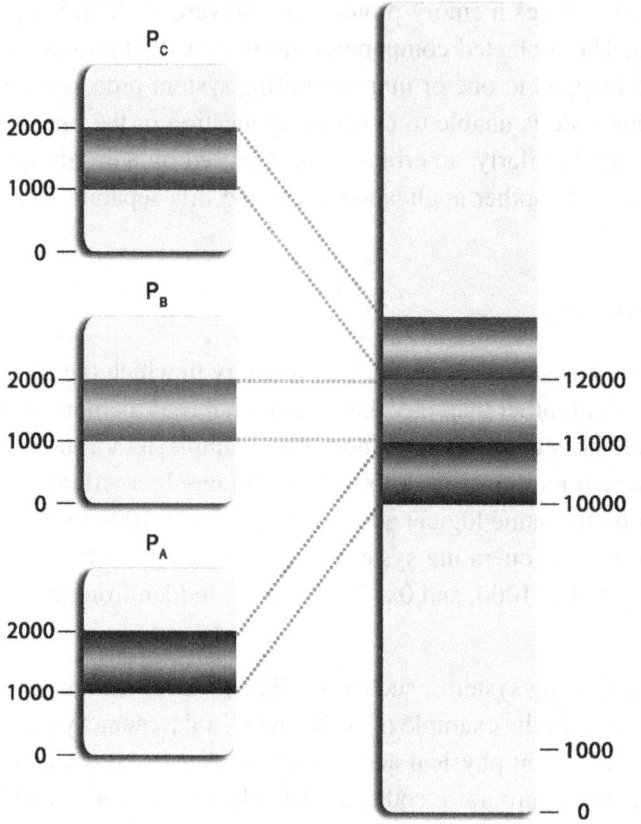

Figure 2.4:
Three virtual-memory-protected processes.

ROM), so unmapping a page effectively causes that chunk of RAM, which is typically limited in an embedded system, to be lost. Memory protection, including this kind of stack overflow detection, is also very useful during the development of an application. Programming errors often generate exceptions that are immediately detected and easily traceable to the source code. Without memory protection, bugs can cause subtle corruptions that are extremely difficult to track down. In fact, on some microprocessors, RAM is often located at physical address 0, so even a NULL pointer dereference can go undetected!

2.4.2.2 Location Obfuscation

Another security benefit of virtual memory is that it hides the physical locations of code and data, making it harder for an attacker to discern internal operation or state. For example, let's suppose that an attacker is able to snoop the physical address bus. By watching address requests, the attacker can determine which process is executing and gain insight into the behavior of that process. In a virtual memory system, the reference to a physical address cannot be mapped back to a process without having first compromised the operating system's internal data structures.

2.4.3 Fault Recovery

The fact is that there are bugs in software, and failures will occur. Most programmers have probably seen statistics detailing the average number of bugs per line of production source code. And as applications become more complex, performing more functions for a technology-hungry world, the number of actual bugs in fielded systems will continue to increase.

> **Key Point**
>
> System designers must plan for failures and employ fault recovery techniques.

Of course, fault recovery is application dependent: the response to a statistics gathering program generating incorrect data is going to differ from the response to a failed network connection. But the operating system must provide some fundamental features that enable the system designer to build in fault tolerance and high availability.

When a process faults (e.g., due to an invalid memory access), the kernel must provide some mechanism whereby notification can be sent to an agent that is in charge of performing some type of fault recovery action. This supervisor process should be able to run in its own address space since the data in the address space containing the faulted process may be corrupted. The kernel must provide a mechanism enabling a supervisor process to close down the faulted process. The kernel must provide a mechanism for restarting an application. The kernel must provide an event logging mechanism so that the cause of a failure can be determined by later examining everything that happened in the system, such as kernel service calls, process context switches, and interrupts, prior to the fault. The kernel must provide a software watchdog capability whereby a supervisor process can be notified when a periodic process does not execute its expected code sequence; this is important because some failures may not directly cause a hardware exception.

The kernel must also protect itself against improper service calls. Many kernels pass the actual pointer to a newly created kernel object, such as a semaphore, back to the application as a handle and then dereference this pointer when passed into subsequent kernel service calls made by the application. The application can pass in an invalid pointer with disastrous results. No kernel service call must ever be permitted to take down the kernel. The operating system should employ opaque descriptors for application references to kernel objects. The kernel must validate all service call parameters.

2.4.4 Guaranteed Resources

> **Key Point**
>
> Despite memory protection and virtual memory, malicious code can still take down a critical application by starving it of resources.

Most operating systems employ a central store for memory resources. As protected processes ask for new resources (e.g., heap memory, threads), the operating system allocates these resources from the central store. An errant or malicious application can request too many resources, causing the central store to be depleted and critical applications to fail when they attempt to obtain their required resources. A similar denial-of-service attack can occur with CPU time. A malicious process can spawn multiple "confederate" processes that soak up the CPU time, keeping critical applications from accomplishing their tasks.

It is possible to architect an operating system such that these critical resources are partitioned according to individual application requirements. An infected process can exhaust its own quota of resources, but it cannot possibly affect the quota of resources held by critical applications. The difference between the central store and quota approaches to memory allocation is depicted in Figure 2.5.

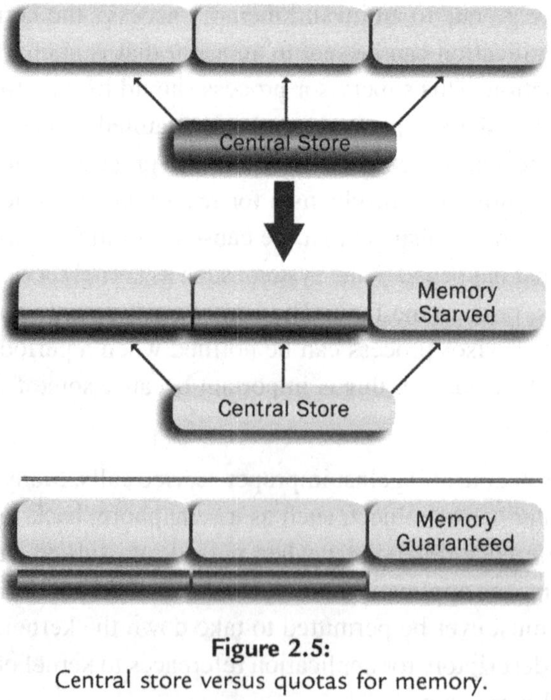

Figure 2.5:
Central store versus quotas for memory.

In the top portion of the diagram, two of the applications make memory allocation requests until the entire central store is exhausted, starving the third application. In the lower portion of the diagram, each application has a quota that cannot be affected by the other applications.

It is important to understand that guaranteed resource partitioning is not practical to retrofit to an operating system not designed for it. For example, it is relatively easy to retrofit a basic quota system for memory, files, and other resources. In fact, UNIX and Linux provide such a system, called *rlimit*, which was first added to Berkeley Software Distribution (BSD) UNIX in version 4.2, circa 1983. The problem with a bolt-on approach like rlimit is that it fails to account for resources allocated internally by the kernel on behalf of its processes. For example, the kernel must use physical memory for internal storage of object control blocks, scheduling lists, virtual memory page tables, device driver buffers, and other resources. These resources are strewn throughout the kernel, and UNIX rlimit fails to attribute their allocation to any particular process. Thus, each of these shared resources becomes a potential denial-of-service vulnerability.

A simple denial-of-service exploit of an rlimit-equipped UNIX operating system is the *fork bomb*: spawn a large number of processes to exhaust the kernel's internal process table. Even if the system enforces per-user process count limits, malware that infiltrates the accounts of multiple users can still bring a system to its knees, preventing other applications from being launched. Most users of UNIX or Linux systems have at one time or another experienced the frustrating message "No more processes."

Furthermore, the process table is only one of the shared resources affected by a fork bomb. A fork operation is expensive in terms of CPU and memory bandwidth, and the mere act of executing a large number of forks concurrently is enough to keep critical applications from making progress.

A secure partitioning operating system never dynamically allocates resources from a shared pool. For example, let's suppose a process loads a dynamic link library (DLL). The DLL is mapped into the process's virtual address space, requiring new logical to physical page mappings that must be managed by the kernel. This management may require storage for internal accounting; that memory must be attributed to and subtracted from the quota of the process that originated the service request. Secure partitioning is the cornerstone of a MILS-compliant separation kernel.

It is also important to note that using a secure partitioning microkernel does not necessarily imply that any of the microkernel's hosted processes providing their own services, such as network communications or file management, will also provide resource guarantees to their clients. Embedded systems designers will likely be faced with a choice of process servers, some of which provide secure partitioning and some that do not. For example, a microkernel may host a UNIX-compatible file system that does not enforce secure partitioning of its internal data structures or of the physical file system media. The UNIX-compatible file system may be useful for portability and compatibility with other computer systems but could be the source of denial-of-service problems if the file system's services are accessible by both critical and malicious processes.

Once again, a secure microkernel can help: multiple instances of the UNIX-compatible file system can be included in the overall system architecture; each instance serves an isolated security domain (see Figure 2.6). Critical applications in one domain are protected from denial-of-service attacks from malicious software in a separate domain (e.g., a domain connected to the Internet). Multi-instance servers need not be expensive in terms of memory resources; the secure microkernel should offer an ability to share the static, read-only portions of the file server (including all its executable code) across domains.

Another potential drawback of a quota system is its static nature. Some complex systems may have varying resource requirements during their lifetimes and may not have enough memory for the union of all the maximum quotas. To address this, a partitioning operating system must provide the ability for system designers to create a trustworthy resource manager that can export resources to, and reclaim resources from, applications on demand. The system designer must be able to specify which critical applications can connect to the resource manager. Malicious software imported into new processes are denied access to the resource manager and governed by quotas.

Alternatively, the system designer can use multiple resource managers, one for each security domain. The applications within a security domain are free to share its domain-specific resources that cannot be hijacked by any applications running in a distinct security domain.

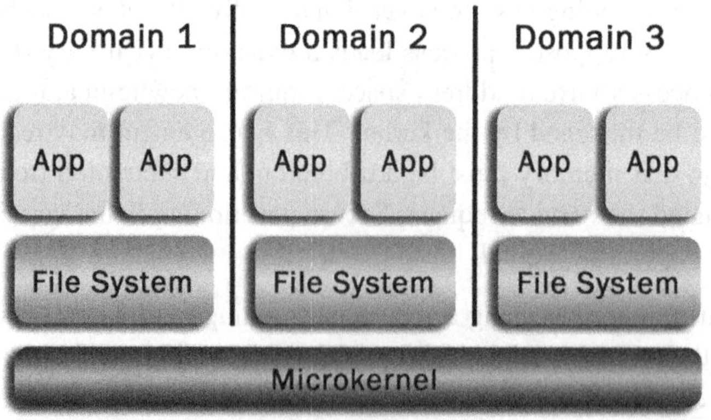

Figure 2.6:
Multiple-domain file system architecture.

Key Point

Partitioning, coupled with the ability to specify resource managers and multiple instance servers, enables system designers to obtain flexibility where needed and security where it is essential.

2.4.5 Virtual Device Drivers

Key Point

Device drivers are frequently the cause of system reliability problems and the target of hackers; thus, device drivers are some of the most important components of the operating system to isolate and protect.

Monolithic operating systems commonly allow users and processes to dynamically install device drivers into the kernel. A faulty device driver can suffer a buffer overflow attack where malware overwrites the runtime stack to install code into the kernel. *Virtual device drivers* prevent these types of attacks because an infiltrated device driver can harm only the process containing the driver, not the kernel itself. To facilitate the development of virtual device drivers, the operating system needs to provide a flexible mechanism for I/O control to the virtual driver process. The virtual driver, however, must be provided access only to the specific device resources that the driver needs to achieve its intended function. Figure 2.7 shows a couple of examples of virtual device drivers, an interrupt-driven communications driver

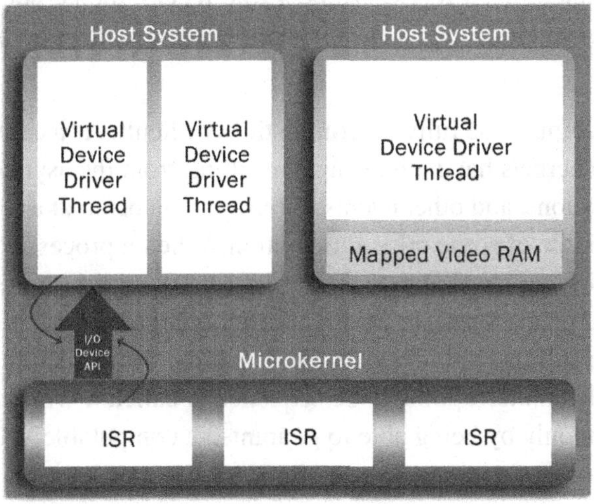

Figure 2.7:
Virtual device drivers.

consisting of two threads, and a graphics device driver that only writes to mapped video RAM without requiring any interrupts.

2.4.6 Impact of Determinism

Operating systems perform a wide range of services on behalf of their client processes and the overall embedded system. Operating systems schedule workloads across one or more CPU cores, allocate memory, shuffle messages between processes, control access to I/O, manage power consumption, monitor system health, and so on. There is much leeway in terms of how the operating system is architected to prioritize all these activities. Most general-purpose operating systems are architected to provide good, or fair, responsiveness to all of their clients. While processes may be assigned numeric priorities, they are interpreted as guidelines that do not interfere with the ability of the rest of the system to make progress.

A *real-time embedded system* is one whose inability to respond within a certain maximum time frame constitutes failure. For example, an aircraft fly-by-wire control computer must read and process a plethora of sensors at fixed regular intervals. If the control application is unable to read and process sensor readings in a timely manner, safe flight is endangered. The B-2 bomber is a great example: its flying wing structure is inherently unstable; the fly-by-wire system overcomes this instability with continuous split-second control adjustments, independent of the pilot, based on input received from sensors. So-called soft real time simply refers to an inability to make absolute guarantees (in other words, not real time).

A real-time operating system (RTOS) is able to guarantee absolute worst-case execution times (WCET) for services, such as handling peripheral interrupts. Thus, a real-time embedded system must be controlled by an RTOS. To be a true RTOS, the kernel's operation must be completely deterministic: the WCET for every kernel service must be bounded and computable.

Building operating systems to be fully deterministic is difficult. As a case in point, let's look at how operating system kernels handle interrupts resulting from the asynchronous arrival of I/O messages, timer expirations, and other events. Interrupts can occur at any time, including when the kernel is in the middle of some critical operation. When a process executes a system call, and the kernel is modifying internal data structures to accomplish the request, interrupt preemption is disabled by the kernel to ensure that another process cannot be context switched in to execute a system call that could access these internal data structures while they are in an inconsistent state. This uninterruptible code sequence is called a *critical section.* The kernel can claim determinism only by being able to guarantee a computable WCET for every critical section.

Even simple, self-described real-time operating systems can disable interrupt processing in hundreds of locations. In one study of a real-time operating system's interrupt latency,[6] researchers estimated that the kernel contained approximately 1,200 critical sections, of which the researchers were able to statically compute the WCET for approximately half, leaving approximately 600 disabling sequences that have no known statically computed WCET. Of the computable sequences, the researchers discovered triply nested loops with estimated WCET of more than 25,000 cycles.

The RTOS analyzed in this study is far less complicated than a general-purpose operating system such as Windows or Linux. Despite significant efforts to reduce typical latencies for many operations, the ability of these monolithic kernels to guarantee determinism remains infeasible.

Why does any of this matter for security? The most obvious impact of determinism is the ability of a real-time embedded system to assure the integrity and availability of its function by avoiding missed deadlines that can lead to system failures. However, only a relatively small subset of all embedded systems has hard real-time deadlines.

Key Point

Determinism is required to enforce secure time partitioning.

Security-critical applications must not be starved of execution time by malicious or errant applications. If a malicious application discovers a weakness in which a certain operating system service call takes an arbitrarily long time to execute in certain circumstances, then this service call can be abused by the malicious application to delay critical applications. Furthermore, variations in service call execution times can be used as a *covert* or *side timing channel*, examples of sophisticated security attacks.

A side channel is a mechanism by which a malicious entity can infer sensitive information about a critical process simply by observing physical variations caused by that process. Observable physical variations include timing (timing channels), power consumption, electromagnetic emanation, and temperature. Side channel weaknesses, threats, and countermeasures are frequently analyzed in cryptosystems, which we discuss further in Chapter 4.

A covert channel also exploits physical variations but assumes two or more colluders use these effects to create an unauthorized communications channel for themselves. A covert timing channel uses variations in timing as the means for communication. Suppose that two isolated but colluding applications, Alice and Bob, are each given 10 milliseconds of execution time in a repeating 20-millisecond schedule (see Figure 2.8). It is the operating system's job to ensure that Alice and Bob get exactly 10 milliseconds of execution time. The kernel will set a timer to expire every 10 milliseconds and context switch Alice and Bob when the timer interrupt fires.

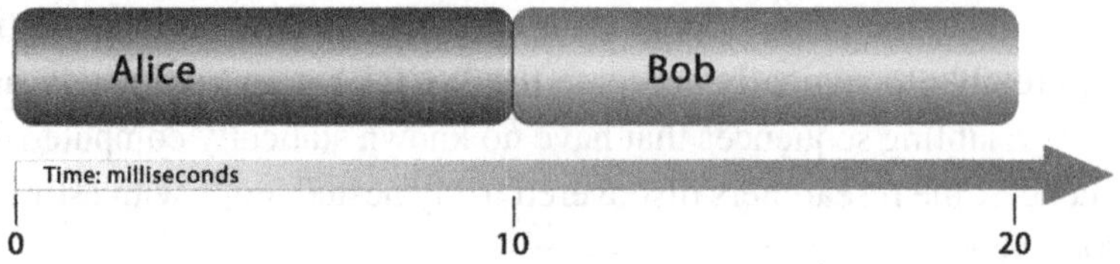

Figure 2.8:
Simple partition schedule.

If Alice is able to temporarily increase its execution time above 10 milliseconds by delaying the context switch, then Bob will discern its loss of execution time. Thus, Alice can transmit a bit of information to Bob for each transit through the schedule: one for a 10-millisecond execution window and zero for a sub-10-millisecond window. This is a high-bandwidth covert channel: a 128-bit private encryption key is disclosed in under three seconds.

Alice can use an operating system service call as the means by which to delay the context switch. The service call acts as a critical section delaying the scheduler from switching to Bob. Thus, Alice can execute a system call near the end of its execution window to accomplish this delay (see Figure 2.9).

The obvious countermeasure to this timing attack is to place a time buffer between Alice and Bob equal to the length of the longest possible service call. Thus, regardless of what system call it attempts, Alice cannot delay the start of Bob's time partition.

This is why determinism is security relevant. If Alice can find a system call that, under certain circumstances, executes longer than expected, then Alice can use that system call as a covert timing channel. Not all embedded systems need to worry about the threat of timing channels, but those that do should obtain a WCET analysis from the operating system vendor. In addition, some security evaluations and certifications require a rigorous analysis of covert channels from all applicable physical sources within the embedded system. By detecting and measuring covert channel bandwidth, designers can determine the need for and/or feasibility of specific countermeasures.

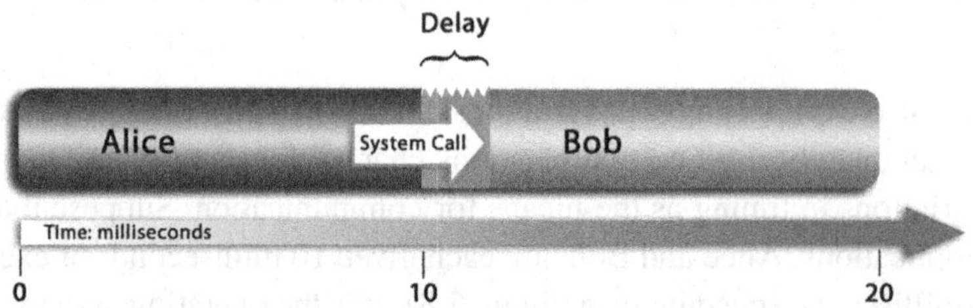

Figure 2.9:
Malicious use of operating system service calls to perpetrate a timing attack between partitions.

2.4.7 Secure Scheduling

Embedded systems developers familiar with real-time operating systems will note that the scheduling approach described in the preceding section is unlike the typical priority-based preemptive scheduler that uses developer-assigned priorities for Alice and Bob. Suppose Alice is assigned priority 100, higher than Bob's priority of 50. As long as Alice continues to execute, Alice runs ahead of Bob (assuming a single processing core). Priority-based scheduling fails to provide secure time partitioning: Alice can easily starve Bob of execution time by spinning, and Alice and Bob can modulate their CPU utilization to trivially form a high-bandwidth covert timing channel.

The execution window approach described in the previous section is known as *partition scheduling*. Depending on the implementation, a time partition may include a single process or a collection of related processes that can safely share a portion of CPU time. The operating system may permit the system designer to use numeric priorities or other scheduling parameters to control how the time window is spread across the partition's constituent processes or threads (see Figure 2.10).

It may also be possible to assign certain trusted processes into more than one execution window.

There are numerous forms of partition scheduling; from a security perspective, the important consideration is that untrusted processes not be able to modify the system-defined schedule. Secure time partitioning is a critical feature of separation kernels; not all microkernels provide secure time partitioning.

The problem inherent in typical RTOS thread schedulers is that they are ignorant of the process or address space in which the threads reside. Suppose that Alice executes in a statistics

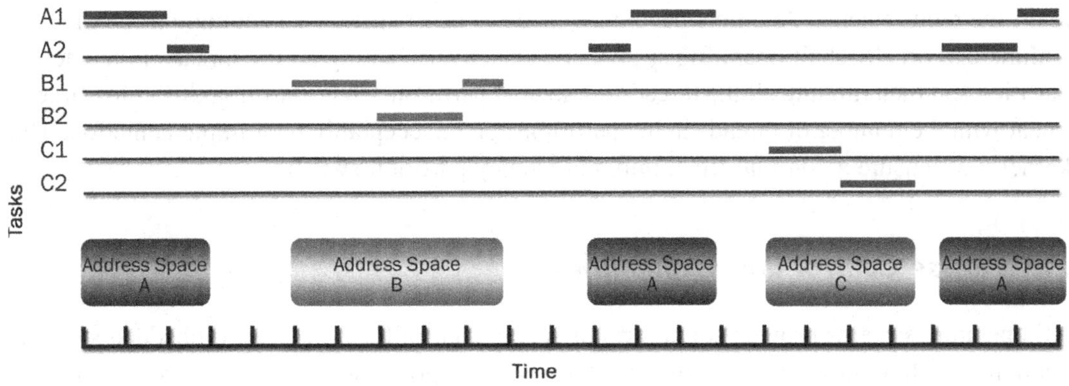

Figure 2.10:
Partition scheduling.

gathering process while critical thread Bob executes in a call processing process. The two applications are partitioned and protected in the space domain, but not in the time domain. Designers of secure systems require the ability to guarantee that the runtime characteristics of the statistics gathering application cannot possibly affect the runtime characteristics of the call processing system. Thread schedulers simply cannot make this guarantee. Let's consider a situation in which Bob normally gets all the runtime it needs by making it higher priority than Alice or any of the other threads in the statistics gathering application. Due to a bug or poor design or improper testing, Bob may lower its own priority (the ability to do so is available with practically all thread schedulers), causing the thread in the statistics gathering application to gain control of the processor's runtime. Similarly, Alice may raise its priority above the priority of Bob with the same effect. The only way to guarantee that the threads in different criticality address spaces cannot affect each other is to provide an address space level, or partition, scheduler.

Designers of safety critical software have known this requirement for a long time. The partition-scheduling concept is a major part of ARINC Specification 653, an Avionics Application Software Standard Interface. The ARINC 653 partition scheduler executes time partitions according to a timeline established by the system designer. Each address space is provided one or more windows of execution within the repeating timeline. During each window, all the threads in the other partitions are not runnable; only the threads within the currently partition are runnable (and typically are scheduled according to the standard thread scheduling rules). When the call processing application's window is active, its processing resource is guaranteed; the statistics gathering application cannot run and take away processing time from the critical application. Although not specified in ARINC 653, a prudent addition to the implementation is to provide the concept of a background partition. When no runnable threads exist within the active partition, the partition scheduler should be able to run background threads, if any, in the background partition instead of idling. A sample background thread might be a low-priority diagnostic agent that runs occasionally but does not have hard real-time requirements. Attempts have been made to add partition scheduling on top of commercial off-the-shelf operating systems by selectively halting all the threads in the active partition and then running all the threads in the next partition. Thus, partition-switching time is linear with the number of threads in the partitions: an unacceptably poor implementation. The kernel must ensure a constant time, minimal latency partition switch.

2.5 Access Control and Capabilities

All operating systems provide some form of access control for processes. Embedded systems contain a plethora of resources: communications, files, processes, various devices, memory, and more. An important goal of security design is to ensure that applications are able to access the resources they need and are disallowed to access resources they do not need.

Key Point

Many security problems are caused by poor access control architecture and/or implementation within the operating system or improper use of access control facilities by the embedded system designer.

At a coarse level, access control policies can be divided into two classes: *discretionary access control* (DAC) and *mandatory access control* (MAC). Both have their place in secure embedded systems design.

An example of a DAC is a UNIX file: a process or thread can, at its sole discretion, modify the permissions on a file it owns, thereby permitting access to the file by another process in the system. DACs are useful for some kinds of objects in some kinds of systems. But an operating system must go one big step further for security and provide MAC of critical system objects. A MAC is managed by the system and cannot be modified by users or processes. For example, let's consider a communications device managed by a call processing application. The system designer must be able to configure the system such that the call processing application, and only the call processing application, has access to this device. Another application in the system cannot dynamically request and obtain access to this device. And the call processing program cannot dynamically provide access to the device to any other application in the system. The access control is enforced by the kernel, is non-bypassable by application code, and is thus mandatory. Mandatory access control provides guarantees. Discretionary access controls are only as effective as the applications using them.

The combination of mandatory and discretionary access controls is often required to implement sophisticated embedded systems. For example, mandatory access controls may be used to divide system resources across multiple security domains such that no application within a security domain may access unauthorized resources. However, within a security domain, discretionary access controls may be used to provide a more flexible sharing of resources.

DACs and MACs can come in many flavors; access control decisions may be based on a practically infinite combination of subject (user/process) and object (device) attributes, roles, and states.

2.5.1 Case Study: Secure Web Browser

Access control enforces the principle of least authority over the processes and resources in an embedded system. When high-profile security vulnerabilities caused by software flaws in common operating systems such as Windows are published, we often make the false assumption that the software vulnerability deserves all the blame. In truth, it is the system's lack of access authority restriction that often is the culprit. A software flaw, properly contained, may have no security impact whatsoever. Sadly, vulnerabilities in an application are often exploited by attackers to gain access to resources that the application should never have had in the first place.

Let's look at the example of a typical web browser, such as Firefox or Chrome. The browser is a user-interactive application and therefore needs access to the graphical display and keyboard and mouse inputs (both of which may be provided via access to a system window manager). The browser needs access to the file system to save files downloaded from the network or to load files for browsing. The browser needs executable access to its own private application code and data. The browser needs to be able to launch and/or communicate with a large variety of plug-ins, such as HTML5 and JPEG renderers. Browser plug-ins are usually implemented using dynamic link libraries (DLLs). DLLs allow an application to gain dynamic, programmatic access to external functionality by loading that functionality directly into the calling application's address space. DLLs are easy to configure, launch, and distribute.

Unfortunately, DLLs—at least the way they are implemented on most operating systems—are the quintessential example of poor access control privilege management. DLLs execute within the calling application's environment and therefore usually have the same access control permissions as has the user. Thus, a plug-in that the user does not trust can launch other programs, access any of the user's files, access any of the browser's memory and other resources, access the networks, and so on. Furthermore, many browser plug-ins are large, sophisticated software programs, riddled with bugs.

A buffer overflow vulnerability in a plug-in provides the potential for an attacker's code to be executed within the user's environment. Such attacks occur regularly in the wild. Popular applications that have been exploited include Adobe Flash, Adobe Acrobat, and JPEG renderers. A common attack vector is to craft a malicious Flash, PDF, or JPEG file that, when browsed over the web, causes a buffer overflow during the rendering process and malicious code to be injected as a result.

Building a secure browser immune to such problems requires a lot more than the use of access control, but let's start there (in Chapter 3 we discuss more about how to build secure software). Why should the JPEG renderer have access to a file system, network, or any other application resources beyond that needed to decode JPEG? Let's assume for now that the JPEG renderer could be implemented as a process whose access to computer resources could be regulated independently of the user or the user's browser application. The input to the JPEG process is a stream of bits representing a JPEG compressed input file. The output of the JPEG process is the stream of bits representing the decompressed image. The only resources required by the JPEG process is a little RAM, some processing time, and a communication mechanism by which the browser provides the compressed input and receives the decompressed output.

The JPEG process should be disallowed access to anything else—no files, networks, or even an ability to launch other programs.

Now let's assume that the JPEG process is chock full of bugs, including the same sorts of buffer overflow flaws that are the bane to web users the world over. Flaws in the JPEG application may enable a maliciously crafted JPEG image to overflow the JPEG process's runtime stack, injecting malicious code. The code may enable the attacker to corrupt, disable, or even completely subjugate the JPEG process to its bidding. However, the attacker cannot access the user's files; launch other programs that could cause damage; or access any other resources beyond the aforementioned, limited scope of the JPEG process itself. Access control engineering dramatically limits damage resulting from the buffer overflow.

Key Point

The biggest reason buffer overflows are so damaging is that insecure operating systems and software development techniques promote the use of ambient authority.

As seen in the insecure web browser example, the plug-in's authority is ambient because it is inherited from the user's encompassing environment. For convenience, developers architect the browser so that a plug-in can easily perform functions and access resources on behalf of the user. It is this excessive authority that results in widespread damage from what should be innocuous programming flaws.

2.5.2 Granularity versus Simplicity of Access Controls

Key Point

One of the biggest challenges embedded system designers face with respect to access control security policy is finding the proper balance between the granularity of policy and the maintainability and assurance of policy.

A policy that is overly simple may provide excessive privilege to processes across the system. A policy that is overly complex may, ironically, introduce security vulnerabilities. The policy specification is part of the TCB, and like the source code within the TCB, must be crafted with care.

The modern Linux access control policy framework is called SELinux (Security-Enhanced Linux). A USENIX research paper reported that the Linux 2.4.19 SELinux sample policy consists of more than 50,000 policy statements and 100,000 permission assignments across 700 subject types.[7] This kind of enormous specification is simply impractical to assure and maintain. The researchers state, "We believe that size and complexity of the SELinux example policy make it impractical to expect that typical administrators can customize it to ensure protection of their Trusted Computing Base (TCB) and to satisfy their site's security goals on this TCB." The excessive complexity of SELinux policies should not be construed as a criticism of the access control framework upon which the policy is built; rather, it is a product of the complexity of the Linux TCB itself. Nevertheless, embedded designers must be cognizant of the need to implement manageable, provable policies.

To further explore this important topic, let's look at access control for file systems. The simplest access control policy for a file system is to allow unfettered file system access to any process that deals at all with files. Any process that does not need file services is disallowed any access to the file system. Suppose an embedded system has processes Alice and Bob, and Alice uses a file system for system event logging and Bob is a real-time process that has no need for files. The access control policy is simple:

Process	File System Access
Alice	Yes
Bob	No

Now suppose that a third process, Charlie, requires file system access, but never to the same files used by Alice. If the file system security policy includes directory-level granularity, then the designer can specify isolated sandbox directories, A and C, in which Alice and Charlie can freely create, read, and write their private files:

Process	File System Access	A	C
Alice	Yes	Yes	No
Bob	No	No	No
Charlie	Yes	No	Yes

Now suppose that a fourth process, David, is a health monitor that periodically reads the status files written by Alice. This implies that the security policy must provide not only file-level granularity, but also individual access-type permissions granularity for files:

Process	File System Access	A	C	A/Status Read	A/Status Write
Alice	Yes	Yes	No	No	Yes
Bob	No	No	No	No	No
Charlie	Yes	No	Yes	No	No
David	Yes	Yes	No	Yes	No

Now suppose that Alice is trusted to append status messages to the Status file, but the system designer does not want to give Alice the right to delete or truncate the Status file. This is not unusual; audit logging is often considered a sensitive operation in which access to the audit log is strictly controlled to reduce the possibility of corruption or tampering. The access control parameters may be extended to include any specific operation type that can be performed on files:

Process	File System Access	A	C	A/Status Read	A/Status Write	A/Status Append	A/Status Create	A/Status Trunc	A/Status Seek	A/Status Delete
Alice	Yes	Yes	No	No	Yes	Yes	No	No	No	No
Bob	No	No	No	No	No	No	No	No	No	No
Charlie	Yes	No	Yes	No	No	No	No	No	No	No
David	Yes	Yes	No	Yes	No	No	No	No	Yes	No

These tables imply a flat policy encoded with a single bit (yes/no) for each access right. Instead, the access control system may be implemented hierarchically such that access to the file system is required before any requests of the file system can be made. Then, access to a specific directory is checked before any requests for files within the directory can be made. And so forth. Similarly, instead of binary configuration, the access control structure may be implemented as lists, lists of tuples, and so on; when access to a file is requested, the system may check a list of valid process and operation tuples.

These tables cover only file system access. In an embedded system, the types of subjects and objects over which an access control policy may need to preside include communications devices, communications channels to other processes, specific types of memory, semaphores, timers, and so on. It is easy to see how an access control policy can quickly become too complicated to effectively manage.

Ultimately, it is the system designer who is charged with the responsibility of finding the happy medium for security policy sophistication. The goal should be to find the simplest policy that provides the necessary privilege limitation across the system. In practice, most resources will have a single owner for the life of the resource. Ideally, the security policy for such resources should be encoded with a single entry, whether a bit in a table, XML element, or whatever representation the operating system and/or resource manager employs.

2.5.3 Whitelists versus Blacklists

Simplicity of access control policy is critical in gaining assurance that the policy is complete and correct. Given an object, O, and N processes in the system—P_1, P_2, P_3, …, P_N, of which $M < N$ require access to O—there are two ways to specify an access control list of the M processes authorized to access O. The first method, a *blacklist*, associates with O a list of the N−M processes that are not authorized to access O. In contrast, a *whitelist* associates with

O a list of the M processes that are permitted to access O. When access to O is requested by process P, the security enforcement agent (could be the kernel or a process trusted to manage access to O) checks the access control list to see if P is authorized for O. Alternatively, the access control lists can be tied to subjects instead of objects. In other words, a whitelist for process P includes those objects for which P is authorized.

> **Key Point**
>
> For most objects and resources, whitelists are preferable to blacklists due to the former's tendency toward reduced privilege implementation and design.

Blacklists can introduce vulnerabilities by accidentally permitting excessive authority. Suppose a system event causes a new process, Q, to be created to handle the event. If object O's access is controlled by a blacklist, then Q is automatically provided access to O whether or not Q needs it. In contrast, if access to O is controlled by a whitelist, then Q is not automatically authorized for O. The system designer must take explicit action to add Q to the whitelist for O if Q has a need for it.

Anti-malware packages are essentially a form of blacklist. When a program is introduced into a system, the anti-malware process compares the program to a list of known malicious programs or looks for known malware code patterns. The problem, of course, is that the universe of undiscovered (including newly created) malware passes the test. Increasingly, security-minded designers are looking to whitelists to solve this problem. The embedded system can be shipped with a list of known good programs that can be executed. Any program that fails to match something on the whitelist is rejected. The comparison may include a hash check to verify the integrity of whitelisted programs.

Alternatively, the embedded system can enforce a policy that only loads trusted programs, where trust is established through digital signatures. The system owner must cryptographically sign a loadable program a priori, and the embedded loader verifies the signature before accepting the program's adoption into the system. The signature represents the owner's assertion that this specific collection of bits is safe to deploy. Of course, this mechanism implies that the system owner (or an authority delegated to make such decisions) does not sign a fraudulent or malicious program. Digital signature technology for authentication is discussed in Chapter 4.

Blacklists should be used only for situations in which there is no practical alternative. A notable example is the use of revocation lists for communications endpoints on large, rapidly evolving networks (e.g., the Internet). When a client wants to verify the validity of a server, it is simply impractical to check against a whitelist of valid servers. When a server certificate is deemed invalid prior to its natural expiration date (a relatively infrequent occurrence), the certificate is placed on a revocation list that is checked during communication session establishment.

2.5.4 Confused Deputy Problem

Earlier in this chapter, we studied the concept of a web browser made more secure using access control engineering. The goal was to limit the authority of browser plug-ins relative to the default ambient authority of the user that is automatically delegated to the plug-in when it is launched. Excess privilege through implicit delegation is the cause of another common access control problem, called the *confused deputy*, as coined by Norman Hardy.[8]

The confused deputy is an application serving some client applications. The server holds multiple authorities, each of which is used to access resources for disparate purposes. For example, the server has the authority to access an audit log (with a predetermined pathname within the file system) that records the server's actions. In addition, the server has the authority to write an output file whose name is specified by a client in its request to the server. The server becomes a confused deputy when the client specifies an output file whose pathname matches that of the server's audit log. Because it has the authority to write its own audit log, the server fails to detect that the authority is being misused, and the audit log is corrupted. The server was never designed to handle the possible erroneous (or malicious) attempt by the client to specify an output file that happens to match the name of the audit log.

The confused deputy is an example of *privilege escalation*: even though the client should have the right only to invoke the server to process the client's output file, the confused deputy enables the client to gain an illicit authority to overwrite the server's audit log.

While there are several possible approaches to solve this specific test case, an effective general solution is to bind an object with a specific type of privilege and force the subjects (clients and servers) to deal only with these object-privilege tuples. In the preceding example, the server would be configured with the "audit log" authority that provides access to the audit log, and the client would be configured with a user-specific authority that provides access to the user's output file. The client passes this authority reference into the server upon service invocation. The server uses this authority to write the user's output file. Because the operating system enforces all object references to be made through these privilege tuples, there simply is no defined mechanism for a client to refer to the audit log authority. A properly constrained client would never be imbued with this privilege. The beauty of this approach is that the server need not be encumbered with extra code to track distinct access permissions for each required file access.

2.5.5 Capabilities versus Access Control Lists

The concept of an object reference that provides implicit access to the object is called an object *capability*. With traditional access control lists, an object is referenced with a descriptor (e.g., UNIX file descriptor), and separate metadata must be maintained to track the access rights for that object.

Key Point

A capability acts as both the mechanism for access as well as the access right: there simply is no way for a subject to access an object if the subject does not possess the object capability.

Capabilities are simple and efficient and promote least privilege secure design. In microkernel-based operating systems that employ capabilities, IPC endpoints are often the most important form of capability because IPC is used by clients to request services, either from the kernel or from user-mode service providers such as file systems and network stacks. If a process owns an IPC connection to the file system, then this connection provides the privilege to request services from the file system. If a process does not possess an IPC connection to the file system, then the process is unable to request any file system services (see Figure 2.11).

The IPC capability is also completely scalable in terms of the granularity of access it can control. In the preceding example, the file system's capability granularity is coarse: a process either has complete access or not. An alternative would be to use separate capabilities for each file system volume. Clients with a volume capability are provided access to only that specific volume and its constituent directories and files, and the file server can use these capabilities to partition the media resources across clients. For finer-grained control, the file system could use capabilities for each individual directory and file or even for individual access permissions (read, write, execute, etc.) for each file. While increased granularity provides for reduced privilege, the choice of capability management scheme often has profound effects upon system complexity. And as we have discussed, such excessive complexity may adversely impact system security.

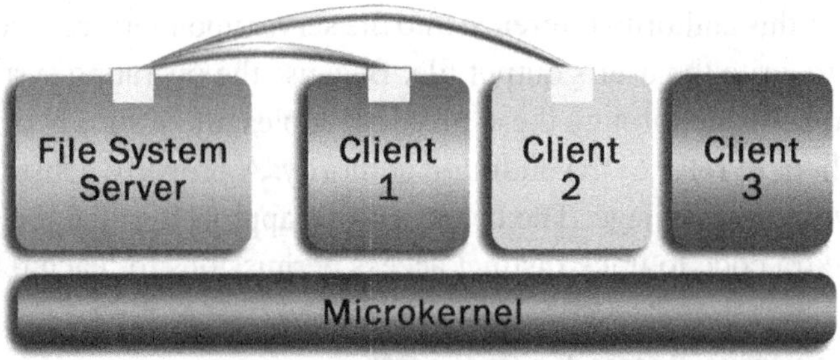

Figure 2.11:
Clients 1 and 2 each own a capability for file system services; client 3 is unable to request any file system services.

In most cases, the operating system provider will provide a default capability scheme for any middleware servers, such as file systems, that are provided along with the operating system. For user-created application servers, IPC capabilities can be used as the designer sees fit. The operating system provider will often offer consultative services to help ensure the best capability model for least privilege relative to complexity.

2.5.5.1 Case Study: MLS Resource Manager

It is not uncommon for a capability-based operating system to support access control lists in addition to capabilities or in combination with capabilities for some services. For example, let's consider a multi-level secure (MLS) resource management application that manages a set of memory pages at varying security levels. A client process is permitted read access to a page if and only if the client's security level dominates (is greater than or equal to) the page's security level. A client is permitted write access to a page if and only if the client's security level is dominated by (is less than or equal to) the page's security level. We also make the assumption that the system is configured such that clients are able to communicate only with other clients at the same security level or with the MLS resource manager; a client is therefore unable to transfer capabilities to clients at higher or lower security levels.

To start, each client that is allowed to make requests of the MLS resource manager is allocated a communications capability to it (see Figure 2.12).

When a client requests read access to a page (e.g., by address or name, depending on how the system is implemented), the resource manager consults its internal access control list of managed pages. In this example, the access control list consists of tuples of a page capability and its associated security level label. Other policies, such as role-based access control, would require additional attributes within the access control tables.

If the client's security level dominates the page's security level, then the resource manager approves the request and uses its communications capability to the client to transfer to it a new capability providing read access to the requested memory page (see Figure 2.13).

The client can now use the newly received capability to perform as many direct reads of various page contents as necessary. The capability provides superior efficiency relative to a pure access control implementation in which each individual access to page content would need to be validated against an access control list by the resource manager.

This example demonstrates that practically any access control policy can be implemented on top of capabilities. Furthermore, this example illustrates how capabilities can be used to provide both mandatory and discretionary access controls. The capability model is, at its root, a mandatory access control system, because all object ownership and access relationships are enforced by the microkernel. The system designer can create a completely static ownership relationship between processes and their own objects. In such a design, no discretionary access

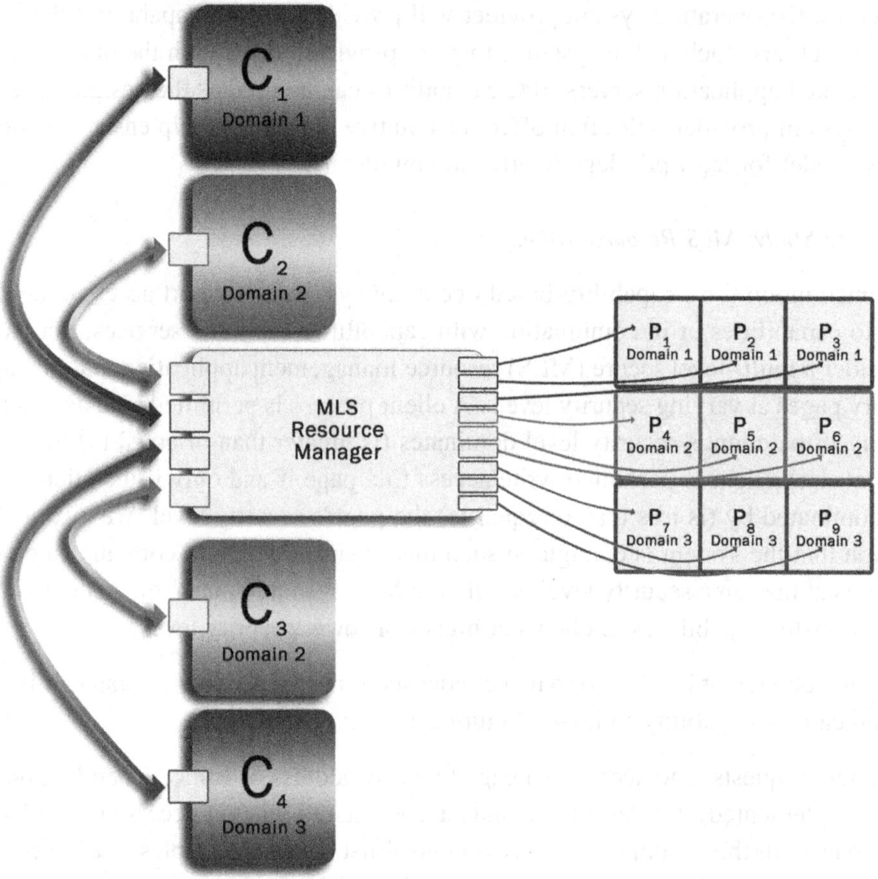

Figure 2.12:
Clients with capabilities to MLS resource manager for requesting memory page capabilities.

controls are present. However, if more flexibility is needed, a trusted server can be designed and imbued with capabilities for which it can manage its own client access control policy (along with capabilities to communicate with those clients). Thus, the capability model can support arbitrary discretionary access control systems on top of, or in addition to, mandatory system-specified controls.

Let's now take this case study a step further and consider the scenario in which the memory page resource manager is tasked with distributing its pages to other servers that will manage a subset of the master resource manager's pages. For example, the system may implement a dynamic link library management service. When a client wishes to load a DLL, the master resource manager provides the DLL manager read/write access to a set of pages that will fit the contents of the DLL. The DLL manager will then provide read-only capabilities of these pages to clients.

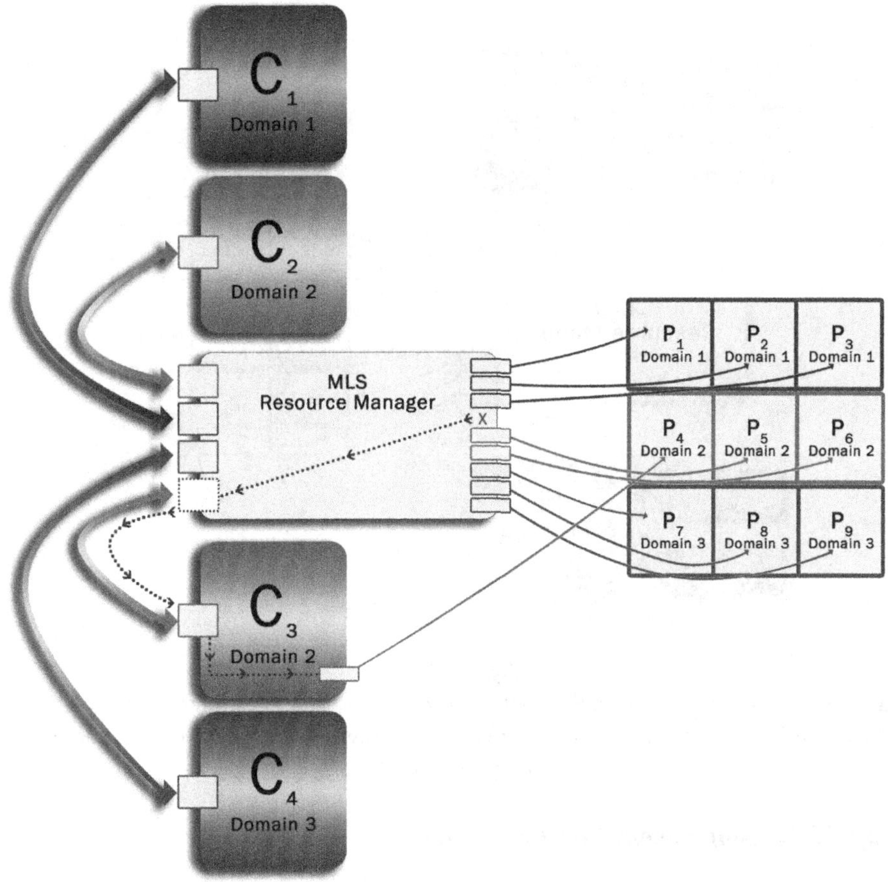

Figure 2.13:
Resource manager transferring a memory page capability to a client.

Of course, the DLL manager may also implement an independent security policy for access to specific DLLs. Separate DLL managers can even be instantiated to manage individual DLLs or sets of DLLs (see Figure 2.14). For example, the system may utilize a distinct DLL manager instance for each security domain. Each DLL manager is trusted to properly distribute its pages to clients, but the DLL manager's overall rights are strictly limited to only those sets of pages provided by the master resource manager. In this manner, capabilities can flow across multiple intermediaries, providing excellent flexibility in system design, while at the same time providing authority minimization for improved security. This combination of flexible, privilege-limited capability distribution occurs often in sophisticated systems. Another example is access to shared network buffers as they flow across the various layers in a communications stack: from network interface driver, to the TCP/IP layer, to TCP/IP clients such as individual processes making socket API calls, HTTP clients/servers, and FTP clients/servers.

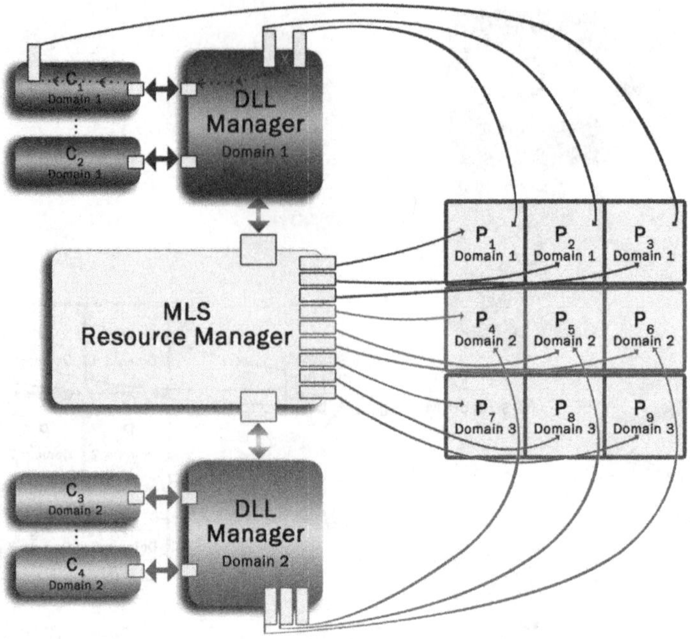

Figure 2.14:
Separate capability domain example: each DLL manager manages memory page capabilities for clients within a security domain on behalf of a higher-level multi-level secure memory page manager.

2.5.6 Capability Confinement and Revocation

One potential advantage that access control lists have over capabilities is the colocation of policy with the designation of authority embodied within the policy. Access control requests come into a server, and the server also contains the access control list for making the access control decision. The access control list is the single repository of policy information relating to that service. If a client's access privilege to a resource within a server must be revoked, then the system designer knows that removing the client from the server's access list will suffice.

In contrast, capabilities do not reside in a single place. Indeed, the ability to distribute capabilities is an important flexibility advantage over access control lists. But what if a server does not want a capability to extend beyond the client to whom access is granted? How can the server or the system designer be assured that this capability is never commuted to other clients? And if a client's authority must be revoked, how is this accomplished when the capability has been distributed beyond the server or even beyond the original client?

In our confused deputy example, the server is entrusted to properly manage the user's output file capability. If, at some point, the system security administrator decides that the server can no longer be trusted, the user is out of luck; the delegated capability is sitting there for the server to abuse. Furthermore, since the server can communicate with other clients, it can maliciously transfer the client's output file capability to other clients.

> **Key Point**
>
> Distributed capability systems point to a need for the operating system to provide a means by which capabilities can be confined within privilege domains as well as a means for revocation.

In a typical UNIX system, processes can easily communicate with other processes using pipes, files, sockets, and other mechanisms. File descriptors can be shared almost as easily as raw data. Because the default policy is to allow communication, a capability system added to UNIX would lack the desirable confinement property. In contrast, confinement is implicitly provided by the microkernel's system of protected processes and ownership of capabilities within each process. Newly created processes lack any capabilities. This forces the system designer to carefully consider *confinement domains*. If Alice lacks communication capability to any other process, then Alice's capabilities are confined to Alice. If Alice and Bob can communicate capabilities between each other, but neither has a connection to any other process, then Alice and Bob form a confinement domain for the union of their capabilities. In addition, the microkernel may provide mechanisms by which processes can communicate (e.g., message passing, synchronization primitives) but cannot transfer capabilities.

To illustrate the need for revocation, let's reconsider the earlier case study of a memory page manager and its subservient DLL managers. Suppose that the master resource manager decides that it must reclaim the memory pages assigned to one of the DLL managers. Also suppose that the DLL manager has distributed read capabilities for these pages to a large set of its clients to be able to provide shared library services. Revocation now implies a requirement to remove the read capabilities from all the affected client processes as well as the read/write capabilities from the DLL manager itself. Furthermore, the master resource manager has no knowledge of which clients were afforded read access. Thus, a scheme that attempts to record all distributions of a capability for future revocation is simply impractical.

The microkernel capability system must provide a simple mechanism for a capability's original owner (in this case, the master resource manager) to revoke that capability, regardless of how it has been propagated across client processes. One solution to this problem is reminiscent of how a C++ function passes read access to its defined objects to other functions. Instead of passing the original object, the function passes a reference to the object. Similarly, a process can distribute references to capabilities. These references are used as access descriptors in the same way as the original object. However, when the owning process issues a revoke operation on the original object, all outstanding references are automatically disabled (see Figure 2.15). The linkages between the references and the original object are stored by the microkernel.

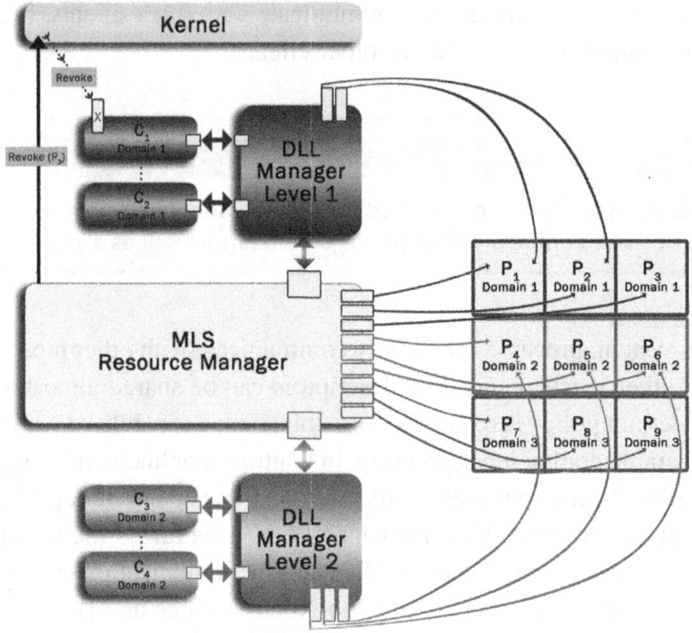

Figure 2.15:
Capability revocation example: a high-level resource manager efficiently revokes/disables memory page capability that has been distributed to a second-level server and then onto its client.

The developer only needs to know to use revocable references instead of transferring the original object.

2.5.7 Secure Design Using Capabilities

Capabilities provide a compellingly simple way to promote secure design habits. Because the capability embodies both the designation of a resource as well as its access authority, software architecture diagrams simultaneously depict the functional interactions as well as the access rights between subjects and objects. In contrast, a purely access-control-list-based system's architectural design showing the connectivity between subjects and objects would fail to provide a complete picture of security policy with respect to object access privileges. The access control lists of each resource manager in the system would need to be added to the picture.

To illustrate, let's revisit our earlier example of the DLL manager that owns a capability to a writeable memory region containing library code and the DLL manager's clients that are provided read-only access to this memory region if and only if they are so authorized by the system designer. Capability-based operating system vendors often provide a graphical tool for defining the interactions between processes and the resources to which they require access.

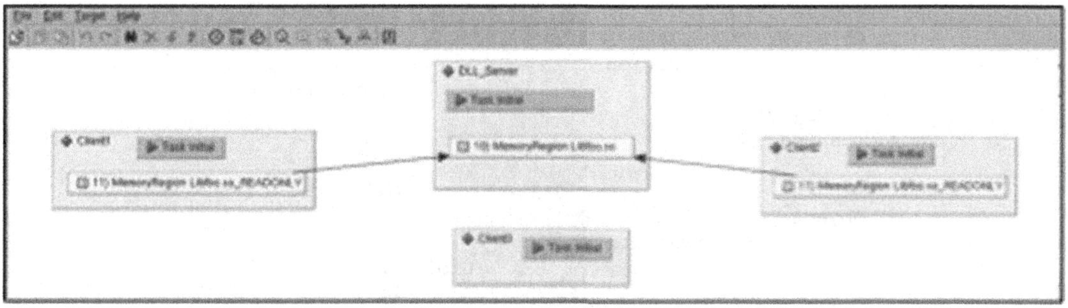

Figure 2.16:
Secure design with capabilities.

Figure 2.16 shows one such tool, *Integrate*, used with the INTEGRITY operating system from Green Hills Software. In this screenshot, the DLL server owns the read/write capability to the memory region corresponding to a library, *Libfoo.so*. The "MemoryRegion Libfoo.so" box within the DLL_Server process box reflects this ownership. Two other processes in the system, *Client1* and *Client2*, are defined to own a read-only capability referencing the memory region. Arrows connecting the authorized Clients' MemoryRegions back to the original object reflect this relationship. Note that process *Client3* lacks such a capability and thus has no authority or mechanism to access the shared library. Thus, a simple architectural diagram conveys the functionality and authority of clients to access a shared library provided by a server.

The Integrate tool stores this information in an ASCII format file (see Figure 2.17) so that system designers who prefer can examine or post-process the security relationships within their favorite editors or utilities. This file is read in directly by the INTEGRITY build system and incorporated into the final image binary itself.

Object capabilities represent the state of the art in operating system access control design. Capabilities are powerful and efficient. They promote least privilege, component-based software architectures. And capabilities are easy for both developers and security administrators to understand. Capabilities can be used to manage a small set of resources with single, exclusive owners all the way up to the most sophisticated policy logic schemes.

2.6 Hypervisors and System Virtualization

The operating system has long played a critical role in embedded systems. A prime historical purpose of the operating system is to simplify the life of electronic product developers, freeing them to focus on differentiation. The operating system fulfills this mission by abstracting the hardware resources—RAM and storage, connectivity peripherals such as USB and Ethernet, and human interface devices such as touchscreens. The abstraction is presented to the developer in the form of convenient APIs and mechanisms for interacting with the hardware

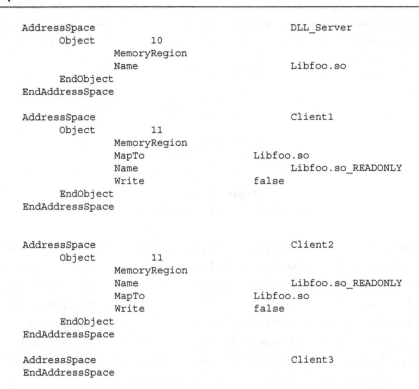

```
AddressSpace                                DLL_Server
      Object        10
               MemoryRegion
               Name                         Libfoo.so
      EndObject
EndAddressSpace

AddressSpace                                Client1
      Object        11
               MemoryRegion
               MapTo                  Libfoo.so
               Name                         Libfoo.so_READONLY
               Write                  false
      EndObject
EndAddressSpace


AddressSpace                                Client2
      Object        11
               MemoryRegion
               Name                         Libfoo.so_READONLY
               MapTo                  Libfoo.so
               Write                  false
      EndObject
EndAddressSpace

AddressSpace                                Client3
EndAddressSpace
```

Figure 2.17:
ASCII format system design representation.

and managing application workloads. The operating system is coupled with a development environment—compilers, debuggers, editors, performance analyzers, and so on—that help engineers build their powerful applications quickly and to take maximum advantage of the operating system.

Of course, this environment has grown dramatically up the stack. Instead of just providing a TCP/IP stack, an embedded operating system must sometimes provide a full suite of application-level protocols and services such as FTP and web servers. Instead of a simple graphics library, the operating system may need to provide sophisticated multimedia audio and 3D-graphics frameworks. Furthermore, as embedded Systems-on-Chip (SoCs) become more capable, application and real-time workloads are being consolidated. The example discussed briefly in Chapter 1 of an automotive infotainment system integrating rear-view camera capability is representative of this trend: the operating system must provide a powerful applications environment while responding instantly to real-time events and protecting sensitive communications interfaces from corruption.

Because of its central role, the operating system has sometimes been the battleground between electronics manufacturers who aim not only to differentiate but also to protect the uniqueness of and investment in their innovations. The smartphone market is an obvious example: silicon vendors, consumer electronics manufacturers, and service providers at all levels want to control that human-machine interface, which acts as the portal for revenue, loyalty, and brand recognition.

The trend toward consolidation has posed a significant challenge to embedded operating system suppliers who must navigate the myriad of stacks, interfaces, standards, and software packages. This complexity has also fueled a trend toward open source models to reap the benefits of a massive, distributed developer base. Linux is the primary success story. However, while Linux has succeeded in gaining dramatic market share, it has also suffered from tremendous fragmentation. What embedded systems developers have realized is that they must customize and extend Linux to be able to obtain differentiation and platform control. In essence, these Linux-based efforts have become the new do-it-yourself proprietary operating system for a collection of stakeholders.

The key problem is that the typical operating system abstractions—files, devices, network communication, graphics, and threads—have begun to reach their limit of utility. Application developers and electronics suppliers who become too dependent on one operating system abstraction environment can find themselves in dire straights if the operating system fails to meet emerging requirements, runs into licensing or IP rights headwinds, or is simply surpassed by another operating system in the market.

Clearly, embedded systems developers need a platform for innovation and differentiation that is flexible enough to accommodate the inevitable rapid evolution in hardware and software technology. Imagine a platform that can run two completely different operating systems at the same time on the same microprocessor. Imagine a platform that can run the most sophisticated consumer multimedia operating system such as Android while still running a hard real-time communications stack and security-critical functions completely protected beyond the reach of Android—all on the same SoC. Imagine a platform that can meet the aforementioned automotive consolidation requirement: a multimedia automotive infotainment system that also attains a cold boot time in milliseconds to support instant-on rear-view camera (on the same screen as the main infotainment operating system) and communications over real-time automotive networks (e.g., Controller Area Network—CAN)—all on the same SoC to save on size, weight, power, and cost.

Key Point

The new level of abstraction needed to cope with increasingly sophisticated, consolidated electronic systems is the operating system itself, not just the computer's hardware resources.

Developers need the flexibility to run any operating system, not just any application, with ease, on the same hardware. The answer to the operating system dilemma is the embedded hypervisor, implementing *system virtualization*. System virtualization enables the hosting of multiple virtual computer systems on a single physical computer system. These virtual computer systems are often called *virtual machines*. The ability to virtualize an entire computer system enables multiple independent operating systems to execute concurrently on the shared hardware platform. These operating systems are often referred to as "guest" operating systems because they are permitted to share the physical hardware as guests of the software layer that owns it. This layer is called a *hypervisor*. Unlike enterprise hypervisors, the embedded hypervisor is designed specifically for embedded and mobile electronics.

The motivations for system virtualization technology in the data center are well known, including resource optimization and improved service availability. But virtualization technology has broader applications throughout the embedded world, including security-enabled mobile devices, embedded virtual security appliances, trusted financial transactions, workload consolidation, and more. This vision is made possible, in part, due to hardware-assisted virtualization technology that now scales down to embedded and mobile devices. The remainder of this chapter provides an overview of the evolution of embedded hypervisor architectures, including both software and hardware trends, and how they affect the security of system virtualization. We also discuss a range of compelling applications for secure virtualization across communities of interest.

2.6.1 Introduction to System Virtualization

Computer system virtualization was first introduced in mainframes during the 1960s and '70s. Although virtualization remained a largely untapped facility during the '80s and '90s, computer scientists have long understood many of the applications of system virtualization, including the ability to run distinct and legacy operating systems on a single hardware platform.

At the start of the millennium, VMware proved the practicality of full system virtualization, hosting unmodified, general-purpose, "guest" operating systems such as Windows, on common Intel Architecture (IA)-based hardware platforms.

In 2005, Intel launched its Virtualization Technology (Intel VT), which both simplified and accelerated virtualization. Consequently, a number of virtualization software products have emerged, alternatively called virtual machine monitors or hypervisors, with varying characteristics and goals. Similar hardware-assists for system virtualization have emerged in other popular embedded CPU architectures, including ARM and Power.

While virtualization may be best known for its application in data center server consolidation and provisioning, the technology has proliferated across desktop and laptop-class systems, and has most recently found its way into mobile and embedded environments.

Key Point

The availability of system virtualization technology across a wide range of computing platforms provides developers and technologists with the ultimate open platform: the ability to run any flavor of operating system in any combination, creating an unprecedented flexibility for deployment and usage.

Yet embedded systems often have drastically different resource and security constraints as compared to server computing. We also focus on the impact of hypervisor architecture upon these constraints.

2.6.2 Applications of System Virtualization

Mainframe virtualization was driven by some of the same applications found in today's enterprise systems. Initially, virtualization was used for time sharing, similar to the improved hardware utilization driving modern data center server consolidation. Another important usage involved testing and exploring new operating system architectures. Virtualization was also used to maintain backward compatibility of legacy versions of operating systems.

2.6.3 Environment Sandboxing

Implicit in the concept of consolidation is the premise that independent virtual machines are kept securely separated from each other (also referred to as virtual machine isolation). The ability to guarantee separation is highly dependent on the robustness of the underlying hypervisor software. As we discussed earlier and will soon expand upon, researchers have found flaws in commercial hypervisors that violate this separation assumption. Nevertheless, an important theoretical application of virtual machine compartmentalization is to isolate untrusted software. For example, a web browser connected to the Internet can be sandboxed in a virtual machine so that Internet-borne malware or browser vulnerabilities are unable to infiltrate or otherwise adversely impact the user's primary operating system environment.

2.6.4 Virtual Security Appliances

Another example, the virtual security appliance, does the opposite: sandbox, or separate, trusted software away from the embedded system's primary operating system environment. Let's consider anti-virus software that runs on a mobile device. A few years ago, the "Metal Gear" Trojan was able to propagate itself across Symbian operating system-based mobile phones by disabling their anti-malware software. Virtualization can solve this problem by placing the anti-malware software into a separate virtual machine (see Figure 2.18).

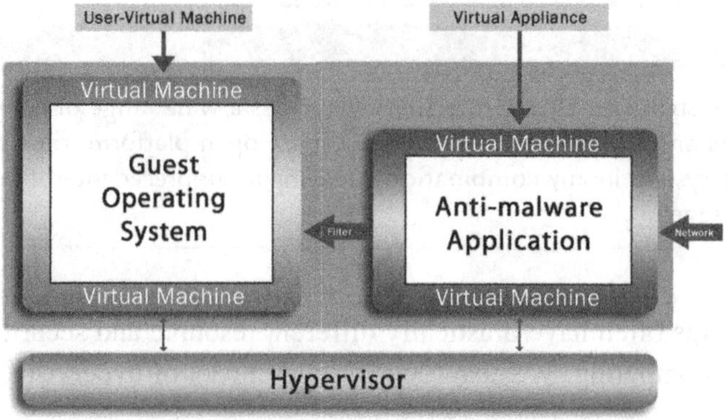

Figure 2.18:
Improved security using isolated virtual appliances.

The virtual appliance can analyze data going into and out of the primary application environment or hook into the platform operating system for demand-driven processing.

2.6.5 Hypervisor Architectures

Hypervisors are found in a variety of flavors. Some are open source; others are proprietary. Some use thin hypervisors augmented with specialized guest operating systems. Others use a monolithic hypervisor that is fully self-contained. In this section, we compare and contrast the available technologies, with an emphasis on security impact. Note that this section compares and contrasts so-called Type-1 hypervisors that run on bare metal. Type-2 hypervisors run atop a general-purpose operating system, such as Windows or Linux, that provides I/O and other services on behalf of the hypervisor (see Figure 2.19).

> **Key Point**
>
> Because they can be no more secure than their underlying general-purpose host operating systems (which are well known to be vulnerable), Type-2 hypervisors are not suitable for mission-critical deployments and have historically been avoided in such environments.

Thus, Type-2 technology is omitted from the following discussion.

2.6.5.1 Monolithic Hypervisor

Hypervisor architectures most often employ a monolithic architecture, as shown in Figure 2.20. Similar to monolithic operating systems, the monolithic hypervisor requires a large body of operating software, including device drivers and middleware, to support the execution of one or more guest environments. In addition, the monolithic architecture often uses a single instance of the virtualization component to support multiple guest environments. Thus, a single flaw in

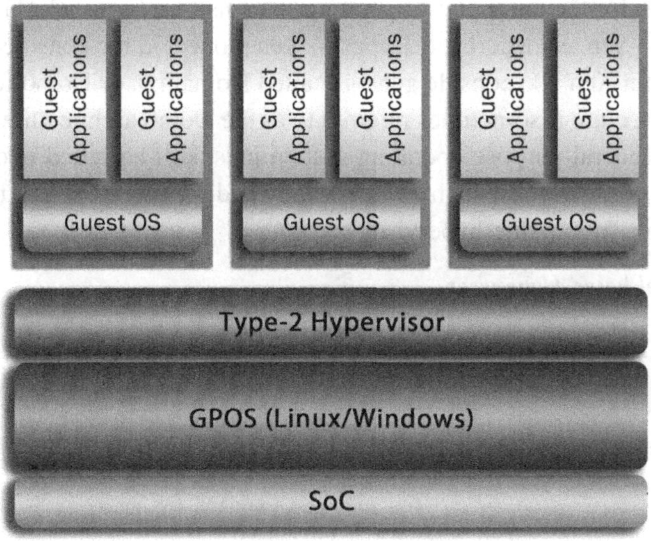

Figure 2.19:
Type-2 hypervisor architecture.

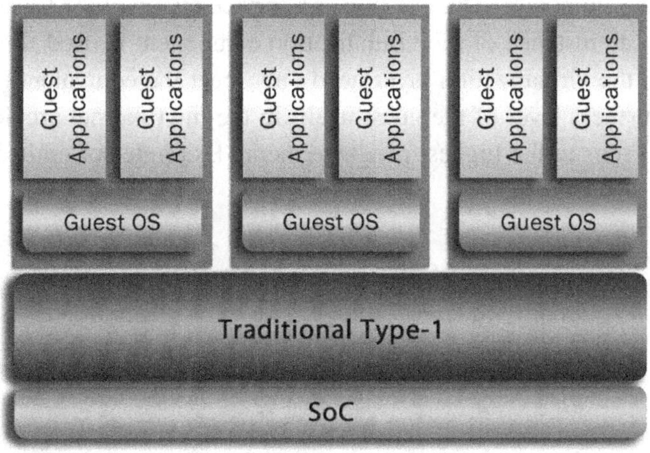

Figure 2.20:
Traditional monolithic type-1 hypervisor architecture.

the hypervisor may result in a compromise of the fundamental guest environment separation intended by virtualization in the first place.

2.6.5.2 Console Guest Hypervisor

An alternative approach uses a trimmed-down hypervisor that runs in the microprocessor's most privileged mode but employs a special guest operating system partition to handle the I/O control and services for the other guest operating systems (see Figure 2.21). Examples of this architecture include Xen and Microsoft Hyper-V. Xen pioneered the console guest approach in the enterprise; within Xen, the console guest is called Domain 0, or Dom0 for short. Thus, the console guest architecture is sometimes referred to as the Dom0 architecture. With the console guest approach, a general-purpose operating system must still be relied upon for system security. A typical console guest such as Linux may add far more code to the virtualization layer than found in a monolithic hypervisor.

2.6.5.3 Microkernel-based Hypervisor

> **Key Point**
>
> The microkernel-based hypervisor, a Type-1 architecture, is designed specifically to provide robust separation between guest environments.

Figure 2.22 shows the microkernel-based hypervisor architecture. Because the microkernel is a thin, bare-metal layer, the microkernel-based hypervisor is considered a Type-1 architecture.

This architecture adds computer virtualization as a service on top of the trusted microkernel. In some cases, a separate instance of the virtualization component is used for each guest environment. Thus, the virtualization layer need only meet the equivalent (and, typically, relatively low) robustness level of the guest itself. In the microkernel architecture, only the trusted microkernel runs in the highest privilege mode. Examples of embedded hypervisors

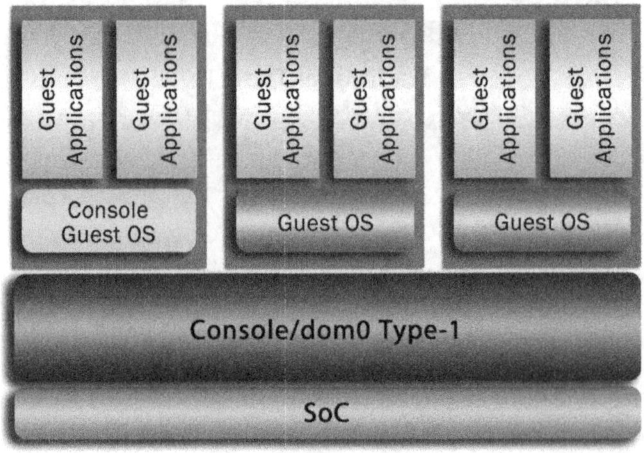

Figure 2.21:
Console guest or Dom0 hypervisor architecture.

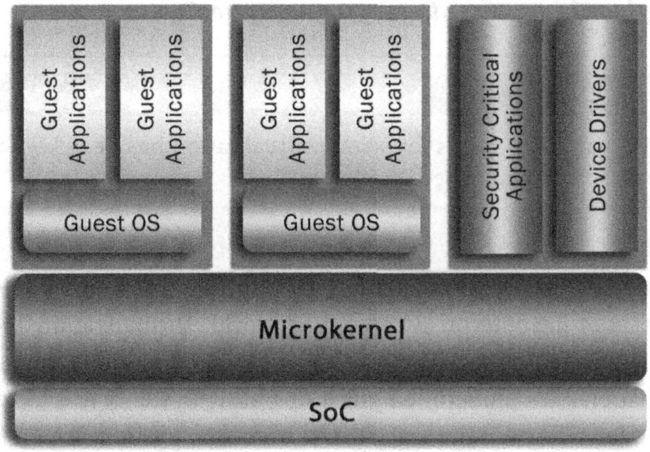

Figure 2.22:
Microkernel-based Type-1 hypervisor architecture.

using the microkernel approach include the INTEGRITY Multivisor from Green Hills Software and some variants of the open standard L4 microkernel.

2.6.6 Paravirtualization

System virtualization can be implemented with full virtualization or *paravirtualization*, a term first coined in the 2001 Denali project.[9] With full virtualization, unmodified guest operating systems are supported. With paravirtualization, the guest operating system is modified to improve the ability of the underlying hypervisor to achieve its intended function.

Paravirtualization is sometimes able to improve performance. For example, device drivers in the guest operating system can be modified to make direct use of the I/O hardware instead of requiring I/O accesses to be trapped and emulated by the hypervisor. Paravirtualization may be required on CPU architectures that lack hardware virtualization acceleration features.

Key Point

The key advantage to full virtualization over paravirtualization is the ability to use unmodified versions of guest operating systems that have a proven fielded pedigree and do not require the maintenance associated with custom modifications.

This maintenance saving is especially important in enterprises that use a variety of operating systems and/or regularly upgrade to new operating system versions and patch releases.

2.6.7 Leveraging Hardware Assists for Virtualization

> **Key Point**
>
> The addition of CPU hardware assists for system virtualization has been key to the practical application of hypervisors in embedded systems.

Intel VT, first released in 2005, has been a key factor in the growing adoption of full virtualization throughout the enterprise-computing world. Virtualization Technology for x86 (VT-x) provides a number of hypervisor assistance capabilities, including a true hardware hypervisor mode that enables unmodified guest operating systems to execute with reduced privilege. For example, VT-x will prevent a guest operating system from referencing physical memory beyond what has been allocated to the guest's virtual machine. In addition, VT-x enables selective exception injection so that hypervisor-defined classes of exceptions can be handled directly by the guest operating system without incurring the overhead of hypervisor software interposing. While VT technology became popular in the server class Intel chipsets, the same VT-x technology is now also available in Intel Atom embedded and mobile processors.

In 2009, the Power Architecture governance body, Power.org, added virtualization to the embedded specification within the Power Architecture version 2.06 Instruction Set Architecture (ISA). At the time of this writing, Freescale Semiconductor is the only embedded microprocessor vendor to have released products, including the QorIQ P4080 and P5020 multicore network processors, supporting this embedded virtualization specification.

In 2010, ARM Ltd. announced the addition of hardware virtualization extensions to the ARM architecture as well as the first ARM core, the Cortex A15, to implement them. Publicly announced licensees planning to create embedded SoCs based on Cortex A15 include Texas Instruments, Nvidia, Samsung, and ST-Ericsson.

Prior to the advent of these hardware virtualization extensions, full virtualization was possible only using dynamic binary translation and instruction rewriting techniques that were exceedingly complex and unable to perform close enough to native speed to be practical in embedded systems. For example, a 2005-era x86-based desktop running Green Hills Software's pre-VT virtualization technology was able to support no more than two simultaneous full-motion audio/video clips (each in a separate virtual machine) without dropping frames. With Green Hills Software's VT-x-based implementation on similar class desktops, only the total RAM available to host multiple virtual machines generally limits the number of simultaneous clips. General x86 virtualization benchmarks showed an approximate doubling of performance using VT-x relative to the pre-VT platforms. In addition, the virtualization software layer was simplified due to the VT-x capabilities.

2.6.7.1 ARM TrustZone

Key Point

An often overlooked and undervalued virtualization capability in modern ARM microprocessors is ARM TrustZone.

TrustZone enables a specialized, hardware-based form of system virtualization. TrustZone provides two zones: a "normal" zone and a "trust" or "secure" zone. With TrustZone, the multimedia operating system (for example, what the user typically sees on a smartphone) runs in the normal zone while security-critical software runs in the secure zone. Although secure zone supervisor mode software is able to access the normal zone's memory, the reverse is not possible (see Figure 2.23). Thus, the normal zone acts as a virtual machine under control of a hypervisor running in the trust zone. However, unlike other hardware virtualization technologies such as Intel VT, the normal zone guest operating system incurs no execution overhead relative to running without TrustZone. Thus, TrustZone removes the performance (and arguably the largest) barrier to the adoption of system virtualization in resource-constrained embedded devices.

TrustZone is a capability inherent in modern ARM applications processor cores, including the ARM1176, Cortex A5, Cortex A8, Cortex A9, and Cortex A15. However, it is important to note that not all SoCs using these cores fully enable TrustZone. The chip manufacturer must permit secure zone partitioning of memory and I/O peripheral interrupts throughout the SoC complex. Furthermore, the chip provider must open the secure zone for third-party trusted operating systems and applications. Examples of TrustZone-enabled mobile SoCs are the Freescale i.MX53 (Cortex A8) and the Texas Instruments OMAP 4430 (Cortex A9).

Trusted software might include cryptographic algorithms, network security protocols (such as SSL/TLS) and keying material; digital rights management (DRM) software; virtual keypad for

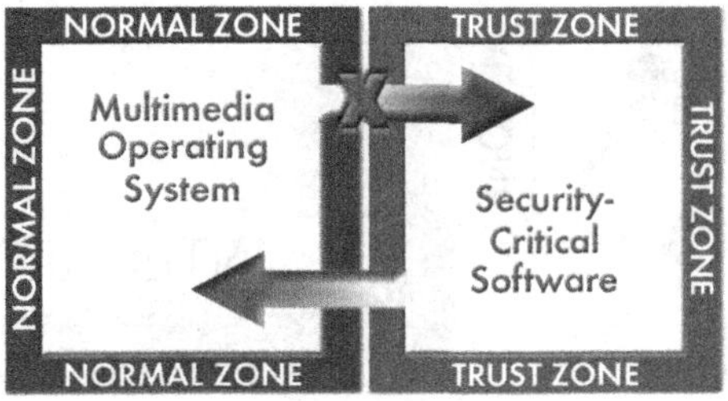

Figure 2.23:
ARM TrustZone.

trusted path credential input; mobile payment subsystems; electronic identity data; and anything else that a service provider, mobile device manufacturer, and/or mobile SoC supplier deems worthy of protecting from the user environment.

In addition to improving security, TrustZone can reduce the cost and time to market for mobile devices that require certification for use in banking and other critical industries. With TrustZone, the bank (or certification authority) can limit certification scope to the secure zone and avoid the complexity (if not infeasibility) of certifying the multimedia operating system environment.

A secure zone operating system can further reduce the cost and certification time, for two main reasons. First, because the certified operating system is already trusted, with its design and testing artifacts available to the certification authority, the cost and time of certifying the secure zone operating environment is avoided.

Second, because the secure zone is a complete logical ARM core, the secure operating system is able to use its memory management unit (MMU) partitioning capabilities to further divide the secure zone into meta-zones (see Figure 2.24). For example, a bank may require certification of the cryptographic meta-zone used to authenticate and encrypt banking transaction messages, but the bank will not care about certifying a multimedia DRM meta-zone, that, while critical for the overall device, is not used in banking transactions and guaranteed by the secure operating system not to interfere.

TrustZone-enabled SoCs are able to partition peripherals and interrupts between the secure and normal states. A normal zone general-purpose operating system such as Android cannot access peripherals allocated to the secure zone and will never see the hardware interrupts associated

Figure 2.24:
TrustZone virtualization implementation with meta-zones within the TrustZone.

with those peripherals. In addition, any peripherals allocated to the normal zone are unable to access memory in the normal zone.

2.6.8 Hypervisor Security

Some tout virtualization as a technique in a "layered defense" for system security. The theory postulates that since only the guest operating system is exposed to external threats, an attacker who penetrates the guest will be unable to subvert the rest of the system. In essence, the virtualization software is providing an isolation function similar to the process model provided by most modern operating systems.

However, common enterprise virtualization products have not met high-robustness security requirements and were never designed or intended to meet these levels. Thus, it should come as no surprise that the theory of security via virtualization has no existence proof in common enterprise implementations. Rather, a number of studies of virtualization security and successful subversions of hypervisors have been published.

2.6.8.1 SubVirt

In 2006, Samuel King's SubVirt project demonstrated hypervisor rootkits that subverted both VMware and Microsoft VirtualPC.[10]

2.6.8.2 Blue Pill

The Blue Pill project took hypervisor exploits a step further by demonstrating a malware payload that was itself a hypervisor that could be installed on the fly, beneath a natively running Windows operating system. Secure boot and platform attestation (discussed later in this chapter) are required to prevent hypervisors from being subverted in this manner.

2.6.8.3 Ormandy

Tavis Ormandy performed an empirical study of hypervisor vulnerabilities. Ormandy's team of researchers generated random I/O activity into the hypervisor, attempting to trigger crashes or other anomalous behavior. The project discovered vulnerabilities in QEMU, VMware Workstation and Server, Bochs, and a pair of unnamed proprietary hypervisor products.[11]

2.6.8.4 Xen Owning Trilogy

At the 2008 Black Hat conference, security researcher Joanna Rutkowska and her team presented their findings of a brief research project to locate vulnerabilities in Xen. One hypothesis was that Xen would be less likely to have serious vulnerabilities, as compared to VMware and Microsoft Hyper-V, due to the fact that Xen is an open source technology and therefore benefits from the "many-eyes" exposure of the code base.

Rutkowska's team discovered three different, fully exploitable, vulnerabilities that the researchers used to commandeer the computer via the hypervisor.[12] Ironically, one of these attacks took advantage of a buffer overflow defect in Xen's Flask layer. Flask is a security framework, the same one used in SELinux, which was added to Xen to improve security. This further underscores an important principle: software that has not been designed for and evaluated to high levels of assurance must be assumed to be subvertible by determined and well-resourced entities.

2.6.8.5 VMware's Security Certification and Subsequent Vulnerability Postings

As VMware virtualization deployments in the data center have grown, security experts have voiced concerns about the implications of "VM sprawl" and the ability of virtualization technologies to ensure security. On June 2, 2008, VMware attempted to allay this concern with its announcement that its hypervisor products had achieved a Common Criteria EAL 4+ security certification. VMware's press release claimed that its virtualization products could now be used "for sensitive, government environments that demand the strictest security."

On June 5, just three days later, severe vulnerabilities in the certified VMware hypervisors were posted to the National Vulnerability Database. Among other pitfalls, the vulnerabilities "allow guest operating system users to execute arbitrary code."[13]

VMware's virtualization products have continued to amass severe vulnerabilities—for example, CVE-2009-3732, which enables remote attackers to execute arbitrary code.

Clearly, the risk of an "escape" from the virtual machine layer, exposing all guests, is very real. This is particularly true of hypervisors characterized by monolithic code bases.

> **Key Point**
>
> It is important for developers to understand that use of a hypervisor does not imply highly assured isolation between virtual machines, no more than the use of an operating system with memory protection implies assured process isolation and overall system security.

Chapter 3 discusses the Common Criteria security evaluation standard and its Evaluated Assurance Levels, and what they imply in terms of security assurance.

2.7 I/O Virtualization

> **Key Point**
>
> One of the biggest impacts on security and efficiency in system virtualization is the approach to managing I/O across virtual machines.

This section discusses some of the security pitfalls and emerging trends in embedded I/O virtualization.

2.7.1 The Need for Shared I/O

In any embedded system, invariably there is a need for sharing a limited set of physical I/O peripherals across workloads. The embedded operating system provides abstractions, such as layer two, three, and four sockets, for this purpose. Sockets provide, in essence, a virtual interface for each process requiring use of a shared network interface device. Similarly, in a virtualized system, the hypervisor must take on the role of providing a secure virtual interface for accessing a shared physical I/O device. Arguably the most difficult challenge in embedded virtualization is the task of allocating, protecting, sharing, and ensuring the efficiency of I/O across the virtual machines and applications.

2.7.2 Emulation

The traditional method of I/O virtualization is emulation: all guest operating system accesses to device I/O resources are intercepted, validated, and translated into hypervisor-initiated operations (see Figure 2.25). This method maximizes reliability, security, and shareability. The guest operating system can never corrupt the system through the I/O device because all I/O

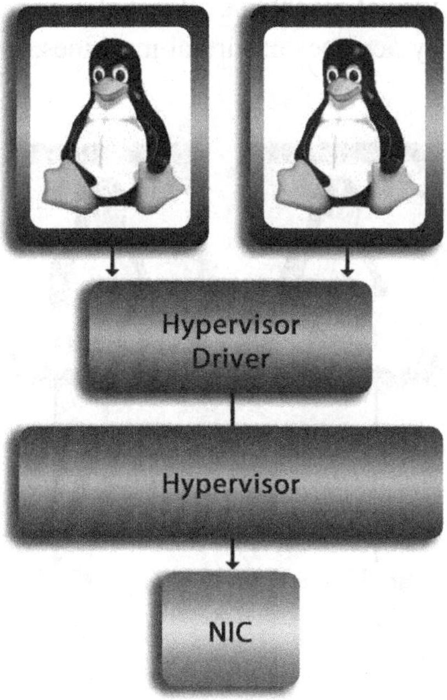

Figure 2.25:
I/O virtualization using emulation.

accesses are protected via the trusted hypervisor device driver. A single device can easily be multiplexed across multiple virtual machines, and if one virtual machine fails, the other virtual machines can continue to utilize the same physical I/O device, maximizing system availability. The biggest drawback is efficiency; the emulation layer causes significant overhead on all I/O operations. In addition, the hypervisor vendor must develop and maintain the device driver independent of the guest operating system drivers.

2.7.3 Pass-through

In contrast, a *pass-through* model (see Figure 2.26) gives a guest operating system direct access to a physical I/O device. Depending on the CPU, the guest driver can either be used without modification or with minimal paravirtualization. A single device can be shared between multiple guests by providing a virtual I/O interface between the guest that owns the physical device and any other guests that require access to that device. For network devices, this virtual interface is often called a virtual switch (layer 2) and is a common feature of most hypervisors. The pass-through model provides improved efficiency but trades off robustness: an improper access by the guest can take down any other guest, or application, or the entire system. This model violates the primary security policy of system virtualization: isolation of virtual environments for safe coexistence of multiple operating system instances on a single computer.

If present, an IOMMU enables a pass-through I/O virtualization model without risking direct memory accesses beyond the virtual machine's allocated memory. As the MMU enables the hypervisor to constrain memory accesses of virtual machines, the IOMMU constrains I/O

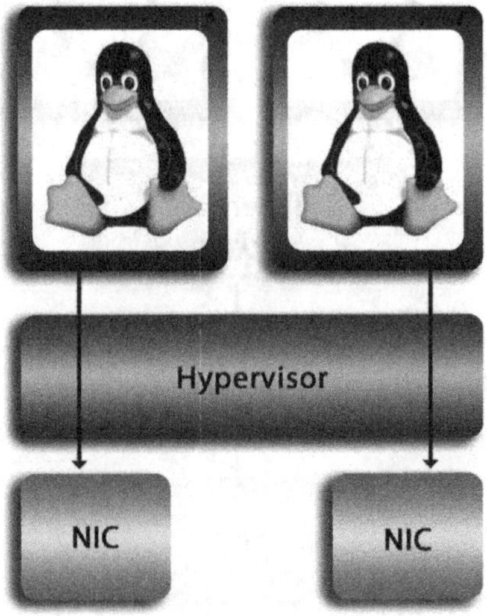

Figure 2.26:
I/O virtualization using pass-through.

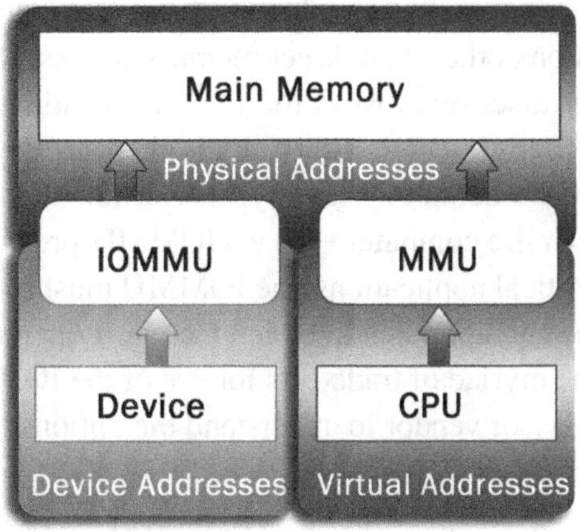

Figure 2.27:
IOMMU is used to sandbox device-related memory accesses.

memory accesses (especially DMA), whether they originate from software running in virtual machines or the external peripherals themselves (see Figure 2.27).

IOMMUs are becoming increasingly common in embedded microprocessors, such as Intel Core, Freescale QorIQ, and ARM Cortex A15. Within Intel processors, the IOMMU is referred to as Intel Virtualization Technology for Directed I/O (Intel VT-d). On Freescale's virtualization-enabled QorIQ processors such as the P4080, the IOMMU is referred to as the Peripheral Access Management Unit (PAMU). On Cortex A15 (and other ARM cores that support the ARM Virtualization Extensions[14]), the IOMMU is not part of the base virtualization specification. Rather, ARM Ltd. has a separate intellectual property offering, called a System MMU (SMMU), which is optionally licensable by ARM semiconductor manufacturers. In addition, the manufacturer may use a custom IOMMU implementation instead of the ARM System MMU. In addition, ARM TrustZone provides a form of IOMMU between the normal and secure zones of an ARM processor; normal zone accesses made by the CPU or by peripherals allocated to the normal zone are protected against accessing memory in the secure zone.

The IOMMU model enables excellent performance efficiency with increased robustness relative to a pass-through model without IOMMU. However, IOMMUs are a relatively new concept.

Key Point

A number of vulnerabilities (ways to circumvent protections) have been discovered in IOMMUs and must be worked around carefully with the assistance of a systems software/hypervisor supplier.

In most vulnerability instances, a faulty or malicious guest is able to harm security via device, bus, or chipset-level operations other than direct memory access. Researchers at Green Hills Software, for example, have discovered ways for a guest operating system to access memory beyond its virtual machine, deny execution service to other virtual machines, install a malicious hypervisor below the default system hypervisor (an attack described later in this chapter), and take down the entire computer—all via IOMMU-protected I/O devices. For high-reliability and/or security-critical applications, the IOMMU must be applied in a different way than the traditional pass-through approach in which the guest operating system has unfettered access to the I/O device. The myriad of trade-offs for use of the IOMMU are beyond the scope of this book; consult a hypervisor vendor to understand the options for use of the IOMMU and I/O virtualization in general.

A major downside of a pass-through approach (with or without the IOMMU) is that it prevents robust sharing of a single I/O device across multiple virtual machines. The virtual machine that is assigned ownership of the pass-through device has exclusive access, and any other virtual machines must depend on the owning virtual machine to forward I/O. If the owning virtual machine is compromised, all virtual machines will be denied servicing for that device.

2.7.4 Shared IOMMU

This deficiency has led to emerging technologies that provide an ability to share a single I/O device across multiple guest operating systems using the IOMMU and hardware partitioning mechanisms built into the device I/O complex (e.g., chipset plus the peripheral itself). One example of shareable, IOMMU-enabled, pass-through devices is Intel processors equipped with Intel Virtualization Technology for Connectivity (Intel VT-c) coupled with PCI-express Ethernet cards implementing Single-Root I/O Virtualization (SR-IOV), a PCI-SIG standard. With such a system, the hardware takes care of providing independent I/O resources, such as multiple packet buffer rings, and some form of quality of execution service among the virtual machines. This mechanism lends itself well to networking devices such as Ethernet, Rapid I/O, and Fibre Channel; however, other approaches are required for secure, independent sharing of peripherals such as graphics cards, keyboards, and serial ports. Nevertheless, it is likely that hardware-enabled, IOMMU-protected, shareable network device technology will grow in popularity across embedded processors.

2.7.5 IOMMUs and Virtual Device Drivers

Earlier in this chapter, we discussed the importance of virtual device drivers in limiting privilege and promoting robust system design. The ideal virtual device driver requires very little device-specific code to reside within the supervisor-mode kernel: the interrupt service routine (ISR) and, in the case of network devices, access to direct memory access (DMA) programming registers. The ISR must be in the kernel (the interrupt vector is executed by the hardware in supervisor mode). The DMA programming is often kept in the kernel because the operation must be trusted: access to DMA registers enables the driver to overwrite any physical memory location, even the kernel itself.

Unfortunately, the virtual driver approach still leaves a bit of device-specific code in the kernel. For improved maintainability and a cleaner architecture, it would be better if the DMA programming could reside in user mode without increasing the driver's privilege. An IOMMU enables the DMA programming to reside in the virtual device driver. Perhaps most importantly, by enabling direct access to memory-mapped device registers and DMA programming, the IOMMU promotes a purer form of virtual device driver architecture without sacrificing performance efficiency.

2.7.6 Secure I/O Virtualization within Microkernels

As discussed earlier, virtual device drivers are commonly employed by microkernel-style operating systems. Microkernel-based hypervisors are also well suited to secure I/O virtualization: instead of the typical monolithic approach of placing device drivers into the hypervisor itself or into a special-purpose Linux guest operating system (the Dom0 method described earlier), the microkernel-based hypervisor uses small, reduced-privilege, native processes for device drivers, I/O multiplexors, health managers, power managers, and other supervisory functions required in a virtualized environment. Each of these applications is provided only the minimum resources required to achieve its intended function, fostering secure embedded system designs. Figure 2.28 shows the system-level architecture of a microkernel-based hypervisor used in a multicore networking application that must securely manage Linux control plane functionality alongside high-throughput, low-latency data plane packet processing within virtual device drivers.

Without virtualization, the preceding application could be implemented with a dual Linux/ RTOS configuration in which the control and data plane operating systems are statically bound

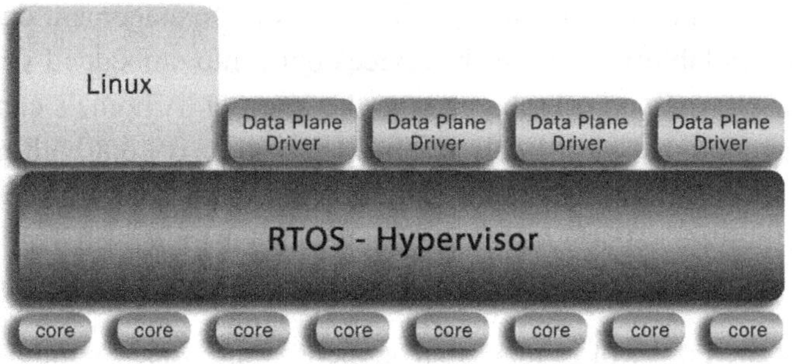

Figure 2.28:
Virtual device drivers in microkernel-based system virtualization architecture.

to a set of independent cores. This is referred to an Asymmetric Multiprocessing (AMP) approach. One advantage of virtualization over an AMP division of labor is the flexibility of changing the allocation of control and data plane workloads to cores. For example, in a normal mode of operation, the architect may want to use only a single core for control and all other cores for data processing. However, the system can be placed into management mode in which Linux needs four cores (SMP) while the data processing is temporarily limited. The virtualization layer can handle the reallocation of cores seamlessly under the hood, something that a static AMP system cannot support.

Security can also be improved by adding applications, or even separate virtual machines (the virtual appliances concept described earlier in this chapter), which perform a dedicated security function such as anti-malware or firewalling.

Increases in software and system complexity and connectivity are driving the evolution in how embedded systems manage I/O and in the architecture of the operating systems and hypervisors that are responsible for ensuring their security. The combination of a reduced-privilege, component-based designs as well as intelligent I/O virtualization to enable secure consolidation without sacrificing efficiency will remain a focus of systems software suppliers in meeting the flexibility, scalability, and robustness demands of next-generation embedded systems.

2.8 Remote Management

When an embedded system fails in the field, developers (and sometimes government forensics teams) are tasked with determining the cause of failure. A flight recorder is a well-known field diagnostic system: the end product (airplane) is shipped with a built-in diagnostic capability (the black box). Yet a burgeoning class of embedded devices requires field diagnostic and management capabilities. Unlike the black box, which is a purely forensic tool, an active network connection is required in many systems. This connection enables technicians to inspect a fielded system to locate the source of anomalous behavior such as loss of function or performance degradation, install patches or other software upgrades, perform automated audits, change configuration, or execute a plethora of other management duties. Furthermore, with the increasing availability of network services built into embedded systems, device management can be conveniently discharged via the Internet. A home's cable or satellite box has a network connection that most likely has been used to carry out both remote diagnostics and firmware upgrades. Device management functionality has been transformational, increasing product lifetime, reliability, serviceability, and customer satisfaction while reducing maintenance cost and total cost of ownership.

A great example of the power of remote management is the Mars Pathfinder: remote management saved the 1997 mission from disaster when a malfunction was diagnosed down to a software defect that was remedied with a patch installed via radio link from Earth.

2.8.1 Security Implications

The hacker's ambition: locate a vulnerability that, when properly manipulated, provides access into a computer system for nefarious purposes. Over time, remote network exploits have become increasingly sophisticated. In April 2010, IBM security researcher Mark Dowd won acclaim with his 25-page report detailing an astoundingly convoluted set of steps that could be taken to exploit a web access vulnerability, previously believed to be innocuous, in Adobe's ubiquitous Flash program.[15]

> **Key Point**
>
> Remote device management is the answer to the hacker's wildest dreams: the embedded system is imbued not only with Internet access, but also a means to remotely modify and patch software.

No Byzantine attack vector required; just get past the basic operating system controls, and the embedded device becomes a playground of iniquity.

In Chapter 1, we briefly described the VxWorks remote management vulnerability in which an operating system diagnostics port is commonly left open for access by any novice hacker to exploit. This vulnerability is so widespread that it is unlikely the Internet will ever be fully scrubbed of devices containing this flaw. In the case of the VxWorks flaw, the remote diagnostics connection trivially enables a hacker to install malware, even rooting or replacing the operating system itself (see Figure 2.29). A basic defense is to guard the remote management port with strong authentication, using standard network security protocols such as TLS/SSL or IKE/IPsec (both described in Chapter 5). First, the remote embedded system must authenticate the computer used by the remote administrator. This ensures that only known, trusted administrative computers are used to access the embedded system for management purposes. Second, if management commands are invoked by a human operator, then the operator must be strongly authenticated to the management computer locally prior to establishing the remote connection. Once the operator is authenticated, SSL or IPsec will use authenticated encryption to protect the integrity and confidentiality of remote management commands and data.

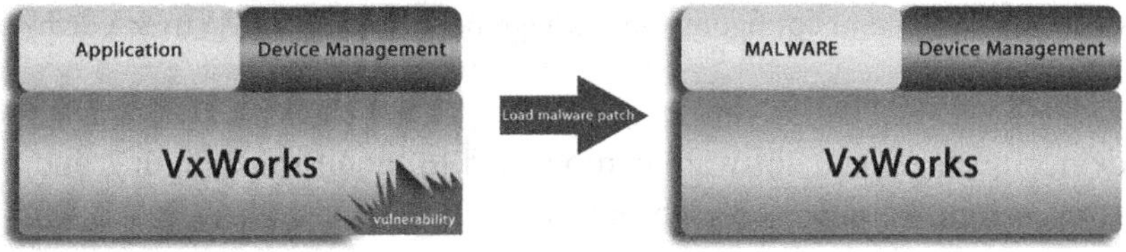

Figure 2.29:
Malware insertion into embedded system via remote management port.

Of course, hacking the management computer may render useless the aforementioned network security protections. For example, malware within the host computer's operating system can piggyback over the encrypted connection to infiltrate the embedded system. Extreme care must be taken to protect the management computer from the Internet or other open network accessible to hackers. Ideally, management workstations are dedicated to their purpose and never connected to the Internet. Using an insecure operator workstation to administer a remote embedded system is akin to putting a padlock on a safe built from cardboard. One example is found in arguably the most famous remote management system in the world: Windows Update. Windows Update is designed to unobtrusively and remotely feed a personal computer with the latest validated security patches. Yet hackers have commandeered this facility to upload unauthorized software.

An SSL connection is only the tip of the iceberg in ensuring that embedded systems are developed properly to enable secure remote management. Insider threats and development flaws can cause an embedded system to be fielded with vulnerabilities that turn a remote management channel into a powerful hacker access method. A secure development process to address these threats is a focus of Chapter 3. Without proper assurance that includes testing at the binary level, secure delivery, and other controls, developers can insert back doors using a wide range of proven techniques. A secure device management solution prevents malicious code insertion by employing high-assurance authentication and digital signatures: unauthorized IT administrators, technicians, janitors, and users cannot circumvent the mandatory access controls imposed by the system.

While developers can employ a high-assurance development process for their own software, how can developers protect against vulnerabilities in third-party operating systems, many of which are shipped in binary form only and lack any security provenance, pedigree, or indemnification? Since the operating system often provides the network security capabilities described earlier, this is obviously a critical issue.

Key Point

System virtualization can provide an effective solution to the problem of incorporating or retrofitting secure remote management to embedded systems: the legacy operating environment is uplifted into a virtual machine, securely isolated from remote management functions, such as connection authentication and configuration management, provided by a trustworthy hypervisor.

In fact, the device management software can be used to monitor, configure, and patch the legacy operating system kernel itself (see Figure 2.30).

In many cases, a secure device management solution involves consulting services to ensure that the appropriate set of security components is integrated and deployed in a robust and

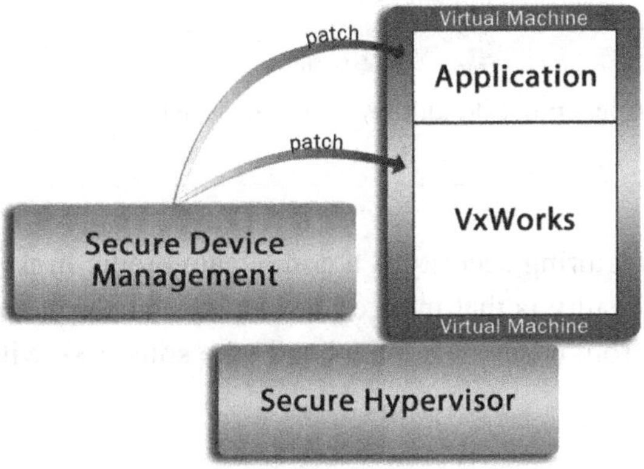

Figure 2.30:
Device management architecture for legacy systems.

cost-effective manner into end devices. Given the increasing financial, safety, and security risks associated with remote access, many embedded and mobile device makers are rethinking their device management strategy. System virtualization is one potentially powerful approach to addressing the secure remote management challenge.

2.9 Assuring Integrity of the TCB

2.9.1 Trusted Hardware and Supply Chain

We now turn to a critical systems-level security threat that can render impotent even the most perfectly implemented, absolutely secure software. Embedded system security requires integrity of the TCB (trusted operating system and any trusted applications and middleware, including data protection protocols discussed in Chapter 5).

For example, a virtual memory operating system depends on the fact that no unauthorized entities can access memory that the operating system allocates to a specific process using the microprocessor memory management and protection hardware. If the operating system is booted up on malicious hardware that exposes certain memory locations (e.g., by sending the contents of that memory over an attached Ethernet interface), then obviously the system security policies enforced by the operating system are moot.

Attacks on hardware components, subsystems, and peripherals are not far fetched. Numerous reports of counterfeit and subverted hardware have occurred over the years. In 2007, some Maxtor/Seagate hard drives were found to have a pre-installed (by the contract manufacturer) virus that would send data stored on the drive to malicious websites.[16] A hidden "kill switch" inserted into a field-programmable gate array (FPGA) during its design was allegedly used to disable radar systems during a military attack.[17]

Key Point

Embedded system developers must do all they can to ensure the trustworthiness of the hardware supply chain.

Supply chain and manufacturing security is a complicated topic in itself and is beyond the scope of this book. The reality is that most embedded systems organizations do the best they can by purchasing from reliable suppliers and take some risk with respect to trusting hardware.

2.9.2 Secure Boot

Key Point

Beyond trusted hardware, we must ensure that the trusted firmware/software cannot be subverted during the boot process.

Platform firmware attacks are far less difficult and expensive to perpetrate than supply chain attacks and represent an important threat that all embedded systems developers must consider. The act of establishing a secure initial state is often referred to as *secure boot.*

If the hardware and boot loader have the capability to load the system firmware (operating system, hypervisor, entire TCB) from an alternative device, such as USB, rather than the intended, trusted device (e.g., Flash), then an attacker with access to the system can boot an evil operating system that may act like the trusted operating system but with malicious behavior, such as disabling network authentication services or adding backdoor logins. Alternatively, an evil hypervisor can be booted, and the hypervisor can then launch the trusted operating system within a virtual machine. The evil hypervisor has complete access to RAM and hence can silently observe the trusted environment, stealing encryption keys or modifying the system security policy. King et al., provide a good example of this attack in a paper that describes SubVirt, a malware hypervisor.[10] Another infamous attack, called the Blue Pill, extended the SubVirt approach to create a permanent rootkit that could easily be launched on the fly using weaknesses in the factory-installed Windows operating system.[18]

2.9.3 Static versus Dynamic Root of Trust

In most embedded systems, a chain of firmware must be executed to establish the secure initial state in which the TCB is up and running and controlling system security. Most commonly, the

CPU first executes a small boot loader, burned into ROM at manufacturing time. Secure boot depends on a hardware-based root of trust; in this case, we depend on the fact that the ROM cannot be modified post-production. The ROM loader will often boot a more functional second-level boot loader residing in internal Flash. For example, many ARM-based embedded systems use the popular *u-boot* boot loader. This boot loader will often boot the primary operating system or hypervisor that in turn boots its higher-level applications.

The typical secure boot method is to verify the authenticity of each component in this boot chain. If any link in the chain is broken, the secure initial state is compromised. The first-stage ROM loader must also have a pre-burned cryptographic key used to verify the digital signature of the next-level boot loader. This key may be integrated into the ROM loader image itself, installed using a one-time programmable fuse, or stored in a local TPM that may provide enhanced tamper protection. The hardware root of trust must include this initial verification key. Chapter 4 describes the concepts of digital signature, usually implemented with public key cryptography.

The signature key is used to verify the authenticity of the second stage component in the boot chain. The known good signature must therefore also be stored in the hardware-protected area. The verification of the second-level component covers its executable image as well as the known good signature and signature verification key of the third stage, if any. The chain of verification can be indefinitely long. It is not uncommon for some sophisticated embedded and mobile computing systems to have surprisingly long chains or even trees of verified components that make up the TCB. Figure 2.31 depicts a sample three-level secure boot sequence. When the verification chain begins at system reset and includes all firmware that executes prior to the establishment of the runtime steady state, this is referred to as a *static root of trust*.

In contrast, a dynamic root of trust allows an already-running system (which may not be in a known secure state) to perform a measurement of the TCB chain and then partially reset the computer resources such that only this dynamic chain contributes to the secure initial state. Dynamic root of trust requires specialized hardware, such as Intel's Trusted Execution Technology (TXT),[19] available on some (at the time of this writing, higher-end) embedded Intel Architecture-based chipsets. The primary impetus behind dynamic root of trust is to remove large boot-time components, which must run to initialize a computer, from the TCB. On Intel Architecture-based systems, the BIOS is often an extremely large piece of software that is used to initialize the system. Because it is a large, monolithic piece of software, the BIOS may (and has in some cases been proven to) contain vulnerabilities that can be exploited. By performing the dynamic reset (also sometimes referred to as *late launch*) after the BIOS has initialized the hardware, removing all privilege from the BIOS execution environment, the system in theory has reduced its TCB and improved the probability of a secure initial state. Unfortunately, several weaknesses,[20] both in hardware and software, that implement the late

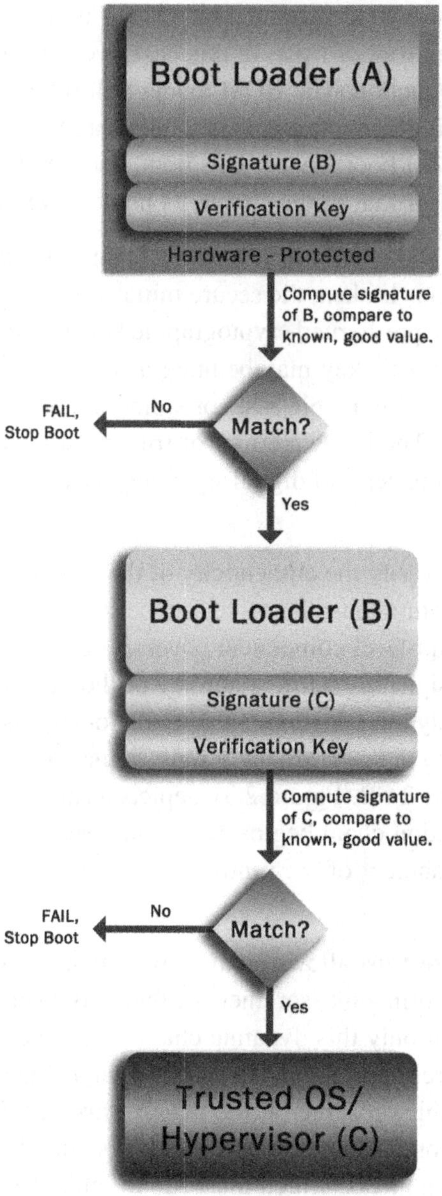

Figure 2.31:
Sample static root of trust secure boot sequence.

launch mechanism, have been found by researchers, bringing into question the ability to achieve a high level of trust in complicated boot environments. Furthermore, while the Trusted Computing Group (TCG) has standardized TPM interfaces,[21] implementations have yet to extend far beyond Intel Architecture-based computing environments.

Key Point

The good news for secure boot is that most embedded and mobile computing systems rely on simple boot loaders that lend themselves well to the static root of trust approach that can be implemented without specialized hardware.

Even a PC-based system can incorporate a custom, secure BIOS developed by embedded software security experts, if the embedded system's security is worth that investment. The readers should consult a systems software vendor to understand the available options.

Another good example of the importance of secure boot is provided in the discussion of data-at-rest protection in Chapter 5.

2.9.4 Remote Attestation

Secure boot provides embedded systems developers with confidence that the deployed product is resistant to low-level, boot-time firmware attacks. Nevertheless, a risk may persist in which sophisticated attackers can compromise the secure boot process. Furthermore, an attacker may be able to replace wholesale the deployed product with a malicious impersonation. For example, a smart meter can be ripped off the telephone pole and replaced with a rogue smart meter that looks the same but covertly sends private energy accounting information to a malicious website. Therefore, even with secure boot, users and administrators may require assurance that a deployed product is actively running the known-good TCB.

When embedded systems are connected to management networks, *remote attestation* can be used to provide this important security function. Once again, the TCG has standardized a mechanism for TCG-compliant systems to perform remote attestation using TPM-based measurements. Network access can be denied when a connecting client fails to provide proper attestation. Within TCG, this function is called Trusted Network Connect (TNC).[22] However, a simple, hardware-independent approach can be used for any embedded system.

Let's assume that the embedded system can communicate to the remote attestation server using a secure channel, such as IKE/IPsec or SSL (both discussed in Chapter 5). The initial session establishment uses public key cryptography. In particular, the static private key representing the identity of the remote embedded system is used to sign data that is then authenticated by the attester. As long as this private key and the client side of the secure connection protocol software are included in the TCB validated during secure boot, the attester has assurance that the embedded system is running some known-good firmware. Therefore, the mere act of a successful IKE or SSL session establishment can be used for remote attestation. An improvement to this approach, providing assurance that the embedded system is running a specific set of trusted firmware components, is to have the client transmit the complete set of digital signatures corresponding to the TCB chain to the attester that stores the known good set of signatures locally. This is more difficult to implement because the signatures must be computed and saved at manufacturing time, before the embedded product is deployed to the field.

2.10 Key Points

1. The operating system bears a tremendous burden in achieving safety and security.
2. The foundation of a MILS-based embedded system is the separation kernel, a small microkernel that implements a limited set of critical functional security policies, including data isolation, information flow control, damage limitation, and periods processing.
3. A separation kernel is considered a reference monitor when the kernel's MILS policy enforcement mechanisms are always invoked, tamper-proof, and evaluatable.
4. The MILS architecture requires the use of security-enforcing components whose functional requirements meet a high level of assurance.
5. Microkernel operating systems provide a better architecture for security than monolithic operating systems.
6. Memory protection is a fundamental requirement for robust embedded systems.
7. Virtual memory provides additional security features, including guard pages and location obfuscation, on top of basic memory protection.
8. System designers must plan for failures and employ fault recovery techniques.
9. Despite memory protection and virtual memory, malicious code can still take down a critical application by starving it of resources.
10. Partitioning, coupled with the ability to specify resource managers and multiple instance servers, enables system designers to obtain flexibility where needed and security where it is essential.
11. Device drivers are frequently the cause of system reliability problems and the target of hackers; thus, device drivers are some of the most important components of the operating system to isolate and protect.
12. Determinism is required to enforce secure time partitioning.
13. Many security problems are caused by poor access control architecture and/or implementation within the operating system or improper use of access control facilities by the embedded system designer.
14. The biggest reason why buffer overflows are so damaging is that insecure operating systems and software development techniques promote the use of ambient authority.
15. One of the biggest challenges embedded system designers face with respect to access control security policy is finding the proper balance between the granularity of policy and the maintainability and assurance of policy.
16. For most objects and resources, whitelists are preferable to blacklists due to the former's tendency toward reduced privilege implementation and design.

17. A capability acts as the mechanism for access as well as the access right: there simply is no way for a subject to access an object if the subject does not possess the object capability.
18. Distributed capability systems point to a need for the operating system to provide a means by which capabilities can be confined within privilege domains as well as a means for revocation.
19. The new level of abstraction needed to cope with increasingly sophisticated, consolidated electronic systems is the operating system itself, not just the computer's hardware resources.
20. The availability of system virtualization technology across a wide range of computing platforms provides developers and technologists with the ultimate open platform: the ability to run any flavor of operating system in any combination, creating an unprecedented flexibility for deployment and usage.
21. Because they can be no more secure than their underlying general-purpose host operating systems (which are well known to be vulnerable), Type-2 hypervisors are not suitable for mission-critical deployments and have historically been avoided in such environments.
22. The microkernel-based hypervisor, a Type-1 architecture, is designed specifically to provide robust separation between guest environments.
23. The key advantage to full virtualization over paravirtualization is the ability to use unmodified versions of guest operating systems that have a proven fielded pedigree and do not require the maintenance associated with custom modifications.
24. The addition of CPU hardware assists for system virtualization has been key to the practical application of hypervisors in embedded systems.
25. An often overlooked and undervalued virtualization capability in modern ARM microprocessors is ARM TrustZone.
26. It is important for developers to understand that use of a hypervisor does not imply highly assured isolation between virtual machines, no more than the use of an operating system with memory protection implies assured process isolation and overall system security.
27. One of the biggest impacts on security and efficiency in system virtualization is the approach to managing I/O across virtual machines.
28. A number of vulnerabilities (ways to circumvent protections) have been discovered in IOMMUs and must be worked around carefully with the assistance of a systems software/hypervisor supplier.
29. Remote device management is the answer to the hacker's wildest dreams: the embedded system is imbued not only with Internet access, but also a means to remotely modify and patch software.
30. System virtualization can provide an effective solution to the problem of incorporating or retrofitting secure remote management to embedded systems: the legacy operating

environment is uplifted into a virtual machine, securely isolated from remote management functions, such as connection authentication and configuration management, provided by a trustworthy hypervisor.

31. Embedded system developers must do all they can to ensure the trustworthiness of the hardware supply chain.
32. Beyond trusted hardware, we must ensure that the trusted firmware/software cannot be subverted during the boot process.
33. The good news for secure boot is that most embedded and mobile computing systems rely on simple boot loaders that lend themselves well to the static root of trust approach that can be implemented without specialized hardware.

2.11 Bibliography and Notes

1. Andersen JP. *Computer Security Planning Study*, http://csrc.nist.gov/publications/history/ande72.pdf; 1972.
2. Liedtke J. Toward Real Microkernels. *Communications of the ACM* September 1996;**39**(9):70–7.
3. DiBona C, Stone S, Ockman M, editors. *Appendix A, Open Sources: Voices from the Open Source Revolution*. Sebastopol, CA: O'Reilly; 1999.
4. Microsoft Security Advisory (2639658). *Vulnerability in TrueType Font Parsing Could Allow Elevation of Privilege*, http://technet.microsoft.com/en-us/security/advisory/2639658; November 3, 2011.
5. US-CERT/NIST National Vulnerability Database. Vulnerability Summary for CVE-2011-3402, http://web.nvd.nist.gov/view/vuln/detail?vulnId=CVE-2011-3402.
6. Carlsson M. Worst Case Execution Time Analysis, Case Study on Interrupt Latency, for the OSE Real-Time Operating System, http://www.astec.uu.se/publications/2002/Carlsson-WCET_Exjobb.pdf.
7. Jaeger T, Sailer R, Zhang X. *Analyzing Integrity Protection in the SELinux Example Policy*. Washington, DC: Proceedings of the 12th USENIX Security Symposium; August 4–8, 2003.
8. Hardy N. "The Confused Deputy," *ACM SIGOPS Operating Systems Review* October 1988;**22**(4).
9. Whitaker A, et al. *Denali: Lightweight Virtual Machines for Distributed and Networked Applications*. Boston, MA: USENIX Annual Technical Conference; June 10–15, 2002.
10. King S, Peter C, Yi-Min W, Chad V, Helen JW, Jacob RL. *SubVirt: Implementing Malware with Virtual Machines*. Berkeley, CA: IEEE Symposium on Security and Privacy; May 21–24, 2006.
11. Ormandy T. *An Empirical Study into the Security Exposure to Hosts of Hostile Virtualized Environments*, http://taviso.decsystem.org/virtsec.pdf; 2006.
12. Rutkowska J, Tereshkin A, Wojtczuk R. *"Detecting and Preventing the Xen Hypervisor Subversions," "Bluepilling the Xen Hypervisor," "Subverting the Xen Hypervisor,"*. Las Vegas, NV: Black Hat USA; August 7, 2008.
13. CVE-2008-2100. National Vulnerability Database. http://web.nvd.nist.gov/view/vuln/detail?vulnId=CVE-2008-2100.
14. Mijat R, Nightingale A. *Virtualization Is Coming to a Platform Near You*. White Paper. ARM Ltd; 2010.
15. Dowd M. *Application-Specific Attacks: Leveraging the ActionScript Virtual Machine*, http://documents.iss.net/whitepapers/IBM_X-Force_WP_final.pdf; April 2008.
16. Maxtor Basics Personal Storage 3200 (PS 3200) virus [205131]. Seagate Knowledge Base, online URL, http://seagate.custkb.com/seagate/crm/selfservice/search.jsp?DocId=205131.
17. Adee S. The Hunt for the Kill Switch. *IEEE Spectrum Magazine*, http://spectrum.ieee.org/semiconductors/design/the-hunt-for-the-kill-switch/; May 2008.
18. Rutkowska J. *Subverting Vista Kernel for Fun and Profit*. Las Vegas, NV: Black Hat Briefings; August 3, 2006.
19. Intel Trusted Execution Technology (Intel TXT). *Software Development Guide/Measured Launched Environment Developer's Guide*; March 2011.

20. Wojtczuk R, Rutkowska J. *Attacking Intel Trusted Execution Technology.* Washington, DC: Black Hat DC; February 18–19, 2009.
21. Trusted Computing Group (TCP) Trusted Platform Module (TPM) Main Specification. *Level 2, Version 1.2, Revision 116 (in three parts, "Design Principles," "Structures of the TPM," "Commands")*; March 1, 2011.
22. Trusted Computing Group (TCG) Trusted Network Connection (TNC) Architecture for Interoperability, Specification Version 1.4, Revision 4, May 18, 2009.
23. Sharon G. *Security Expert: U.S. Companies Unprepared for Cyber Terror.* eSecurity Planet; July 19, 2002; online URL: http://www.esecurityplanet.com/trends/article.php/1429851/Security-Expert-US-Companies-Unprepared-For-Cyber-Terror.htm

CHAPTER 3

Secure Embedded Software Development

Chapter Outline

Embedded Systems Security. DOI: 10.1016/B978-0-12-386886-2.00003-5

3.1 Introduction to PHASE—Principles of High-Assurance Software Engineering

In Chapter 1, we discussed the trend of increased software complexity and its adverse impact on security and safety. As yet one more example, a Swiss National Railways centralized traffic control system was recently reported to contain four million lines of code.[1] Complexity strains traditional reliability techniques, such as code reviews, and implies a growing necessity for a comprehensive approach to software assurance.

> **Key Point**
>
> Software assurance refers to the level of confidence that the software end user and other relevant stakeholders (e.g., certifiers) have that the security policies and functions claimed by that software are actually fulfilled.

Simply meeting functional requirements does not achieve the assurance required for security-critical embedded systems. The purpose of this chapter is to discuss proven techniques for dramatically increasing the assurance of embedded software. Increased assurance decreases vulnerabilities, improving security, safety, and reliability.

We begin the discussion of secure embedded software development with an introduction to PHASE, a methodology that can break the vicious cycle of unreliable software.

> **Key Point**
>
> PHASE—Principles of High Assurance Software Engineering—prescribes a set of five principles to be used in the creation of ultra-reliable software and systems:
>
> 1. Minimal implementation
> 2. Component architecture
> 3. Least privilege
> 4. Secure development process
> 5. Independent expert validation
>
> Each principle is described, with examples, in the following sections.

3.2 Minimal Implementation

> **Key Point**
>
> It is much harder to create simple, elegant solutions to problems than complex, convoluted ones.

Most software developers do not work in an environment in which producing the absolute minimal possible solution to a problem is an unwavering requirement. Spaghetti code is the source of vulnerabilities that run rampant in software and provide the avenue of exploitation for hackers.

As an example, let's consider an HTML 1.1-compliant web server. Engineers at Green Hills Software developed a high-assurance web server (HAWS) that used state-driven protocol processing instead of the typical error-prone string parsing and manipulation. The result: a few hundred lines of perfect code instead of the tens of thousands of lines found in many commercial web servers. The Green Hills web server runs on the high-assurance INTEGRITY operating system. In 2008, a website running on this platform was deployed on the Internet, and Netragard, a leading white hat hacker organization, was invited to perform a vulnerability assessment of the website. Netragard CTO Adriel Desautels reported that the website had "no attack surface whatsoever." HAWS is covered in more depth as a case study at the end of this chapter.

As another example, let's consider file systems. Engineers at Green Hills Software developed a high-assurance journaling file system, called PJFS, using a few thousand carefully crafted lines of code. The file system achieves excellent performance, provides guaranteed media storage quotas for clients (important in safety-critical contexts), and employs transactional journaling to assure the integrity of file system data and metadata (and instant reboot time) in the event of sudden power loss. In contrast, commercial journaling file systems typically exceed 100,000 source lines, with plenty of software flaws.

3.3 Component Architecture

Key Point

An important software robustness principle is to compose large software systems from small components, each of which is easily maintained by, ideally, a single engineer who understands every single line of code.

It is imperative to use well-defined, documented interfaces between components. These interfaces serve as a contract between component owners and must be created carefully to minimize churn that forces a cascade of implementation, testing, and integration changes. If a modification to an interface is required, component owners whose components use these interfaces must agree, involving common management to resolve disagreements if necessary.

Key Point

An important corollary to the component architecture principle is that safety and/or security-enforcing functionality should be placed into separate components so that critical operations are protected from compromise by non-critical portions of the system.

It is not enough to isolate security functions into their own components, however. Each security-critical component must, to the greatest extent practicable, be designed or refactored to remove any functionality that is not part of its security-enforcing function.

One of the key reasons why overly complex software is difficult to manage is that such a piece of software is almost always worked on by multiple developers, often at different times over the life of the product. Because the software is too complex for a single person to comprehend, features and defect resolutions alike are addressed by guesswork and patchwork. Flaws are often left uncorrected, and new flaws are added while the developer attempts to correct other problems.

Componentization also provides the capability for the system designer to make customer-specific changes in a methodical way. By focusing on customer and market requirements, the designer can make changes by swapping out a small subset of components as opposed to the larger part of the software baseline. This minimizes the task of regression testing by decreasing the impact to the overall system. When designers keep this attitude of componentization and interface definition in mind, improvements can be made over time with low risk.

> **Key Point**
>
> Componentization provides many benefits, including improved testability, auditability, data isolation, and damage limitation.

Componentization can prevent a failure in one component from devolving into a system failure. Componentization can also dramatically reduce development cost and certification cost, if applicable, by enabling developers to apply a lower development process rigor on non-critical components while raising the level of assurance for the critical pieces, which are often a relatively small percentage of the entire system.

Dividing a system into components requires that they have well-defined interfaces. Instead of modifying the same shared piece of code, developers must define simple, clear interfaces for components and only use a component's well-documented (or at least well-understood) interface to communicate with other components. Componentization enables developers to work more independently and therefore more efficiently, minimizing time spent in meetings where developers attempt to explain behavior of their software. Re-factoring a large software project in this manner can be time consuming. However, once this is accomplished, all future development will be more easily managed.

> **Key Point**
>
> Ensure that no single software partition is larger than a single developer can fully comprehend.

Each partition must have a well-known partition manager. One way to ensure that developers understand who owns which partitions is to maintain an easily accessible partition manager list that is modified only by appropriate management personnel. The partition manager is the only person authorized to make modifications to the partition or to give another developer the right to make a modification. By having clear ownership of every single line of code in the project, developers are not tempted to edit code that they are not appropriately qualified to handle.

Key Point

Ensure all developers know who the component managers are.

Component managers develop, over time, a comprehensive understanding of their owned partitions, ensuring that future modifications are done with complete knowledge of the ramifications of modifying any software within the partition.

3.3.1 Runtime Componentization

Usually, the embodiment of a component in the target computer system is a single executable program. Examples of components include Windows .EXE applications and POSIX/UNIX processes. Thus, complex software made up of multiple components should always be used in conjunction with an operating system that employs memory protection to prevent corruption of one component's memory space by another partition. Inter-component communication is typically accomplished with standard operating system message-passing constructs.

Different embedded operating systems (and microprocessors) have varying capabilities in terms of enforcing strict separation between components. For example, a small, real-time operating system may not make use of a computer's memory management unit at all; multiple software applications cannot be protected from each other, and the operating system itself is at risk from flaws in application code. These flat memory model operating systems are not suitable for complex, partitioned software systems. General-purpose desktop operating systems such as Linux and Windows employ basic memory protection, in which partitions can be assigned processes that are protected from corruption by the memory management unit, but do not make hard guarantees about availability of memory or CPU time resources.

For secure systems, the embedded operating system must provide strict partitioning of applications in both time and space. A damaged application cannot exhaust system memory, operating system resources, or CPU time because the faulty software is strictly limited to an assigned quota of critical resources. The quota affects literally all memory in use, including heap memory for the C/C++ runtime, memory used for process control blocks and other operating system objects, and processes' runtime stack memory. In addition, the partitioning policies provide strict quotas of execution time and strict control over access to system resources such as I/O devices and files. A more rigorous partitioning of applications at the operating system level ensures that the benefits of partition management policies used in the development process are realized during runtime.

> **Key Point**
>
> If possible, use an operating system that employs true application partitioning.

3.3.2 A Note on Processes versus Threads

When developers factor embedded software into components, a natural embodiment of the runtime component is the *thread*. Threads are flows of execution that share a single address space with other threads. In most modern operating systems, an address space has at least one default thread, and the address space with this single thread is often called a *process*. Because they are easy to create and the way most embedded programmers learn to employ concurrency, threads often get overused. Furthermore, embedded systems developers often have the mistaken impression that a proliferation of processes will exhaust too many system resources relative to threads. While threads are certainly lighter weight than a full-blown process, the distinction has become increasingly less important in modern embedded systems. Another reason for thread overuse can be attributed to the fact that the original real-time operating systems created in the 1980s and early 1990s did not support memory-protected processes at all. Developers became accustomed to threads, and their legacy lives on.

> **Key Point**
>
> Contrary to popular belief, designers should strive for a one-to-one ratio between threads and processes.

In other words, each memory-protected component should contain a minimum number of threads. The key reason is that multi-threaded processes are often the cause of subtle synchronization problems that result in memory corruption, deadlock, and other faults. The use of virtual memory processes forces developers to create well-defined inter-process communication interfaces between components. Each component can be independent unit tested by exercising these interfaces. This thread-less component philosophy avoids some of the nastiest vulnerabilities that plague embedded software.

3.4 Least Privilege

> **Key Point**
>
> Components must be given access to only those resources (e.g., communication pathways, I/O devices, system services, information) that are absolutely required.

Access control must be mandatory for critical system resources and information. As discussed in Chapter 2, insecure systems typically allow any program to access the file system, launch other programs, and manipulate system devices.

For example, browser buffer overflow vulnerabilities may enable an attacker to access any file because the web browser has the privilege to access the entire file system. There is no reason why a web browser should have unfettered access to the entire file system. The web browser either should have write access only to a dedicated, browser-specific directory (out of which the user can carefully decide what can leave the sandbox), or all browser write requests can require user approval via a dialog box. Read access can be limited to a whitelist of files known to be required for browser operation. The web browser's runtime stack should not be executable.

Every component of the entire system should be designed with least privilege in mind, and it is always best to start with no privilege and work up to what is needed rather than start with the kitchen sink and whittle away privileges.

For another example, let's consider the common operating system function of launching a new process. One original UNIX method, *fork()*, creates a duplicate of the parent process, giving the child all the same privileges (e.g., access to file descriptors, memory) as the parent. The developer then must close descriptors and otherwise try to limit the child's capabilities. This requires an unrealistic prescient knowledge of all system resources accessible to a process. Thus, errors in the use of this interface have often led to serious vulnerabilities.

In a secure operating system, the default process creation mechanism establishes a child without access to any of the parent's capabilities for memory, devices, or other resources. The creator can systematically provide capabilities to the child, building up a strictly limited privilege process. The child must also obtain its physical memory resources from its parent, ensuring that a process cannot drain the system or otherwise affect other critical processes with a fork bomb.

3.5 Secure Development Process

> **Key Point**
>
> For critical safety- and security-enforcing components, the software development process must meet a much higher level of assurance than is used for general-purpose components.

The embedded systems developer unfamiliar with the secure development process should study proven high-assurance development standards that are used to certify critical embedded systems. Two noteworthy standards are DO-178B Level A (a standard for ensuring the safety of flight-critical systems in commercial aircraft) and ISO/IEC 15408 (Common Criteria) EAL6/7 or equivalent. A high-assurance development process will cover numerous controls, including configuration management, coding standards, testing, formal design, formal proof of critical functionality, and so on.

Let's consider a case in which a rogue programmer has the capability of installing into aircraft engine software a condition such that the engine will shut down at a certain time and date. This software may reside in all the engines for a particular class of aircraft. One aspect of a secure development process is having separate system and software teams developing the redundant aircraft engine control systems. In this way, systemic or rogue errors create by one team are mitigated by the entirely distinct development path. Independence of systems, software, and testing teams in accordance with standards also contributes to this secure development process.

It can be educational to examine the key differences between a high-assurance development process and a low-assurance development process. This chapter includes a comparison of classes of security standards within Common Criteria, the international standard for evaluating the security of IT systems.

3.5.1 Change Management

A software project may be robust and reliable at the time of its first release, only to endure change rot over ensuing years as new features, not part of the original design, are hacked in, causing the code to become difficult to understand, maintain, and test. Time-to-market demands exacerbate the problem, influencing developers to make hasty changes to the detriment of reliability.

> **Key Point**
>
> An extremely important aspect of maintaining secure software over the long term is to utilize an effective change management regimen.

Some fundamentals of quality change management, such as the employment of configuration management systems to control and record changes and manage code branches and releases, are assumed and not covered in this book.

3.5.2 Peer Reviews

A critical aspect of effective change management is the use of peer code reviews. A common peer code review sequence consists of the code author developing a presentation describing the code change followed by a face-to-face meeting with one or more developers and development managers involved in the project. The developer presents the software design in question, and the others try to poke holes in the code. These meetings can be extremely painful and time consuming. Audience members sometimes feel compelled to nitpick every line of code to demonstrate their prowess.

> **Key Point**
>
> Use asynchronous code reviews with e-mail correspondence or carefully controlled live meetings.

Componentization drastically reduces the time required for code reviews since the code experts are almost always the ones modifying their own components. Debates regarding design decisions are usually avoided. In addition, we advocate limited face-to-face peer reviews along with correspondence reviews when practicable. The partition manager applies the changes to a local copy of the software, selects a suitable peer to review the changes, and then makes a review request via e-mail. Soon after, but at a time convenient for the reviewer, the reviewer reviews the code differences at her desk and then sends back comments via e-mail to the author. When the author receives an e-mail indication that the change is approved, the software is committed to the configuration management system. If a reviewer rejects a modification, the author must correct any discovered flaws or, if she disagrees with the assessment, appeal to the common manager to referee. The configuration management system must provide for the ability to specify the reviewer's user identification as part of the commit comment. For example, CVS allows a script to be run during a commit; the script is provided the commit comment that is parsed for the user identification. If valid user identification is not found, the configuration management system rejects the commit. Thus, the configuration management system can be used to automate the enforcement of a code review policy. Without such an automated system, there can be no guarantee that a developer will not commit a change that has not been properly vetted.

> **Key Point**
>
> Use the configuration management system to automate enforcement of peer reviews for every modification to critical code.

Recording the reviewer's identification in the configuration management system also provides an electronic paper trail for security certification auditors.

Another advantage of partitioning is the ability to minimize process requirements across the system. In any large software project, there is a continuum of criticality among the various pieces of code. By way of example, let's consider an excimer laser system used in semiconductor manufacturing. The laser itself is controlled by a highly critical, real-time software application. If this application faults, the laser in turn may fail, destroying the semiconductor. In addition, the system contains a communications application that uses CORBA over TCP/IP to receive commands and to send diagnostic data over a network. If the communications application fails, then the system may become unavailable or diagnostic data may be lost, but there is no possibility for the laser to malfunction. If both applications were built into a single, monolithic system in which all code executes in the same memory space, then the entire software content must be developed at the highest levels of quality and reliability. If the applications are partitioned, however, the non-critical communications application development can be subjected to a lower level of rigor, saving time to market and development cost.

> **Key Point**
>
> Apply a level of process rigor, including code reviews and other controls, that is commensurate with the criticality level of the component.

Obviously, we do not advocate a free-for-all on components that are not considered critical; management should use judgment regarding which controls to apply to various software teams. When the process controls in non-critical applications are reduced, time to market for the overall system can be improved without jeopardizing reliability where it counts.

3.5.2.1 Security-Oriented Peer Review

Most peer reviews are spent looking for coding bugs, design flaws, and violations of coding standards. While these activities contribute to more reliable and hence secure software, most embedded software organizations do not perform reviews based specifically on security analysis.

When a developer presents a new design or piece of software, the reviewers should consider security-relevant characteristics. For example:

Least privilege: Can the software be refactored such that the least critical components are provided the least amount of privilege in terms of access to resources? Reducing privilege of a component decreases its attack surface and reduces its assurance requirements, improving efficiency in development and certification (if applicable).

Attack potential: Think in terms of an attacker—system-resident (malware) or external (network-borne): where are the access points and weaknesses in the system, and how might an attacker attempt to compromise them? As in poker, success requires putting one's self in the opponent's frame of reference and training to think like one's opponent. Over time, developers with this mindset become proficient at predicting attack potential and therefore can place controls to prevent security failures.

Sophisticated attacks: Even if the code under review is not protecting the power grid, let's consider advanced security concerns such as side and covert channels, transmission security, and DMA corruption via system peripherals. Developers trained to consider sophisticated attack threats will be better prepared to handle the components that demand high robustness against such threats.

Key Point

By making security a part of peer reviews, management will create a culture of security focus throughout the development team.

In fact, because peer reviews account for a significant portion of group interaction in a development organization, they are an ideal venue for engendering the kind of vigilance needed to build secure embedded systems.

3.5.3 Development Tool Security

An *Easter egg is* an intentionally undocumented message, joke, or capability inserted into a program by the program's developers, as an added challenge to the user or simply just for fun. Easter eggs are commonly found in video games. The Linux packaging tool *apt-get* has this bovine egg:

```
> apt-get moo
          (__)
          (oo)
    /------\/
   / |    ||
  *  /\---/\
     ~~   ~~
...."Have you mooed today?"...
```

Cute. Funny. But what if a developer aims to insert something malicious? How can an organization be protected from this insider threat? How can the organization ensure that malware is not inserted by third-party middleware or the compiler used to build the software?

Key Point

Developers and users require assured bit provenance: confidence that every single binary bit of production software originates from its corresponding known-good version of source code.

This is a critical aspect of software security that many embedded systems developers never consider. High-assurance security and safety standards, such as DO-178B Level A (aircraft safety) and Common Criteria Evaluated Assurance Level 7 (IT security), require the ability to re-create the exact bits of the final production software from the configuration management system. Ancillary items, not just product source code, must be configuration managed. For example, any support files (e.g., scripts) used for the creation of production images must be strictly controlled. And the tool chain used to build the software must also be covered. Failure to rigorously control the entire development system can lead to serious vulnerabilities, both inadvertent and malicious.

3.5.3.1 Case Study: The Thompson Hack

One such subversion was performed by Ken Thompson and reported famously in his Turing Award acceptance speech. Thompson inserted a back door into UNIX by subverting the compiler used to build UNIX. Thompson's modification caused the UNIX *login* password verification to match on a string of Thompson's choosing in addition to the normal database validation. In essence, Thompson changed the UNIX login program that used to look something like this:

```
int login(unsigned int uid, char *password)
{
  if (strcmp(pass_dbase(uid), password) == 0)
    return true; // password match, login ok
  else
    return false; // password mismatch, login fail
}
```

into something that looks like this:

```
int login(unsigned int uid, char *password)
{
  if (strcmp(pass_dbase(uid), password) == 0 ||
      strcmp("ken_thompson", password) == 0))
    return true; // password match, login ok
  else
    return false; // password mismatch, login fail
}
```

However, changing the UNIX source code would be too easy to detect, so Thompson modified the compiler to insert the back door. With compiler insertion, examination of the UNIX source code would not be sufficient to detect the bug. The compiler Trojan would be a code fragment that examines the internal syntax tree of the compiled program looking for the specific *login* password check code sequence and replacing it with the back door:

```
if (!strcmp(function_name(), "login")) {
  if (OBJ_TYPE(obj) == IF_STATEMENT &&
      OBJ_TYPE(obj->left) == FUNCTION &&
      !strcmp(OBJ_NAME(obj->left), "strcmp")) {
    Object func = GET_ARG(1, obj->left);
```

```
    if (OBJ_TYPE(func) == FUNCTION) &&
        !strcmp(OBJ_NAME(func),"pass_dbase")) {
      // insert back door
      obj = MAKEOBJ(ORCMP, obj,
          MAKEOBJ(FUNCTION, "strcmp",
          MAKEOBJ(STRING, "ken_thompson"),
          GET_ARG(2, obj->left);
    }
  }
}
```

If the compiler is configuration managed and/or peer reviewed, Thompson's change might be detected by inspection. But if the compiler source code is not under configuration management or is very complicated, the Trojan could go unnoticed for some time. Also, who would think to question code committed by the esteemed Ken Thompson? One lesson learned is that those with the most trust can cause the most damage—another argument for enforcing least privilege mentality throughout the engineering department.

Assuming that the Trojan in the compiler might be detected, Thompson took his attack a step further and taught the compiler to add the Trojan into itself (two levels of indirection). In other words, the preceding compiler Trojan was inserted not into the source code of the compiler, but rather into the object code of the compiler when compiling itself. The Trojan is now said to be *self-reproducing*. While this may sound sophisticated, it really is not difficult once a developer has a basic understanding of how the compiler works (which of course, Ken Thompson did): simply locate the appropriate compiler phase and insert the preceding code fragment into the target's syntax tree when the target is the compiler.

There are ways to detect this more advanced attack. Code inspection is again an obvious method. Another approach is to build the compiler from source (under configuration management), build from the same source again with this new compiler, and require that the two built compiler binaries are bit-for-bit identical. This binary tool comparison method, called *bootstrapping*, is a cheap and effective method to detect some classes of development tool vulnerabilities. With Thompson's Trojan just inserted into the compiler source code, the first binary will not contain the Trojan code, but the second one will, causing the bootstrap test to fail. Of course, this approach works only if the compiler and the compiler's compiler have the same target processor back end—that is, the compiler is *self-hosting*. Since most UNIX systems have self-hosting compilers, this generational test is effective.

However, to cover his tracks even further, Thompson removed the compiler's Trojan source code, leaving only a subverted compiler binary that was installed as the default system compiler. Subsequent bootstrapping tests would fail to detect the subversion since both the first- and second-generation compilers contain the Trojan binary code.

This attack shows how sophisticated attackers can thwart even good development tool security. Ideally, we would like to formally prove correspondence between a tool's source code and its resulting compiled object code. A practical alternative is to require a bootstrap every time a default tool chain component is replaced. Performing the bootstrap test for every new compiler will generate a chain of trust that would have prevented Thompson's subversion if this process had been in place prior to his attack.

Modified condition/decision coverage (MC/DC) validation of the UNIX *login* program would also have detected the Thompson hack since the comparison to Thompson's backdoor password string will never succeed in normal testing. MC/DC testing is discussed further later in this chapter. However, a good defense-in-depth strategy should not assume that testing would find all forms of development tool subversion.

Of course, the configuration management system must be protected from tampering, either via remote network attack or by physical attack of the computers that house the configuration system. Some organizations may go as far as to require that active configuration management databases be kept in a secure vault, accessible only by authorized security personnel.

If an organization is not thinking about development security, now is the time to start.

3.5.4 Secure Coding

Many aspects of a secure embedded product development apply directly to the software authoring process. The following sections discuss a number of important secure coding techniques. As approximately 80% of embedded systems employ C or C++, much of the advice is directly applicable to these or similar programming languages. However, much of the advice is generic to any language. This book does not attempt a significant comparison of high-level programming languages. First, the vast majority of embedded systems organizations are faced with large legacy code bases written in a particular set of programming languages and engineering staffs trained in the those languages and their associated tool sets; picking a new language is simply not an available luxury. Second, there is much more to the efficacy of a programming language besides issues relating to security, and therefore any reasonable programming language comparison would range far beyond this book's subject matter. For readers interested in such comparisons, a plethora of online and printed resources covering this subject are available. A good place to start is the Wikipedia entry, "Comparison of programming languages."[2]

3.5.4.1 Coding Standards

Most safety and quality certification standards and guidance espouse the use of a coding standard that governs how developers write code.[3] Some of them go further to recommend or require specific rules be included in the coding standard. The goal of the coding standard is to increase reliability by promulgating intelligent coding practices. For example, a coding standard may contain rules that help developers avoid dangerous language constructs, limit complexity of functions, and use a consistent syntactical and commenting style. These rules can drastically reduce the occurrence of flaws, make software easier to test, and improve long-term maintainability.

> **Key Point**
>
> Develop and deploy a coding standard that governs software development of all critical components.

It is not uncommon for a coding standard to evolve and improve over time. For example, the development team may discover a new tool that can improve code reliability and recommend that management add a requirement that this tool be used during the development process.

It is also not uncommon to see a coding standard consisting of guidance rules whose enforcement is accomplished primarily with human code reviews. Developing a new coding standard with dozens of rules that must be verified manually is a sure way to reduce developer efficiency, even if it increases the reliability of the code.

> **Key Point**
>
> Maximize the use of automated verification of the coding standard; minimize the use of manually verified coding rules.

Numerous static code analyzers, and some compilers, can automate large portions of a typical secure coding standard. Furthermore, although some coding standard rules are necessarily language-specific, there are some universally or almost universally applicable rules that should be a part of a high-quality coding standard. Assuming they are improving software quality, the best coding standard rules are those whose enforcement can be automated and are applicable to any software project. Some examples are provided in this chapter.

> **Key Point**
>
> Prohibit compiler warnings.

Compilers and other tool chain components (e.g., the linker/loader) often emit warnings, as opposed to halting a build with a fatal error. A warning is an indicator to the developer that a construct may be technically legal but questionable, such as exercising a corner of the language that is not well defined. Such constructs are not infrequently the cause of subtle bugs. To ensure that developers do not intentionally or accidentally ignore warnings, tell the compiler to treat all warnings as errors. Many compilers have such an option.

> **Key Point**
>
> Take advantage of the compiler's strictest language settings for security and reliability.

Compilers also tend to provide a variety of strictness levels in terms of language standard interpretation. In addition, some compilers are capable of warning the developer about constructs that are technically legal but dangerous. For example, the Motor Industry Software Reliability Association (MISRA) has published guidelines for the use of the C language in critical systems,[4] and some compilers can optionally enforce some or all of these guidelines that essentially subset the language by excluding constructs believed to lead to unreliable software. Some MISRA guidelines are advisory and may yield warnings instead of errors; once again, if the MISRA rule is enabled, the compiler should be forced to generate a fatal build error on any non-compliant construct.

The authors are certainly not recommending that all development organizations adopt full MISRA compliance as part of their coding standards. On the contrary, there are good reasons for not adopting the entire standard. Once management decides to enable a MISRA rule checker that will force product builds to fail on non-conformant source code constructs, the developers will immediately need to edit the code to fix the discovered issues. This editing phase brings cost: time spent to change the code, retesting overhead, and risk of adding new flaws during the editing process. Therefore, management must be careful when adopting new coding rules. The following case study demonstrates this need.

3.5.4.2 Case Study: MISRA C:2004 and MISRA C++:2008

Like practically any language-related standard, MISRA has many good rules along with a few rules that are either questionable or simply inappropriate for some classes of users and applications. MISRA 2004, with 141 rules, fixed a few questionable guidelines in the original MISRA 1998 standard. If MISRA is used as part of a coding standard, it may be acceptable to enforce only a subset; however, that subset must be carefully considered and approved by management. It is also important that the MISRA checker (often built directly into the compiler) be able to selectively enable and disable specific rules within individual code modules and functions.

The following is a sampling of some MISRA rules that demonstrate some of the pitfalls of the C programming language and how *selective* use of MISRA will help avoid them:

1 *Rule 7.1: Octal constants (other than zero) and octal escape sequences shall not be used.* The following example demonstrates the utility of this rule:

```
a | = 256;
b | = 128;
c | = 064;
```

The first statement sets the eighth bit of the variable *a*. The second statement sets the seventh bit of variable *b*. However, the third statement does not set the sixth bit of variable *c*. Because the constant *064* begins with a 0, it is interpreted in the C standard as an octal value. Octal 64 is equal to 0x34 in hexadecimal; the statement thus sets the second, fourth, and fifth bits of variable *c*.

Because octal numbers range from zero to seven, developers easily misinterpret them as decimal numbers. MISRA avoids this problem by requiring all constants to be specified as decimal or hexadecimal numbers.

2 Rule 8.1: Functions shall have prototype declarations and the prototype shall be visible at both the function definition and call. The MISRA informative discussion for this rule includes the sound recommendation that function prototypes for external functions be declared in a header file and then included by all source files that contain either the function definition or one of its references. It should be noted that a MISRA checker might only validate that *some* prototype declaration exists for calls to a function. The checker may be unable to validate that all references to a particular function are preceded by the *same* prototype. Mismatched prototypes can cause insidious bugs, which is worse than not having any prototype.

For example, let's consider the following C function definition and code reference, each located in a separate source file:

File1:

```
void read_temp_sensor(float *ret)
{
  *ret = *(float *)0xfeff0;
}
```

File2:

```
float poll_temperature(void)
{
  extern float read_temp_sensor(void);
  return read_temp_sensor();
}
```

The preceding code fragments are perfectly legal ANSI/ISO C. However, this software will fail since the reference and definition of *read_temp_sensor* are incompatible (the former is written to retrieve the return value of the function, and the latter is written to return the value via a reference parameter).

One obviously poor coding practice illuminated in the preceding example is the use of an extern function declaration near the code containing the reference. Although strict ANSI C requires a prototype declaration, the scope of this declaration is not covered by the specification. MISRA rule 8.6, "functions shall be declared at file scope," attempts to prevent this coding pitfall by not allowing function declarations at function code level. However, the following code fragment would pass this MISRA test yet fail in the same manner as the preceding example:

```
  extern float read_temp_sensor(void);
float poll_temperature(void)
{
  return read_temp_sensor();
}
```

While MISRA does not explicitly disallow function declarations outside header files, this restriction is an advisable coding standard addition. Declaring all functions in header files certainly makes this error less likely yet still falls short: the header file containing the declaration may not be used in the source file containing the incompatible definition.

There is really only one way to guarantee that the declaration and definition prototypes match: detect incompatibilities using a program-wide analysis. This analysis could be performed by a static code analyzer or by the full program linker/loader. We describe the linker approach here for illustration of how a high-quality tool chain can be critical to enforcing coding standards. When compiling the aforementioned code fragment, the compiler can insert into its output object file some marker, such as a special symbol in the symbol table or a special relocation entry, that describes the signature of the return type and parameter types used in a function call. When the function definition is compiled, the compiler also outputs the signature for the definition. At link time, when the final executable image is being generated, the linker/loader compares the signature for same-named functions and generates an error if any incompatible signature is detected. This additional checking should add negligible overhead to the build time (the linker already must examine the references of functions to perform relocation) yet guarantees function parameter and return type compatibility and therefore improves reliability and quality of the resulting software.

One major advantage of the link-time checking approach is the ability to encompass libraries (assuming they were compiled with this feature) whose source code may not be available for static analysis.

3 Rule 8.9: An identifier with external linkage shall have exactly one external definition. This rule is analogous to the preceding rule. Mismatched variable definitions can cause vulnerabilities that will not be caught by a standard language compiler. Let's consider the following example in which the variable *temperature* should take on only values between 0 and 255:

File1:

```
#include <stdio.h>
unsigned int temperature;
int main(void)
{
  set_temp();
```

```
  printf("temperature = %d\n", temperature);
  return 0;
}
```

File2:

```
unsigned char temperature;
void set_temp(void)
{
  temperature = 10;
}
```

Without additional error checking beyond the C standard, this program will build without error despite the mismatched definitions of *temperature*. On a big-endian machine with 32-bit *int* type and 8-bit *char* type, this function will execute as follows:

```
  temperature = 167772160
```

As with the preceding example with function prototypes, an inter-module analysis is required to detect this mismatch. And once again, the linker/loader is a sensible tool to provide this checking.

4 Rule 8.11: The static storage class specified shall be used in definitions and declarations of objects and functions that have internal linkage. Two programmers may use variables of the same name for independent purposes in independent modules within the same program. One module's modification of the variable will corrupt the other module's instance and vice versa. Furthermore, global variables may be more visible to attackers (if, for example, the global symbol table for the program is available), opening up opportunities to alter important data with malware. MISRA rule 8.11 is designed to prevent this by enforcing the generally good policy of limiting the scope of declarations to the minimum required.

While MISRA rules 8.9 and 8.11 will prevent many forms of incompatible definition and use errors, they will not prevent all such occurrences. Another example of improper symbolic resolution relates to the unintended use of exported library definitions. Libraries are often used to collect code modules that provide a related set of functions. In fact, the use of libraries to collect reusable software across projects is worthy of mention in a coding standard. For example, most operating systems come with a C library, for example, *libc.so*, that provides support for the C runtime, including string manipulation, memory management, and console input/output functions. A complex software project is likely to include a variety of project-specific libraries. These libraries export functions that can be called by application code. A reliability problem arises due to the fact that library developers and application developers may not accurately predict or define a priori the library's exported interfaces. The library may define globally visible functions intended for use only by other modules within the library. Yet once these functions are added to the global namespace at link time, the linker may resolve references made by applications that were not intended to match the definitions in the library.

For example, let's consider an application that makes use of a *print* function. The application developer envisions the use of a printing library provided by the printer management team. However, the font management team created a library, also used by the application developer, that provides a set of font manipulation functions. The font management team defines a *print* function intended for use by other modules within the font management library. However, if there does not exist a facility for limiting the namespace of libraries (the use of such a facility, if available, should be covered by the coding standard), the font library's *print* function may be inadvertently used by the linker to resolve *print* references made by the application developer, causing the system to fail.

Therefore, this problem may need to be solved by something other than the compiler's front end. One method is to use a tool chain utility program that hides library definitions so that they are used by the linker when resolving intra-library references but ignored when resolving extra-library references. The Windows platform employs user-defined library export files to accomplish this separation.[5] When creating Windows DLLs, developers specify which functions are exported. Functions not included in the export file will not be used to resolve application references.

Some high-level languages, such as C++ and Ada, do a better job of automatically enforcing type consistency and name spacing than other languages such as C. Language choice may well make certain coding standard rules trivial to enforce.

5 Rule 16.2: Functions shall not call themselves, either directly or indirectly. While directly recursive functions are easy to detect, and almost always a bad idea in resource-constrained or safety-critical embedded systems due to the risk of stack overflow, indirect recursion can be far more difficult to detect. Sophisticated applications with complex call graphs and calls through function pointers may contain unnoticed indirect recursion. This is yet another case in which an inter-module analyzer, such as the linker/loader, is required to detect cycles in a program's call graph. Handling all cases of indirect function calls, such as dynamically assigned function pointers, tables of function pointers, and C++ virtual functions, can be extremely difficult for an automated tool due to the ambiguity of potential functions that may be referenced by these pointers. A developer should try out simple test cases with a MISRA checker to see what kinds of limitations it has. If a tool vendor is unable to improve or customize the tool to meet specific needs, the developer should consider other tool choices or adopt stricter coding standard rules for limiting the use of problematic forms of indirect function calls.

MISRA for C++ was released in 2008[6] and, as one would expect, includes significant overlap with MISRA C. However, the MISRA C++ standard includes 228 rules, approximately 50%

more than the MISRA C standard. The additional ground covers rules related to virtual functions, exception handling, namespaces, reference parameters, access to encapsulated class data, and other facets specific to the C++ language.

6 Rule 9-3-2: Member functions shall not return non-const handles to class data. A simple example of a non-compliant class is as follows:

```
#include <stdint.h>
class temperature
{
public:
  int32_t &gettemp(void) { return the_temp; }
private:
  int32_t the_temp;
}
int main(void)
{
  temperature t;
  int32_t &temp_ref = t.gettemp();
  temp_ref = 10;
  return 0;
}
```

One of the major design goals of the C++ language is to promote clean and maintainable interfaces by encouraging the use of information hiding and data encapsulation. A C++ class is usually formed with a combination of internal (private) data and class member functions. The functions provide a documented interface for class clients, enabling class implementers to modify internal data structures and member implementations without affecting client portability. The preceding class member function *gettemp* returns the address of an internal data structure. The direct access of this internal data by the client violates object-oriented principles of C++. An obvious improvement (and MISRA-compliant) implementation of the preceding sample class would be as follows:

```
#include <stdint.h>
class temperature
{
public:
  int32_t gettemp(void) { return the_temp; }
  void  settemp(int32_t t) { the_temp = t; }
private:
  int32_t the_temp;
}
int main(void)
{
  temperature t;
  t.settemp(10);
  return 0;
}
```

If the temperature class owner decides that only eight bits of data are required to store valid temperatures, then she can modify the internal class without affecting the class clients:

```
#include <stdint.h>
class temperature
{
public:
  int32_t gettemp(void) { return the_temp; }
  void  settemp(int32_t t) { the_temp = t; }
private:
  int8_t the_temp;
}
```

The non-compliant implementation requires modification to the client-side code due to the size change.

3.5.4.3 Embedded C++

A number of advanced features in C++, such as multiple inheritance, can result in programming that is error prone, difficult to understand and maintain, and unpredictable or inefficient in footprint and execution speed. Because of these drawbacks, a consortium of semiconductor and developments tools vendors created a C++ subset specification called *Embedded C++* that has been in widespread use for more than a decade.[7]

> **Key Point**
>
> The goal of Embedded C++ is to provide embedded systems developers who come from a C language background with a programming language upgrade that brings the major object-oriented benefits of C++ without some of its risky baggage.

Embedded C++ removes the following features of C++:

- Multiple inheritance
- Virtual base classes
- New-style casts
- Mutable specifiers
- Namespaces
- Runtime type identification (RTTI)
- Exceptions
- Templates

One example of the rationale for Embedded C++ is the difficulty in determining the execution time and footprint of C++ exception handling. When an exception occurs, the compiler-generated exception-handling code invokes a destructor on all automatic objects constructed since the applicable *try* block was executed. The number and execution time of this destructor chain may be extremely difficult to estimate in sophisticated applications. Furthermore, the compiler-generates exception-handling code to unwind the call stack linking the handler to its original *try* block. The additional footprint may be significant and difficult to predict. Because the standard C++ runtime is compiled to support exception handling, this feature adds code bloat to programs that do not even make use of the *try* and *catch* exception-handling mechanisms.[8] For this reason, purpose-built runtime libraries supporting the reduced language subset typically accompany an Embedded C++ tool chain.

Footprint concerns also led C++ templates to be left out of the Embedded C++ standard; in some cases, the compiler may instantiate a large number of functions from a template, leading to unexpected code bloat.

Of course, some of these removed features can be extremely useful. Careful use of templates can avoid unnecessary code bloat while proving simpler, more maintainable source code interfaces.

For this reason, many compilers provide variants of Embedded C++ that enable a development organization to add back features that may be acceptable for security-critical development, especially if those features are used sensibly (such as enforcing some or all of the rules of MISRA C++). For example, Green Hills Software's C++ compiler provides options for allowing the use of templates, exceptions, and other individual features with the Embedded C++ dialect (along with enabling MISRA checking).

3.5.4.4 Complexity Control

Much has been published regarding the benefits of reducing complexity at the function level. Breaking up a software module into smaller functions makes each function easier to understand, maintain, and test.[9] One can think of this as *meta-partitioning*: applying the aforementioned software componentization paradigm at a lower, programmatic, level. A complexity-limitation coding rule is easily enforced at compile time by calculating a complexity metric and generating a compile-time error when the complexity metric is exceeded. Once again, since the compiler is already traversing the code tree, it does not require significant additional build time to apply a simple complexity computation, such as the popular McCabe metric. Because the compiler generates an actual error pointing out the offending function, the developer is unable to accidentally create code that violates the rule.

Key Point

Use automated tools to enforce a complexity metric maximum and ensure that this maximum is meaningful (such as a McCabe value of 20).

Adopting a coding standard rule that allows a McCabe complexity value of 200 is useless; most legacy code base will be compliant despite having spaghetti-like code that is hard to understand, test, and maintain. The selection of a specific maximum complexity value is open to debate. If an existing code base is well modularized, a value may be selected that allows most of the properly partitioned code to compile; future code will be held to the same stringent standard. When the complexity metric is applied to a large code base that has previously not been subjected to such an analysis, it is likely that a small number of large functions will fail the complexity test. Management then needs to weigh the risk of decomposing a large function with the risk of changing the code at all. Modifying a piece of code that, while complex, is well exercised (proven in use) and serves a critical function may reduce reliability by increasing the probability of introducing a flaw. The complexity enforcement tool should provide a capability to allow exceptions to the complexity enforcement rule for specific functions that meet this profile. Exceptions, of course, should always be approved by management and documented as such. The coding standard should not allow exceptions for code that is developed subsequent to the adoption of the coding rule. These types of coding standard policies conform to their spirit while maximizing efficiency, enabling them to be employed effectively in legacy projects.

3.5.4.5 Static Source Code Analysis

Static source code analyzers attempt to find code sequences that, when executed, could result in buffer overflows, resource leaks, or many other security and reliability problems. Source code analyzers are effective at locating a significant class of flaws that are not detected by compilers during standard builds and often go undetected during runtime testing as well.

Most static source code analyzers use the same type of compiler front end that is used to compile code. In fact, ideally, a static source code analyzer should be integrated with the everyday compiler to maximize use and reduce complexity of the tool chain. In addition, integrated checking enables source code parsing to be performed only once instead of twice. The use of a compiler front end is only natural because the analyzer takes advantage of pre-existing compiler dataflow algorithms to perform its bug-finding mission.

A typical compiler will issue warnings and errors for some basic potential code problems, such as violations of the language standard or use of implementation-defined constructs. In contrast, a static source code analyzer performs a full program analysis, finding bugs caused by complex interactions between pieces of code that may not even be in the same source file (see Figure 3.1).

The analyzer determines potential execution paths through code, including paths into and across subroutine calls, and how the values of program objects (such as standalone variables or fields within aggregates) could change across these paths. The objects could reside in memory or in machine registers.

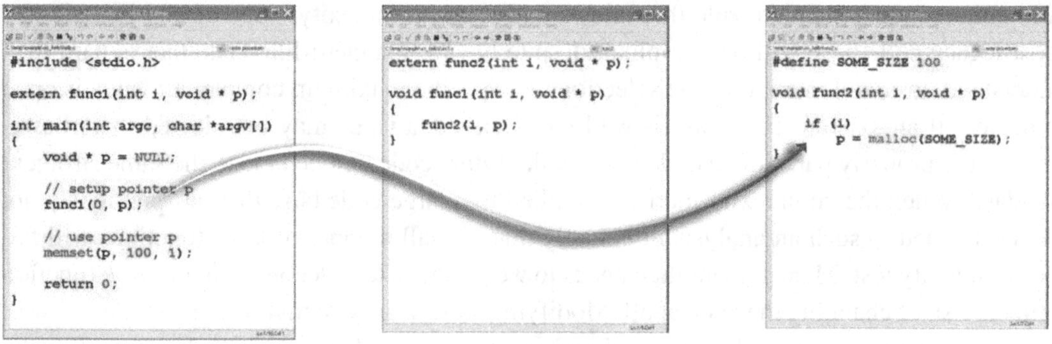

Figure 3.1:
Inter-module static analysis.

The analyzer looks for many types of flaws. It looks for bugs that would normally compile without error or warning. The following is a list of some of the more common errors that a modern static source code analyzer will detect:

- Potential NULL pointer dereferences
- Access beyond an allocated area (e.g., array or dynamically allocated buffer); otherwise known as a buffer overflow
- Writes to potentially read-only memory
- Reads of potentially uninitialized objects
- Resource leaks (e.g., memory leaks and file descriptor leaks)
- Use of memory that has already been deallocated
- Out-of-scope memory usage (e.g., returning the address of an automatic variable from a subroutine)
- Failure to set a return value from a subroutine
- Buffer and array underflows

The analyzer also has knowledge about how many standard runtime library functions behave. For example, the analyzer knows that subroutines such as *free* should be passed pointers to memory allocated by subroutines such as *malloc*. The analyzer uses this information to detect errors in code that calls or uses the result of a call to these functions.

3.5.4.5.1 Limiting False Positives

The analyzer can also be taught about properties of user-defined subroutines. For example, if a custom memory allocation system is used, the analyzer can be taught to look for misuses of this system. By teaching the analyzer about properties of subroutines, users can reduce the number of false positives. A false positive is a potential flaw identified by the analyzer that could not actually occur during program execution. Of course, one of the major design goals of a static source code analyzer is to minimize the number of false positives so that developers can minimize time looking at them. If an analyzer generates too many false positives, it will become irrelevant because engineers will ignore the output. A modern static source code analyzer is much better at limiting false positives than traditional UNIX analyzers such as *lint*. However, since a static analyzer is not able to understand complete program semantics, it is not possible to totally eliminate false positives. In some cases, a flaw found by the analyzer may not result in a fatal program fault, but could point to a questionable construct that should be fixed to improve code clarity. A good example of this is a write to a variable that is never subsequently read.

3.5.4.5.2 Case Study: Open Source Internet Security Applications

To help demonstrate the types of coding errors that can be efficiently detected and prevented using static source code analysis, we consider a case study of three popular, security-critical open source applications—*Apache, OpenSSL,* and *sendmail*—that were analyzed using Green Hills Software's *DoubleCheck* static source code analyzer.

The *Apache* open source hypertext transfer protocol (HTTP) server is the most popular web server in the world, powering a majority of the websites on the Internet. Given the ubiquity of *Apache* and the world's dependence on the Internet, the reliability and security of *Apache* represent an important concern for all of us. A serious flaw in *Apache* could cause widespread inconvenience, financial loss, or worse. The *Apache* web server consists of approximately 200,000 lines of code, 80,000 individual executable statements, and 2,000 functions.

OpenSSL is an open source implementation of Secure Sockets Layer (SSL) and Transport Layer Security (TLS) as well as a comprehensive cryptographic algorithm library. TLS is the modern reimplementation of SSL, although SSL is often used as a general term covering both protocols. TLS and OpenSSL are discussed in more detail in Chapter 5.

SSL forms the basis of much of the secure communication on the Internet. For example, SSL is what enables users to send private credit card information securely from their browsers to an online merchant's remote server. In addition to being intimately involved with data communication, *OpenSSL* contains implementations of a variety of cryptographic algorithms used to secure the data in transit. *OpenSSL* is available for Windows; however, *OpenSSL* is the standard SSL implementation for Linux and UNIX worldwide. In addition, because of its liberal licensing terms (not GPL), *OpenSSL* has been used as a basis for a number of commercial offerings. Like *Apache*, *OpenSSL* is a keystone of worldwide secure Internet communication. Flaws in this software could have widespread deleterious consequences. *OpenSSL* consists of approximately 175,000 lines of code, 85,000 individual executable statements, and 5,000 functions.

Although its use is in decline, *sendmail* is among the most popular electronic mail server software used in the Internet. *Sendmail* has been the de facto electronic mail transfer agent for UNIX (and subsequently, Linux) systems since the early 1980s. Given the dependence on electronic mail, the stability and security of *sendmail* is certainly an important concern for many. The name *sendmail* might lead one to think that this application is not very complicated. Anyone who has ever tried to configure a *sendmail* server knows otherwise. *Sendmail* consists of approximately 70,000 lines of code, 32,000 individual executable statements, and 750 functions.

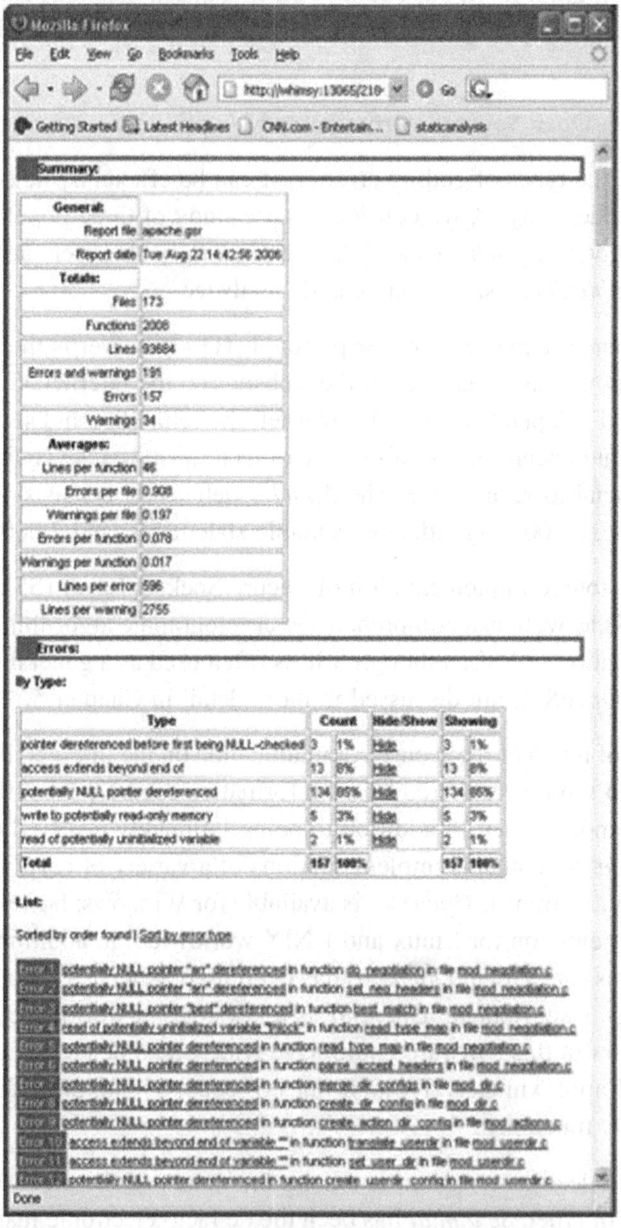

Figure 3.2:
Static source code analyzer summary report.

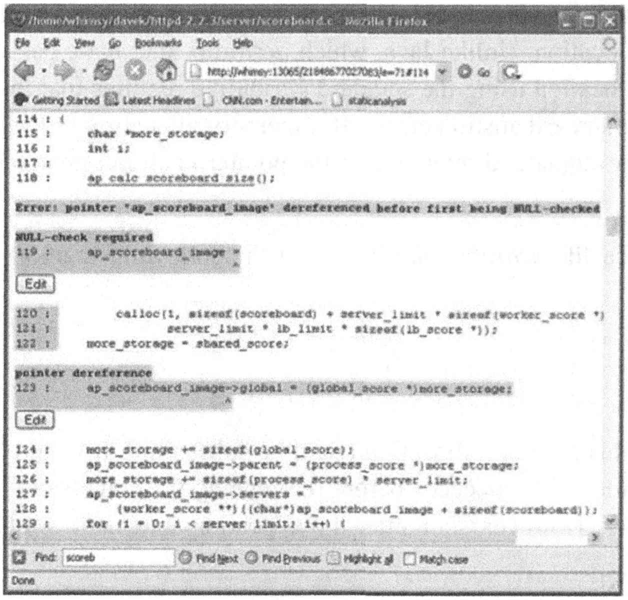

Figure 3.3:
In-context display of source code vulnerability.

3.5.4.5.2.1 Output of a Static Source Code Analyzer Many leading source code analyzers generate an intuitive set of web pages, powered by an integrated web server. The developer can browse high-level summaries of the different flaws found by the analyzer (see Figure 3.2) and then click on hyperlinks to investigate specific problems. Within a specific problem display, the error is displayed inline with the surrounding code, making it easy to understand (see Figure 3.3). Function names and other objects are hyperlinked for convenient browsing of the source code. Since the web pages are running under a web server, the results can easily be shared and browsed by any member of the development team.

The following sections provide examples of actual flaws in *Apache*, *OpenSSL*, and *sendmail* that were discovered by DoubleCheck. The results are grouped by error type, with one or more examples of each error type per section.

3.5.4.5.2.2 Potential NULL Pointer Access By far the most common flaw found by the analyzer in all three suites under testing was potential NULL pointer access. Many cases involved calls to memory allocation subroutines that were followed by accesses of the returned pointer without first checking for a NULL return. This is a robustness issue. Ideally, all memory allocation failures are handled gracefully. If there is temporary memory exhaustion, service may falter but not terminate. This is of particular importance to server programs such as *Apache* and *sendmail*. Algorithms can be introduced that prevent denial of service in overload conditions such as that caused by a malicious attack.

The *Apache* web server, *sendmail*, and *OpenSSL* all make profligate use of C runtime library dynamic memory allocation. Unlike Java, which performs automatic garbage collection, dynamic memory allocation using the standard C runtime requires that the application itself handle potential memory exhaustion errors. If a memory allocation fails and returns a NULL pointer, a subsequent unguarded reference of the pointer is all but guaranteed to cause a fatal crash.

In the *Apache* source file *scoreboard.c*, we have the following memory allocation statement:

```
ap_scoreboard_image = calloc(1,sizeof(scoreboard) +
  server_limit * sizeof(worker_score *) +
  server_limit * lb_limit * sizeof(lb_score *));
```

Clearly, the size of this memory allocation could be substantial. It would be a good idea to make sure that the allocation succeeds before referencing the contents of *ap_scoreboard_image*. However, soon after the allocation statement, we have this use:

```
ap_score_board_image->global = (global_score
  *)more_storage;
```

The dereference is unguarded, making the application susceptible to a fatal crash. Another example from *Apache* can be found in the file *mod_auth_digest.c*:

```
entry = client_list->table[idx];
prev = NULL;
while (entry->next){/* find last entry */
  prev = entry;
  entry = entry->next;
  ...
}
```

The variable *entry* is unconditionally dereferenced at the beginning of the loop. This alone would not cause the analyzer to report an error. At this point in the execution path, the analyzer has no specific evidence or hint that entry could be NULL or otherwise invalid. However, the following statement occurs after the loop:

```
if (entry) {
  ...
}
```

By checking for a NULL entry pointer, the programmer has indicated that entry could be NULL. Tracing backward, the analyzer now sees that the previous dereference to entry at the top of the loop is a possible NULL reference.

The following similar example was detected in the *sendmail* application, in the file *queue.c*, where the code unconditionally dereferences the pointer variable *tempqfp*:

```
errno = sm_io_error(tempqfp);
```

sm_io_error is a macro that resolves to a read of the *tempqfp->f_flags* field. Later in the same function, we have this NULL check:

```
if (tempqfp != NULL)
  sm_io_close(tempqfp, SM_TIME_DEFAULT);
```

In addition, there are no intervening writes to *tempqfp* after the previously noted dereference. The NULL check, of course, implies that *tempqfp* could be NULL; if that were ever the case, the code would fault. If the pointer can never in practice be NULL, then the extra check is unnecessary and misleading. What may seem harmless sloppiness can translate into catastrophic failure under certain conditions.

In *sendmail*, there are many other examples of unguarded pointer dereferences that are either preceded or followed by NULL checks.

The final example in this category comes from *OpenSSL*, in file *ssl_lib.c*:

```
if (s->handshake_func == 0) {
  SSLerr(SSL_F_SSL_SHUTDOWN, SSL_R_UNINITIALIZED);
}
```

Shortly thereafter, we have a NULL check of the pointer *s*:

```
if ((s != NULL) && !SSL_in_init(s))
```

Again, the programmer is telling us that *s* could be NULL, yet the preceding deference is not guarded.

3.5.4.5.2.3 Buffer Underflow A buffer underflow is defined as an attempt to access memory before an allocated buffer or array. Similar to buffer overflow, buffer underflows cause insidious problems due to the unexpected corruption of memory. The following flaw in file *queue.c* in *sendmail* was discovered by static analysis:

```
if ((qd == -1 || qg == -1) &&
  type != 120)
  ...
else {
  switch (type) {
  ...
  case 120:
    if (bitset(QP_SUBXF,
      Queue[qg]->qg_qpaths[qd].qp_subdirs))
        ...
  }
}
```

As you can see, the *if* statement implies that it is possible for *qd* or *qg* to be −1 when *type* is 120. But in the subsequent *switch* statement, always executed when *type* is 120, the *Queue* array is unconditionally indexed through the variable *qg*. If *qg* were −1, this would be an underflow. The program was not studied exhaustively to determine whether *qg* can indeed be −1 when *type* is 120 and hence reach the fault. However, if *qg* can't be −1 when *type* is 120, then the initial *if* check is incorrect, misleading, and/or unnecessary.

Another example of buffer underflow is found in file *ssl_lib.c* in *OpenSSL*:

```
p = buf;
sk = s->session->ciphers;
for (i = 0; i < sk_SSL_CIPHER_num(sk); i++) {
  ...
  *(p++)=':';
}
p[-1] = '\0';
```

The analyzer informs us that the underflow occurs when this code is called from file *s_server.c*. From a look at the call site in *s_server.c*, it is clear that the analyzer has detected that *buf* points to the beginning of a statically allocated buffer. Therefore, in the *ssl_lib.c* code, if there are no ciphers in the cipher stack *sk*, then the access *p[−1]* is an underflow. This demonstrates the need for an inter-module analysis, since there would be no way of knowing what *buf* referenced without examining the caller.

If it is the case that the number of ciphers cannot actually be 0 in practice, then the *for* loop should be converted to a *do* loop to make it clear that the loop must always be executed at least once (ensuring that *p[−1]* does not underflow).

Another problem is a potential buffer overflow. No check is made in *the ssl_lib.c* code to ensure that the number of ciphers does not exceed the size of the *buf* parameter. Instead of relying on convention, a better programming practice would be to pass in the length of *buf* and then add code to check that overflow does not occur.

3.5.4.5.2.4 ***Resource Leak*** In file *speed.c* in *OpenSSL*:

```
fds=malloc(multi*sizeof *fds);
```

fds is a local pointer and is never used to free the allocated memory prior to return from the subroutine. Furthermore, *fds* is not saved in another variable where it could be later freed. Clearly, this is a memory leak. A simple denial-of-service attack on *OpenSSL* would be to invoke or cause to be invoked the *speed* command until all of memory is exhausted.

Many would argue that the code quality of such popular open source applications is expected to be relatively high. As one person put it, "By sharing source code, open source developers make software more robust. Programs get used and tested in a wider variety of contexts than one programmer could generate, and bugs get uncovered that otherwise would not be found."[10] Unfortunately, in a complex software application such as *Apache*, it is simply not feasible for all flaws to be found by manual inspection. In addition to this case study, other commercial static code analyzers have been used successfully on large open source applications, including the Linux operating system, to locate numerous latent security vulnerabilities.

Numerous mechanisms are available to help in the struggle to improve software quality, including improved testing and design paradigms. But automated source code analyzers are one of the most promising technologies.

> **Key Point**
>
> Use of static analysis should be a required part of every security-conscious software organization's development process.

3.5.4.5.3 Which Static Analyzer Should an Organization Use?

The best answer to this question is that a development organization should use multiple tools from different vendors. Empirical use within government software safety and security evaluation teams has demonstrated that a surprising majority of software flaws caught by one static analyzer will not be caught by another tool, and vice versa. Many forms of full-program static analysis are inherently intractable, requiring carefully tuned heuristic algorithms to provide high-quality results.

> **Key Point**
>
> The best coverage for software flaw detection via static analysis requires multiple tools from multiple vendors to be used in concert.

In addition to accuracy, there are large differences in the execution time of static analyzers. A recommended approach is to employ at least one runtime efficient analysis pass during everyday software builds executed by individual developers and relegate the remainder of the available tools to offline execution that can asynchronously notify the development team of discovered flaws. Since some compilers include built-in full program static analysis, development teams should consider enabling this feature as a compile option for all builds.

The U.S. Food and Drug Administration's Center for Device and Radiological Health (CDRH) uses static source code analyzers as a forensics tool to help locate causes of medical device failures.[11] In some cases, several different static analyzers are used in concert. Similarly, the U.S. National Security Agency (NSA) uses multiple static analyzers to help perform security vulnerability assessments on software.

Development organizations should consider evaluating numerous products for both execution efficiency as well as quality of output on the same code base. Pick a combination of tools that provide excellent flaw detection coverage while offering sufficient execution time efficiency to enable developers to use at least one of them on every compile.

3.5.4.6 Creating a Tailored Organizational Embedded Coding Standard

So far in this chapter we've discussed the importance of having a high-quality coding standard to limit vulnerabilities and therefore improve software reliability and security. We have discussed how the coding standard should require the application of numerous proven code-quality techniques, including static source code analysis, MISRA rule checking, and complexity control. We've also stressed (over and over, and we're not done yet) how important it is to automate as much as possible the implementation of this coding standard; if the developers' everyday tool chain is always enforcing the coding standard, then the software security techniques will become assimilated into the minds of all engineers and managers. In the same way that a professional golfer relies on muscle memory to create a repeatable swing, embedded software developers must think software security every day to ensure that lapses do not occur.

3.5.4.7 Preparing for a One-Time Retrofit Cost

Unfortunately, the task of writing a coding standard that requires all these great ideas and turning on these software analyzers in the tool chain is not as simple or straightforward as it may sound.

> **Key Point**
>
> For organizations new to static analyzers, MISRA, and other automated code-quality enforcement tools discussed in this chapter, managers and their development teams need to prepare themselves for what may be a painful initial deployment.

For example, enabling a static source code analyzer for the first time in a large code base is almost certain to identify hundreds, if not thousands, of problematic code sequences. Some of these will undoubtedly be false positives that must be worked around by modifying the code to mollify the analyzer. Numerous real flaws will be discovered, and the organization will be encouraged by their eradication, even though the review and correction of the identified flaws may require significant resource investment. Nevertheless, experience has shown that once the code base has been made "coding standard clean," keeping it that way will become routine.

3.5.4.8 Allowing for Management-Approved Exceptions to Reduce Regressions

Developers must take extreme care when modifying software that has been around a long time (and possibly fielded for a long time). Some studies have shown that static analyzers and checkers have the potential to harm software security by introducing new flaws when correcting identified problems.[12] This risk is especially high when retrofitting a new coding standard or new analyzer tool to a code base.

In some cases, it may be prudent to disable a check for a particular piece of software rather than take the risk of modifying it. For example, let's suppose management decides to retrofit a new rule limiting all functions to a maximum McCabe complexity value of 20. A large code base may include dozens or even hundreds of functions that initially fail to meet this metric. Some of the offending subroutines may be straightforward to refactor. Others may be difficult and risky. In fact, some of the worst offenders that are begging for a rewrite may be the exact wrong ones to change, especially if they are providing a security-critical function. If the software includes a hand-coded AES cryptographic algorithm that has been painstakingly developed to be side-channel attack-resistant and has been through FIPS certification, perhaps the best approach is to leave this function alone as a documented exception to the coding standard.

Key Point

Management must allow for exceptions to coding standard rules for specific portions of critical software in which the risk and cost of retrofit clearly exceed the benefit.

Obviously, this judgment is subjective, and management must be careful not to fall into the trap of using exceptions as a means to avoid resource expenditure.

3.5.4.9 Language Standards Are Never Perfect

Earlier in this chapter, we discussed how Embedded C++ was created to provide a simpler, more efficient subset dialect of the complete C++ language standard. Then we went on to explain how some organizations might find this subset too limiting. This is a good example of the need to customize automated software quality tools, especially those based on generic open standards.

The MISRA standard provides another good example. A draconian management edict to conform to MISRA is almost certain to be impractical, generate much internal strife, and breed a counterproductive burden throughout the development organization. For example, imagine that one of your embedded software projects uses dynamic memory allocation liberally. Many C++ developers would laugh if asked to follow a standard that precludes all uses of the *new* and *delete* operators. Yet this is exactly what MISRA conformance requires (Rule 18-4-1 of MISRA C++:2008 and rule 20.4 of MISRA C:2004). While it may make sense to enforce this rule for certain hard real-time or security-critical applications that require complete determinism, many other general-purpose applications may benefit from dynamic memory allocation. Management, in consultation with their development teams, must examine each MISRA rule in turn and determine which rules represent safe and practical options. While it is advisable to include this default set in the coding standard, programmers must be prepared to allow for project-specific exceptions as well as management-approved exceptions in any piece of code. Most automated MISRA checking tools will allow for the selection of individual rules within the standard.

Another rule that has been met with dismay in some embedded software organizations is MISRA C:2004 20.9: "The input/output library *<stdio.h>* shall not be used in production code." The rationale argues that a large number of C standard I/O functions include undefined and/or implementation-defined behavior. Furthermore, "It is assumed within this document that they [streams and file I/O] will not normally be needed in embedded systems." One obvious example of an important security-critical function made easier with C standard I/O is the use of *sprintf* to create messages that form audit log records. Without a reasonably conservative use of some C standard I/O functionality, programmers may be forced to create more complex (and therefore more prone to error and security vulnerabilities) audit logging or other security-critical code.

Some MISRA rules are based on sound ideas that may be overly restrictive and in need of customization. A good example is the following:

> *Rule 20.1: "Assignment operators shall not be used in expressions that yield a Boolean value."*

The following code fragment is flagged as nonconforming:

```
if (a = b) { ... }
```

Did the programmer really intend to simultaneously assign and test the value of *a*? Or perhaps the programmer intended to compare *a* to *b* but left out the second equal sign—a rather common flaw. This is a case that clearly represents poorly constructed software that may indicate programmer error. However, the following code sequence is also disallowed by MISRA:

```
if ((a = b) <= c) { ... }
```

This construction is far more reasonable. In this case, the programmer's use of an assignment is more explicit, distinct from the comparison, and hence not likely to be an error. Combining assignments with conditionals in this manner may actually improve readability and maintainability. Unfortunately, there is no language standard that would allow the preceding example while disallowing the previous one. Some software quality-checking tools provide a significant collection of customized options like this. The programmer's best bet is to talk to the tools provider and see what kind of customizability is possible in this realm.

3.5.4.10 Case Study: Green Hills Standard Mode

Green Hills Software has implemented MISRA checking, static analysis, complexity control, and many other automated software quality enforcement techniques into its internally deployed checking tool, the Green Hills compiler and tool chain. While Green Hills has made these capabilities available to its customers, the actual coding standard used within the Green Hills development organization includes a carefully tailored set of automated facilities that represent a comprehensive yet practical software security system. This system has been proven in use for 30 years and used to create software certified to the highest levels of safety and security.

Recently, Green Hills decided to make its tailored set of rules available to users of the Green Hills tool chain using a special build option that specifies the latest and greatest available code-checking options: *–coding_standard=ghstd<year>*. For example, *–coding_standard=ghsstd2010* is the option for the recommended set of rules in a release of the tool chain made in the year 2010. In subsequent releases, users have the option of enabling the compatible set of 2010 rules or trying the newer versions. The rules are defined in a special coding standard profile (.csp) file that can be customized by end users. A development organization can take the *ghsstd2010.csp* file and modify it to create *yourstandard.csp* and then add the following option to build software: *–coding_standard=yourstandard.csp*. As an example, diagnostic number 187 specified in *ghstd2010.csp* enables the previously discussed, more relaxed version of MISRA C:2003 Rule 20.1 and is referred to as "*use of '=' where '==' may have been intended.*"

Let's look at another example of how the customization of automated software quality tools can be beneficial. Let's consider the following MISRA C:2004 rule:

> *Rule 15.3: The final clause of a* switch *statement shall be the* default *clause.*

The rationale is to force the programmer to handle all possible values of the *switch* expression. The following example, however, demonstrates a situation in which this requirement is overkill:

```
typedef enum { red, yello, blue } colors;
void carcolors(colors c)
{
  switch (c) {
    case red:
      printf("red\n");
      break;
    case yellow:
      printf("yellow\n");
      break;
    case blue:
      printf("blue\n");
      break;
  }
}
```

The preceding code is not MISRA-compliant and would be flagged as an error. However, because the *switch* expression is of enumerated type whose range of possible values is covered completely by the statement's *case* clauses, the addition of a *default* clause is unnecessary:

```
typedef enum { red, yellow, blue } colors;
void carcolors(colors c)
{
  switch (c) {
    case red:
      printf("red\n");
      break;
    case yellow:
      printf("yellow\n");
      break;
    case blue:
      printf("blue\n");
      break;
    default:
      printf("Unexpected color!\n");
      break;
  }
}
```

In fact, the addition of the *default* clause in the preceding snippet would add unreachable code (ironically, in violation of MISRA C:2004 Rule 14.1, which prohibits unreachable code) and make the function confusing to readers and maintainers. For these reasons, the *−coding_standard=ghstd2010* option slightly relaxes the MISRA rule and permits the lack of a *default* clause if and only if all possible expression values are covered by the *switch* statement's other *case* clauses.

Another customization example, also involving switch statements, is related to yet another MISRA C:2004:

> *Rule 15.2: "An unconditional* break *statement shall terminate every non-empty* switch *clause."*

The rationale is that programmers sometimes forget to add a *break* statement, causing control flow to fall through to the next *case* clause with unintended consequences. While Rule 15.2 is generally a good one to follow, on occasion a *switch* statement is cleaner when a *case* clause can fall through to avoid duplicating code. However, the developer must indicate her intention—for example, with an appropriate comment just before the fall-through point. To allow for this situation, the *−coding_standard=ghstd2010* option permits a fall-through if and only if the vernacular *FALLTHRU* comment is so included. It is interesting to note that the C# language, by default, does not permit *case* clause fall-through. And C#, like MISRA, does not allow for exceptions to this rule. The following example is not MISRA-compliant but is permitted by the Green Hills default coding standard option:

```
switch (i) {
  case 0:
  handle_zero();
    /* FALLTHRU */
  case 1:
  case 2:
  case 3:
    process(i);
    audit("handle_input", i);
    post_buffer();
    break;
  default:
    audit("Unexpected input", i);
    reset();
    break;
}
```

When a developer is considering how to write this code in a MISRA-compliant manner, the most obvious approach would be to duplicate the code in the 1−3 clause for the 0 clause or create a new function containing the code to be shared by the 0−3 case clauses. However, these alternatives create unnecessary complexity. The *FALLTHRU* comment makes the developer's intentions clear with no extra code. C# has a reasonable workaround using a *goto* statement that is syntactically illegal in C or C++ and not MISRA-compliant:

```
switch (i) {
  case 0:
  handle_zero();
  goto case 1;
  case 1:
  case 2:
  case 3:
    process(i);
    audit("handle_input", i);
    post_buffer();
    break;
  default:
    audit("Unexpected input", i);
    reset();
    break;
}
```

3.5.4.11 Importance of Management Resolve

Change is always difficult, and experience has shown that the best intentions of incorporating important code-quality controls all too often fall by the wayside when the efforts hit snags such as disagreements with rules or retrofit edits, tedious coding workarounds, or product development schedule pressures.

> **Key Point**
>
> It is absolutely critical that management decide the course, invest in the initial retrofit cost and schedule, and then stay the course until a high-quality coding standard is firmly entrenched into the development system and indoctrinated throughout the engineering organization.

Implementing a new process is like entering the playoffs in professional basketball. The winner is often not the best shooting team but rather the team that has the fortitude to play a strong, stubborn defense for every play of every game of the series. Managers need to be stubborn and demand compliance until the project is complete.

3.5.4.12 Case Study: Netrino Embedded C Coding Standard

Netrino is an embedded systems consulting firm whose president, Michael Barr, has authored an embedded software coding standard for C programmers.[13] For development organizations looking to create a coding standard for the first time, *Embedded C Coding Standard* is a worthwhile read and an excellent starting point.

The Netrino coding standard dedicates approximately one-half of its rules to stylistic concerns, ranging from the use of braces, spaces, and alignment to naming conventions and comment style. While developers new to coding standards may be surprised to see this emphasis on arguably subjective content, the importance of stylistic standards should not be underestimated. These rules promote code readability and maintainability that ultimately contributes to software reliability and security. Many software style rules can be automatically enforced with code beautifiers such as *Uncrustify*, reducing peer review workload.

The Netrino coding standard also attempts to indicate when rules should be enforceable with automated tools. For example, one excellent rule, 4.3.c, "*Each source file shall be free of unused include files*" is advisable for any coding standard. Extraneous include files can occasionally harm code efficiency (e.g., a reference within a header file could cause library code to be pulled into a link when it otherwise would not), increase code complexity for no benefit, and make software more difficult to maintain (e.g., a change in the header file may affect the source file including it). The Netrino standard points out that this rule can be automatically enforced with *lint*, a lightweight open source static analyzer. While *lint* (and its brethren, such as *splint*[14]) may require significant investment in terms of configuration to avoid false positives, the tool may be a reasonable addition to the developer's toolbox, alongside more professional static analyzers that specialize in locating sophisticated software bugs while minimizing false positives.

Netrino's standard requires the use of C99 dialect if available. As of this writing, C99 is the latest ratified version of C programming language[15] and adds several desirable features for code quality, such as the ability to use C++ style // comments and the ability to declare *for* loop iteration variables within the initial loop statement. Before allowing C99, however, developers should consider all its additions; features such as variable-length arrays may be highly undesirable (cause unexpectedly high runtime stack use and increase the probability of overflow). They should consult with embedded software specialists like Netrino and/or development tool suppliers to see what level of configuration of C99 may be possible within the tool chain. At the time of this writing, another revision of the C language, called C1X,[16] is in the works. The latest revision of C++, C++11, was released in August 2011.[17] One welcome software reliability addition to C++11 is *static_assert*, compile-time assertions. While these can be implemented in the older versions of C or C++ using typedef trickery, the acceptance into the standard should encourage developers to make good use of them. Unlike runtime assertions, which incur execution overhead (often leading them to be omitted from a final product), static assertions have no impact on program execution. C++ developers are encouraged to explore the C++11 standard and consult with their development tool suppliers for availability of conformant products.

The Netrino standard disallows unconditional jumps: *goto*, *continue*, and the use of a *break* statement to exit a loop. While this is a sensible rule and certainly a good idea for new code, experience has shown that retrofitting this rule to large code bases can be hazardous. Some code bases simply make profligate use of these jumps, especially the *goto* statement. This rule is also one that is likely to be met with vigorous argument over a reasonable use of the *goto* statement—a means to centralize error-handling code that, by design, requires unconditional termination of processing. For example, the following open source Integrated Device Electronics (IDE) device driver function (portions elided for brevity) that handles the many cases of writes to an I/O port uses no fewer than 18 *goto* statements, all of which are used to terminate processing due to a fatal error condition and execute some cleanup work:

```
static void ide_ioport_write(void *opaque, uint32_t addr, uint32_t val)
{
  IDEState *ide_if = opaque;
  IDEState *s;
  int unit, n;
  int lba48 = 0;
#ifdef DEBUG_IDE
  printf("IDE: write addr=0x%x val=0x%02x\n", addr, val);
#endif
  addr &= 7;
  switch(addr) {
    case WIN_READ:
    case WIN_READ_ONCE:
      if (!s->bs)
        goto abort_cmd;
      ide_cmd_lba48_transform(s, lba48);
      s->req_nb_sectors = 1;
      ide_sector_read(s);
      break;
```

```
case WIN_MULTREAD:
  if (!s->mult_sectors)
    goto abort_cmd;
  ide_cmd_lba48_transform(s, lba48);
  s->req_nb_sectors = s->mult_sectors;
  ide_sector_read(s);
  break;
case CFA_WRITE_MULTI_WO_ERASE:
  if (!s->mult_sectors)
    goto abort_cmd;
  ide_cmd_lba48_transform(s, lba48);
  s->error = 0;
  s->status = SEEK_STAT | READY_STAT;
  s->req_nb_sectors = s->mult_sectors;
  n = s->nsector;
  if (n > s->req_nb_sectors)
    n = s->req_nb_sectors;
  ide_transfer_start(s, s->io_buffer, 512 * n,
       ide_sector_write);
  s->media_changed = 1;
  break;
case WIN_READDMA:
case WIN_READDMA_ONCE:
  if (!s->bs)
    goto abort_cmd;
  ide_cmd_lba48_transform(s, lba48);
  ide_sector_read_dma(s);
  break;
case WIN_WRITEDMA:
case WIN_WRITEDMA_ONCE:
  if (!s->bs)
    goto abort_cmd;
  ide_cmd_lba48_transform(s, lba48);
  ide_sector_write_dma(s);
  s->media_changed = 1;
  break;
case WIN_SETFEATURES:
  if (!s->bs)
    goto abort_cmd;
  switch(s->feature) {
  case 0xcc: /* reverting to power-on defaults enable */
  case 0x66: /* reverting to power-on defaults disable */
  case 0x02: /* write cache enable */
  case 0x82: /* write cache disable */
  case 0xaa: /* read look-ahead enable */
  case 0x55: /* read look-ahead disable */
  case 0x05: /* set advanced power management mode */
  case 0x85: /* disable advanced power management mode */
  case 0x42: /* enable Automatic Acoustic Mode */
  case 0xc2: /* disable Automatic Acoustic Mode */
```

```
    s->status = READY_STAT | SEEK_STAT;
    ide_set_irq(s);
    break;
  case 0x03: { /* set transfer mode */
  uint8_t val = s->nsector & 0x07;
  switch (s->nsector >> 3) {
    case 0x00: /* pio default */
    case 0x01: /* pio mode */
     put_le16(s->identify_data + 63,0x07);
     put_le16(s->identify_data + 88,0x3f);
     break;
    case 0x04: /* mdma mode */
     put_le16(s->identify_data + 63,0x07 | (1 << (val +
        8)));
     put_le16(s->identify_data + 88,0x3f);
     break;
    case 0x08: /* udma mode */
     put_le16(s->identify_data + 63,0x07);
     put_le16(s->identify_data + 88,0x3f | (1 << (val +
        8)));
     break;
    default:
     goto abort_cmd;
  }
  s->status = READY_STAT | SEEK_STAT;
  ide_set_irq(s);
  break;
 }
 default:
   goto abort_cmd;
}
break;
case WIN_SRST:
  if (!s->is_cdrom)
    goto abort_cmd;
  ide_set_signature(s);
  s->status = 0x00; /* NOTE: READY is _not_ set */
  s->error = 0x01;
  break;
case WIN_PACKETCMD:
  if (!s->is_cdrom)
    goto abort_cmd;
  /* overlapping commands not supported */
  if (s->feature & 0x02)
    goto abort_cmd;
  s->status = READY_STAT;
  s->atapi_dma = s->feature & 1;
  s->nsector = 1;
  ide_transfer_start(s, s->io_buffer, ATAPI_PACKET_SIZE,
          ide_atapi_cmd);
```

```
    break;
  /* CF-ATA commands */
  case CFA_REQ_EXT_ERROR_CODE:
    if (!s->is_cf)
      goto abort_cmd;
    s->error = 0x09;  /* miscellaneous error */
    s->status = READY_STAT;
    ide_set_irq(s);
    break;
  case CFA_ERASE_SECTORS:
  case CFA_WEAR_LEVEL:
    if (!s->is_cf)
      goto abort_cmd;
    if (val == CFA_WEAR_LEVEL)
      s->nsector = 0;
    if (val == CFA_ERASE_SECTORS)
      s->media_changed = 1;
    s->error = 0x00;
    s->status = READY_STAT;
    ide_set_irq(s);
    break;
  case CFA_TRANSLATE_SECTOR:
    if (!s->is_cf)
      goto abort_cmd;
    s->error = 0x00;
    s->status = READY_STAT;
    memset(s->io_buffer, 0, 0x200);
    s->io_buffer[0x00] = s->hcyl; /* Cyl MSB */
    s->io_buffer[0x01] = s->lcyl; /* Cyl LSB */
    s->io_buffer[0x02] = s->select;    /* Head */
    s->io_buffer[0x03] = s->sector; /* Sector */
    s->io_buffer[0x04] = ide_get_sector(s) >> 16;/* LBA MSB */
    s->io_buffer[0x05] = ide_get_sector(s) >> 8; /* LBA */
    s->io_buffer[0x06] = ide_get_sector(s) >> 0; /* LBA LSB */
    s->io_buffer[0x13] = 0x00; /* Erase flag */
    s->io_buffer[0x18] = 0x00; /* Hot count */
    s->io_buffer[0x19] = 0x00; /* Hot count */
    s->io_buffer[0x1a] = 0x01; /* Hot count */
    ide_transfer_start(s, s->io_buffer, 0x200,
        ide_transfer_stop);
    ide_set_irq(s);
    break;
  case CFA_ACCESS_METADATA_STORAGE:
    if (!s->is_cf)
      goto abort_cmd;
    switch (s->feature) {
    case 0x02: /* Inquiry Metadata Storage */
      ide_cfata_metadata_inquiry(s);
      break;
    case 0x03: /* Read Metadata Storage */
```

```
        ide_cfata_metadata_read(s);
        break;
      case 0x04: /* Write Metadata Storage */
        ide_cfata_metadata_write(s);
        break;
      default:
        goto abort_cmd;
      }
      ide_transfer_start(s, s->io_buffer, 0x200,
          ide_transfer_stop);
      s->status = 0x00; /* NOTE: READY is _not_ set */
      ide_set_irq(s);
      break;
    case IBM_SENSE_CONDITION:
      if (!s->is_cf)
        goto abort_cmd;
      switch (s->feature) {
      case 0x01: /* sense temperature in device */
        s->nsector = 0x50;   /* +20 C */
        break;
      default:
        goto abort_cmd;
      }
      s->status = READY_STAT;
      ide_set_irq(s);
      break;
    default:
    abort_cmd:
      ide_abort_command(s);
      ide_set_irq(s);
      break;
    }
  }
}
```

While the preceding code will not be nominated for any awards, the example does demonstrate an effective use of the *goto* statement. We include this sizeable software fragment in the text to demonstrate that retrofitting some rules can be problematic.

When an organization is retrofitting, an advisable approach is to speculatively enable each rule, in turn, in a local checkout of the project code to ascertain the quantity of violations and estimate the redress work. Management should not abandon a good rule simply because significant engineering is required to implement it. If possible, the organization should allow for individual approved management exceptions. However, when faced with a conceptually sound code-quality rule whose implementation is intractable, the organization also should consider a compromise approach. In the preceding *goto* situation, a compromise rule could be applied:

```
The continue statement shall not be used.
The break shall not be used to terminate a loop.
```

Use of the *goto* statement is disallowed, with the exception of safe, centralized error handling.

The Netrino standard includes naming requirements for functions used as entry points for threads and interrupt service routines (ISRs). These rules are an excellent example of how embedded coding standards often include features not found in general-purpose coding standards.

As mentioned previously, while it stresses the use of automated checks, the Netrino standard includes a reasonable percentage of rules that may require enforcement via peer review.

Key Point

Management should carefully limit the number of coding standard controls that must be manually enforced; if that number is too large, either peer reviews will become overly inefficient or enforcement of those rules will become lax and ultimately irrelevant.

3.5.4.13 Dynamic Code Analysis

Key Point

A secure development process should employ dynamic code analysis in addition to static code analysis.

A simple example demonstrates this need. The following code will be flagged as an error by a static source code analyzer:

```
int *getval(void)
{
  return 0;
}
void foo(void)
{
  int *b = getval();
  *b = 0;
}
```

The pointer *b* is initialized by the return value from a function call that obviously returns a NULL pointer. Then *b*, the NULL pointer, is dereferenced.

However, the following similar code may not be flagged as an error by a static source code analyzer:

```
int fd;
int *getval(void)
{
  int *tmp;
  read(fd, &tmp, sizeof(tmp));
  return tmp;
}
void foo(void)
{
  int *b = getval();
  *b = 0;
}
```

In this example, *b* is also initialized by the return value from a function call. However, the source code provides no indication of potential return values from the function call. In particular, the return value is read from a file. While the file may well contain an invalid pointer, causing this program to crash, many static analyzers will adopt a conservative approach (to minimize false positives) and will not assume anything specific about the externally read data.

Dynamic analysis uses code instrumentation or a simulation environment to perform checks of the code as it executes. For example, an instrumented program will have a check prior to the dereference of *b* which validates that *b* is not NULL. Or a simulator can validate all memory references to check for writes to address 0.

Some compilers have dynamic code analysis instrumentation available as a standard option. The development process should require that these checks be enabled at appropriate stages of development, testing, and integration. For example, the Green Hills Software compiler has the option *−check=memory,* which causes the maximum amount of dynamic analysis instrumentation for various forms of memory errors, including NULL pointer dereferences. The instrumented code performs the check and then calls a diagnostic function, provided by a library that is automatically linked to the program when using this option, which informs the user that a fault occurred as well as the type and location of the error within the source code:

```
> gcc myfile.c -check=memory
> ./a.out
Nil pointer dereference on line 15 in file myfile.c
```

This is one example in which the program likely would have crashed, helping the developer locate the program, even if dynamic analysis were not enabled. However, many other kinds of failures are far more insidious, leading to subtle corruptions that may go completely unnoticed or cause a downstream failure that is extremely difficult to trace back to its root cause. Dynamic analysis detects the fault at its source, turning a thorny bug into a trivial one.

Let's examine a few other examples of dynamic code analysis controls that developers should use during development and testing.

3.5.4.13.1 Buffer Overflow

There are many forms of buffer overflow errors, many of which will not be caught by static analysis because the amount of data being written to a buffer is unknown at build time. The following is a simple example:

```
int an_array[10];
void a_func(int index)
{
  an_array[index] = 0;
}
```

If the parameter passed to *a_func* is a value read from a file or message queue by a caller to *a_func*, most static analyzers will conservatively ignore this array reference. However, if *index* turns out to be a value greater than nine, a dynamic analyzer will catch the fault, as shown here:

```
> gcc myfile.c -check=bounds
> ./a.out
Array index out of bounds on line 50 in file myfile.c
```

3.5.4.13.2 Assignment Bounds

The C and C++ programming languages (especially C) suffer from a lack of strong, compile-time enforced-type safety that languages such as Ada and C# provide. However, quality coding standards as well as the use of static and dynamic analysis can provide reasonable compensation for these language limitations. Integer overflow is one risk of weak typing, as shown in the following example:

```
void assign(unsigned int p)
{
  static volatile unsigned short s;
  s = p;
}
void myfunc(void)
{
  assign(65536);
}
```

This code fragment is perfectly legal ANSI C; the assignment of *p* to *s* is defined to truncate *p*'s value to fit *s*'s type. In typical implementations, an *unsigned short* integer occupies 16 bits of storage, allowing values in the range of 0 to 65,535. However, in the example, a value just beyond this range is passed as the parameter *p*, clearly a programming error. Yet standard compilers will not emit even a warning on the preceding code sequence.

Dynamic analysis can detect assignments of values that are out of range for a type, even if the values are read externally (e.g., from a file). The analyzer build command and output for the preceding example may look as follows:

```
> gcc myfile.c -check=assignbound
> ./a.out
Assignment out of bounds on line 57 in file myfile.c
```

3.5.4.13.3 Missing Case

Most imperative programming languages, such as C, C++, C#, and Java, have a *switch/case* selection control equivalent. It is perfectly legal in these languages to have a *switch* statement whose *case* arms do not cover all possible values of the control expression type. For example:

```
typedef enum { red, yellow, blue, green } colors;
void carcolors(colors c)
{
  switch (c) {
    case red:
      printf("red\n");
      break;
    case yellow:
      printf("yellow\n");
      break;
    case blue:
      printf("blue\n");
      break;
  }
}
```

Despite the legality of the preceding code, some compilers and static analyzers will emit a diagnostic, complaining of a lack of *case* to handle the value of *green* for *switch* control *c*. For example, the open source GCC compiler will emit a warning when passed the *-Wall* option that enables some checks beyond the language standard:

```
> gcc myfile.c -Wall
myfile.c: In function 'carcolors':
myfile.c:64: warning: enumeration value 'blue' not handled in switch
```

Some programmers will include a default *case* arm as a matter of habit to ensure that all possible values of the control variable are handled and avoid such warnings. However, this approach is not always a good idea. A catchall case can lead to unintended consequences in which the *default* handling is not appropriate for all inputs. For this reason, some high-assurance coding standards eschew the use of the *default* case whenever practical and instead promote the use of explicit *cases* for all expected control values.

In the preceding example, the programmer may know that the cars can be only red, yellow, and blue (no green cars). But what if some day green cars are invented? Will the software be updated to reflect this new reality? The preceding *carcolor* function will compile and execute, but the lack of *green* handling could have unintended consequences. Once again, in such cases dynamic analysis can be used as a code-quality enforcement mechanism. If a *switch* statement is passed a value for its control variable that matches no existing case, then the dynamic analyzer will generate a runtime exception:

```
> gcc myfile.c —check=switch
> ./a.out
Case/switch index out of bounds on line 7 in file myfile.c
```

3.5.4.13.4 Stack Overflow

Earlier in this chapter, we discussed MISRA C rule 15.2, which prohibits recursion to avoid runtime stack overflow, and how static detection of cycles in a complicated program's call graph can be difficult due to indirect function calls. Furthermore, programs devoid of recursion can also suffer from stack overflow simply due to a long function call sequence and/or excessive automatic storage usage.

Detecting stack overflow is critical both for reliability and security of embedded systems. Embedded systems are often memory constrained, requiring system designers to carefully allocate and minimize stack usage for all processes and threads. Stack overflows may manifest themselves in subtle corruptions that are difficult to track down during development and testing. Overflow vulnerabilities that go undetected during product development may cause fielded programs to crash. Attackers who become aware of stack overflow vulnerabilities can use them to subvert execution in numerous ways. For example, a stack overflow triggered by crafted input to one thread may overwrite the data in a second thread, causing it to crash or execute malware.

Key Point

Whenever possible, a static analysis tool should be used to check for the largest potential runtime stack memory requirements for a program or for all threads in a multi-threaded program.

Your tool chain provider should include a tool for this purpose. However, because of the aforementioned indirect function call dilemma, maximum potential runtime stack memory requirement cannot always be computed statically.

In Chapter 2, we discussed how virtual memory-capable embedded operating systems could employ guard pages to detect stack overflow at runtime. For developers not using a virtual memory operating system, a second option for dynamic analysis of stack overflow is to instrument the program with overflow checks in the prologue of each function call. This feature is available in some compilers and may not be appropriate for multi-threaded applications. Building a program that overflows its stack would generate an appropriate runtime error, halting execution when the stack pointer first exceeds the bounds of the allocated runtime stack:

```
> gcc myfile.c —check=stack
> ./a.out
Stack overflow
```

If no documented dynamic stack overflow detection option exists in a tool chain or operating system, a developer should consider the following do-it-yourself method that works reasonably well. Most operating systems have a hook for executing a developer-defined function call on every system context-switch as well as a means of reading each thread's stack pointer and the location of the thread's allocated stack segment. The context-switch function can simply compare the stack pointer of the thread about to be executed with the thread's runtime stack bounds. On most computers, stacks grow downward to lower addresses, so a comparison that shows a stack pointer below the bottom of its allocated stack segment would generate an alarm, audit record, and so on. Readers should consult operating system documentation for the common context-switch hook feature.

3.5.4.13.5 Memory Leaks

One of the major reliability benefits touted by the Java language is its avoidance of programmer-controlled dynamic heap memory allocation by using automatic garbage collection. However, many embedded applications use dynamic memory allocation and suffer from vulnerabilities due to improper memory management errors. Many such errors can be prevented via dynamic code analysis.

Memory leaks are one class of memory management error. A memory leak occurs when a function allocates memory but never releases it. If the function is called sporadically, then the loss of memory may be gradual, escaping detection until a system is field deployed. Furthermore, if an attacker is aware of a leaking function, it can focus its attention on causing the function to be executed, draining the system of memory resources and forcing a system failure.

A search of the memory leak vulnerabilities in the NIST's National Vulnerability Database uncovers numerous instances in commercial products, including security appliances. For example, CVE-2010-2836 is a recent high-severity security vulnerability identified in the SSL virtual private network (VPN) feature of Cisco's network appliance operating system called IOS. The vulnerability enables remote attackers to cause a denial of service via memory exhaustion by improperly disconnecting SSL sessions.[18]

Memory leak detection is a form of dynamic analysis that eliminates programmer leak vulnerabilities. Leak detection works by comparing a program's pointer references to the memory management library's outstanding allocations. A program's pointer references may reside in memory-resident data variables, runtime automatic stack storage, or CPU registers. The memory leak detector, therefore, is usually offered as a tightly integrated feature of the developer tool chain (compiler, runtime libraries).

Memory leaks can occur at any time during a program's lifetime. The runtime library can perform its memory leak detection algorithm at sensible call points (such as when memory is allocated or released). In addition, the user can add explicit calls to the memory leak detection algorithm as a sanity check at regular intervals in time or at specific points in the application code. Leak detection can be performed during debugging, during testing, or even in a fielded product.

Ideally, the memory management library is able to record an execution call stack within its allocation database. When the leak detection algorithm identifies a leak, the call stack can be reported to the developer, making it easy to identify the specific allocation that has been left hanging.

A static source code analyzer should detect the simple memory leak error shown next. However, as with the other cases discussed in this section, many forms of leaks are beyond the insight of static analysis and require dynamic leak detection.

```
void leak(void)
{
  char *buf = malloc(100);
  sprintf(buf, “some stuff\n”);
  printf(buf);
}
int main()
{
  leaks();
  __malloc_findleaks();  // call the leak detector
}
```

In the preceding example, the leak function allocates memory pointed to by a local variable and never deallocates the memory. Upon return from the function, therefore, this memory is leaked. A call to the runtime library's leak detector will report the leak:

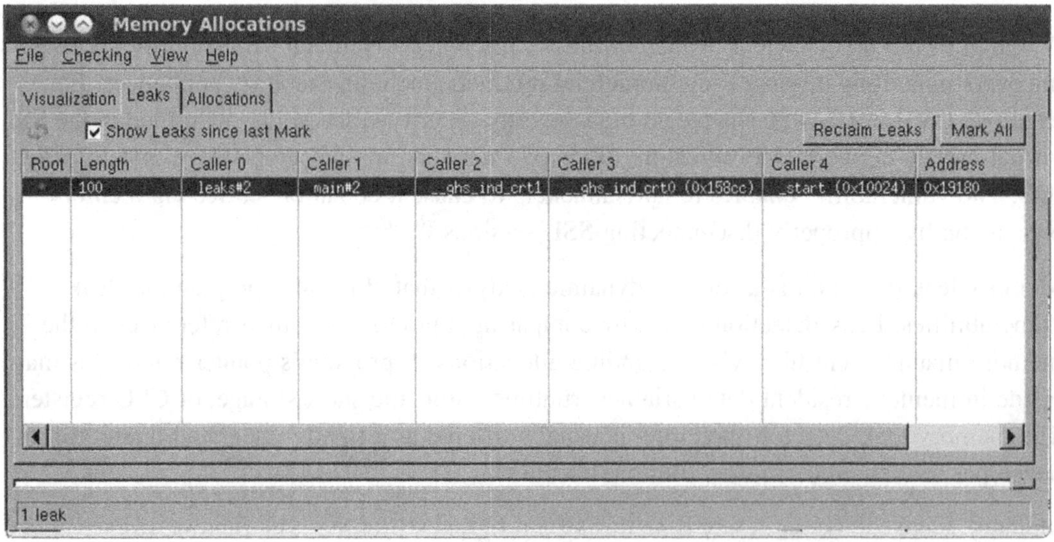

Figure 3.4:
Memory leak detection integrated into software development environment.

```
> gcc myfile.c -check=memory
> ./a.out
Unreferenced memory adr=0x18d40 allocated at 0x103f4 called from 0x1043c then 0x15f18
then 0x10028
```

When integrated with the software development environment, the leak detection report's call stack addresses are mapped to actual source code locations, enabling the developer to more easily locate and understand the leak source (see Figure 3.4).

For the precise name of the leak detection applications programming interface (API) and build-time options used to enable leak detection, readers should consult a tool chain supplier.

3.5.4.13.6 Other Dynamic Memory Allocation Errors

With programmatic control of memory allocation and deallocation, there are many more ways for developers to shoot themselves in the foot. The following simple function shows a couple more examples:

```
void badalloc(void)
{
  char *buf = malloc(100);
  char localbuf[100];
  free(buf);
  free(localbuf);
  free(buf);
}
```

The first call to *free(buf)* is fine; it references a valid allocated buffer. However, the second call to *free(buf)* is invalid, since *buf* has already been deallocated. The call to *free(localbuf)* is also invalid because *localbuf* is a local buffer, not allocated using a corresponding dynamic memory allocation call such as *malloc* or *calloc*. Similar errors in C++ occur with the operators *new* and *delete*. Once again, static analysis can locate the errors in this example, but dynamic analysis will find other memory allocation errors that static checking cannot. For example, the following change will confuse many static analyzers:

```
char localbuf[100];
char *b = localbuf;
void badalloc(void)
{
  free(b);
}
```

Because the variable *b* is now globally defined, a static source code analyzer may assume less knowledge about that to which *b* may point. Dynamic analysis detects the invalid deallocation during program execution:

```
> gcc myfile.c -check=memory
> ./a.out
Attempt to free something not allocated adr=0x18484
```

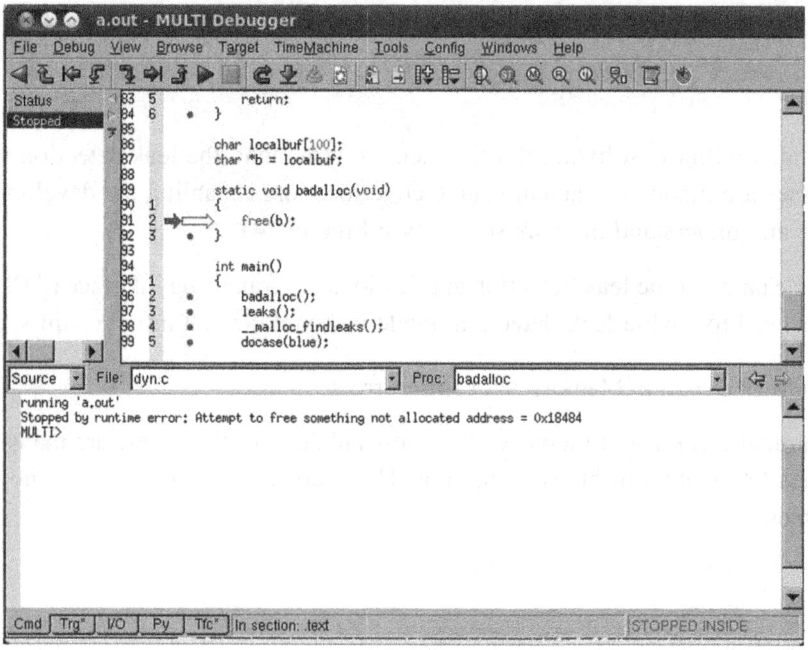

Figure 3.5:
A dynamic analysis error stops the program at the offending line in the debugger, making it easy for the developer to locate and fix common security vulnerabilities.

If dynamic analysis is integrated into the debugger, the preceding failure is even easier for the developer to detect and correct. As shown in Figure 3.5, the debugger is automatically halted when the memory deallocation error occurs, pointing the developer to the exact offending line of code.

It goes without saying that software managers should strongly weigh the diagnostic capability of a compiler and tool chain when selecting such an important tool.

3.5.5 Software Testing and Verification

> **Key Point**
>
> A comprehensive test regimen, including functional, regression, performance, and coverage testing, is well known to be one of the best mechanisms to assure that software is reliable and secure.

Testing is an important component of many high-assurance development standards and guidance documents, such as that promulgated by the U.S. Food and Drug Administration.[19]

In the realm of functional testing, we have fault-based testing, error-based testing, stress testing, white-box testing, and black-box testing.

Two approaches to testing are almost always required to ensure security. First, all software within security-critical components must be covered by some form of functional test. Coverage is verified using code coverage tools. Furthermore, all security-critical software must be traceable to the software's component requirements. Software that fails to trace back to a test and to a requirement is more likely to introduce latent security vulnerabilities.

3.5.5.1 Modified Condition/Decision Coverage

Because code coverage analysis is so important for security, it is worth examining the various levels of coverage testing that can be applied to embedded software. To aid this discussion, we consider the code coverage requirements across the five assurance levels specified in the standard that the U.S. Federal Aviation Administration (FAA) uses to perform safety certification of commercial aircraft. This standard, published by RTCA, is titled *Software Considerations in Airborne Systems and Equipment Certification*, commonly referred to as DO-178B.[20] In fact, DO-178B is the most commonly used software safety standard in the worldwide avionics industry.

The five assurance levels, in increasing level of criticality, of DO-178B are as follows:

Level E: Failure has no impact on flight safety.
Level D: Failure impact is minor, noticeable but not critical to flight safety (e.g., passenger inconvenience).
Level C: Failure impact is major, safety-related but not severe (e.g., passenger discomfort but not injury).
Level B: Failure impact is severe (e.g., passenger injury).
Level A: Failure impact is catastrophic (e.g., aircraft crash).

The structural code coverage requirements corresponding to each assurance level are shown in the following table:

Assurance Level	Structural Coverage Requirement
E	None
D	None
C	Statement
B	Decision
A	Modified Condition/Decision

DO-178B Level C requires *statement coverage*: demonstrating that every program statement has been executed at least once (covered) by the verification test regimen.

Statement coverage is what most developers equate with the more general term *code coverage.*

Level B augments statement coverage with *decision coverage*, a requirement that every decision point in the program has been executed with all possible outcomes. For example, a conditional branch's comparison both succeeds (branch taken) and fails (branch not taken) at least once each.

Finally, *modified condition/decision coverage* (MC/DC) augments decision coverage with a specialized form of *condition coverage* in which each condition within a decision must be shown to have an independent effect on the outcome of that decision.

We use a few simple code examples to illustrate the increasing rigor and security-enforcing quality of each coverage approach:

```
if (a || b || c) {
  <code executed on true decision>
}
```

Statement coverage requires that the *if* statement is executed and that the code within the *if* block (executed on a true decision) is fully executed. As there are no statements corresponding to a false decision, statement coverage would not require any test cases that force the *if* block not to execute.

In contrast, decision coverage would require at least one test to execute the false decision path, even though there is no explicit code associated with that path. This extra coverage is desirable from a security perspective because it indicates that the developer has considered the impact of a false decision, which may have some other side effects. Let's consider this slightly more detailed example:

```
uint32_t divisor = 0;
if (a || b || c) {
  divisor = a | b | c;
}
result /= divisor;
```

The final division statement will fail (divide by zero) on a false decision, but statement coverage testing may never activate this pathway. If an attacker were somehow able to control the decision (e.g., by controlling the values of *a*, *b*, and *c*), then the attacker could cause a denial of service (program crash). Decision coverage testing would have pointed out this problem before it could be fielded.

Condition coverage requires that each condition within the decision be tested with true and false values. The following two test cases will force each of the three conditions to take on both a true and a false value at least once: ($a=1$, $b=1$, $c=1$) and ($a=0$, $b=0$, $c=0$).

While testing a decision's constituent conditions may seem like an improvement over decision coverage, condition coverage is not a superset of decision coverage, as shown in this example:

```
if (a || !b) {
  <code executed on true decision>
} else {
  <code executed on false decision>
}
```

The two test cases, (a=0, b=0) and (a=1, b=1), satisfy condition coverage (both conditions executed with true and false inputs) but neglect to cover the false decision path. Clearly, decision and condition coverage techniques used in concert is desirable.

Multiple condition coverage requires all combinations of conditions. In other words, every row of a decision's truth table must have a corresponding test case. In the earlier test case with conditions a, b, and c, the truth table is as follows:

a	b	c	result
F	F	F	F
F	F	T	T
F	T	F	T
F	T	T	T
T	F	F	T
T	F	T	T
T	T	F	T
T	T	T	T

Thus, multiple condition coverage requires 2^n tests, where n is the number of independent conditions. This approach is viewed as impractical; exhaustive condition testing would simply take too many test cases and too long to execute.

Languages with short-circuiting Boolean operators (e.g., C, C++, Java) reduce the number of required test cases:

a	b	c	result
F	F	F	F
F	F	T	T
F	T	-	T
T	-	-	T

Nevertheless, compound Boolean expressions may yield an impractical explosion in test cases across realistic programs.

MC/DC is the selected compromise for most high-assurance safety and security standards. MC/DC includes both decision and condition coverage. However, in addition, MC/DC requires that each condition be demonstrated to have an independent effect on the decision. This modified condition requirement is accomplished by varying a single condition while holding the remainder constant and verifying that the decision changes. Let's consider the following example:

```
if (a || b || c || d) {
  <code executed on true decision>
}
```

MC/DC requires the following test cases (shown in truth table form):

a	b	c	d	result
F	*F*	*F*	*F*	F
T	F	F	F	T
F	*T*	F	F	T
F	F	*T*	F	T
F	F	F	*T*	T

The italicized values below each condition are the test cases that cover both true and false inputs and result in both a true and false result when all other condition inputs are held constant. For uncoupled conditions, MC/DC requires N+1 test cases, where N is the number of Boolean conditions. This linear growth in test cases makes MC/DC practical to implement, and its effectiveness in locating testing gaps as well as design flaws is well regarded and documented.[21]

Key Point

We strongly recommend the use of MC/DC coverage testing for the most critical components of an embedded system—for example, the operating system kernel, network security protocols, and cryptographic components.

MC/DC may be overkill for some de-privileged applications. The converse is that system designers should not assume that MC/DC coverage implies perfect testing. Limitations of (and improvements to) the traditional MC/DC definition have been reported.[22]

There are numerous other aspects of a program to test for coverage other than code execution flow. For example, if the program has an enumerated type with five possible values, it would be sensible to validate that all five values are at least used somewhere in the program.

Another problem with code coverage testing is the loss of fidelity when translating from source to machine code. In some cases, it is preferable that code coverage testing be performed on the machine code. By doing so, we increase the assurance that malicious code is not instrumented as part of the build process (see the Thompson Hack case study earlier in this chapter). In fact, machine code coverage is required by some high-assurance security and safety certifications. Let's examine the following simple function:

```
int foo(int a, int b, int *arr, int n)
{
  int i;
  for (i = 0; i < n; i++) {
    arr[i] += a / b;
  }
  return i;
}
```

The loop body contains a divide operation in which numerator and denominator are both loop invariant. A compiler would like to hoist this divide outside the loop to improve performance:

```
int foo(int a, int b, int *arr, int n)
{
  int i;
  int tmp = a / b;
  for (i = 0; i < n; i++) {
    arr[i] += tmp;
  }
  return i;
}
```

However, this optimization is disallowed because it changes the function's semantics. In the pre-optimized version, the divide may never be executed (if the loop itself is never executed, i.e., argument *n* is zero). In the optimized version, the divide is unconditionally executed. If the argument *b* is zero, then the optimization could theoretically induce a program crash that might not otherwise occur.

But compilers are smart! Most compilers will hoist the divide but introduce a guard against zero division:

```
int foo(int a, int b, int *arr, int n)
{
  int i, tmp;
  if (b != 0)
    tmp = a / b;
  for (i = 0; i < n; i++) {
    arr[i] += tmp;
  }
  return i;
}
```

The preceding code shows the compiler optimization visualized as source code changes. Of course, the compiler performs this optimization as it is generating the machine code. Therefore, the machine code contains a new decision, *b != 0*, that was not present in the source code. A code coverage tool that operates on source code alone would fail to cover this additional decision.

If a code coverage tool is capable of providing decision coverage but not MC/DC, the source can be modified to remove compound conditions. With singleton conditions throughout, MC/DC reduces to decision or branch coverage:

```
if (a || b) {...}
```

modified to

```
if (a)
  if (b) {... }
```

While this approach may seem heavy-handed, it can be acceptable because the amount of critical code requiring MC/DC level testing is often relatively small.

> **Key Point**
>
> A lack of complete test coverage almost always points to a lack of other important validation (e.g., functional testing), design flaws, or simply latent code that has unknown and potentially security-relevant impact.

It is left as an exercise for readers to explore other various types of testing and their relative advantages. Beyond the important example of coverage testing, our guidance is more concerned with the integration of testing methodology into the development process to maximize its value.

Organizations that do not follow a rigorous development process often resort to ad hoc testing that is often an afterthought when most of the software has already been written. Organizations that follow a rigorous process often focus testing during a release process, again after much of the software has been written.

> **Key Point**
>
> The testing system should be running 24x7.

If a testing system is run only on demand, occasionally, or only during a release process, then errors that can be detected by the testing system tend to go unnoticed for an unnecessarily long period of time. When a flaw is discovered, the developer has a much harder time trying to remediate it than if the flaw was introduced the previous day. In some cases, the developer may have moved on to another project, if not another company, leaving someone else to try to learn the code and fix the flaw. Fixing flaws discovered by the testing system should be prioritized higher than anything other than emergency customer support issues;

keeping the system running cleanly at all times guarantees that test system failures are almost always new failures that have not been examined by anyone else and need immediate attention.

> **Key Point**
>
> The testing system should run on actively developed products as well as currently shipping products.

This key point is a corollary to the previous key point but important enough to be stressed: when a testing system is used throughout the development process, developers are forced to keep the product in a working state at all times. Software projects that move to rigorous test only after a code freeze are subjected to test phases that last longer overall because developers must wrestle with problems inserted throughout months of development time. When a product is always working, a code freeze leads directly to final quality assurance testing, saving time to market. If a developer cannot develop code in a manner that prevents the product from failing, then a private branch can be used as long as it is not allowed to live too long; integrating old code branches that have drifted far from the trunk often causes unforeseen conflicts that affect the efficiency of the entire development team.

> **Key Point**
>
> The testing system should be able to effectively test a software project in less than one night.

A testing system that takes too long to run tends to become underutilized if not completely ignored. Developers should be able to quickly validate a change overnight before committing it to the project. In addition, the automated tests running 24x7 on dedicated testing compute farms can detect flaws very quickly so they can be corrected while the understanding of the recently added code is still fresh in the developer's mind. It is reasonable to have more tests that can run in one night; however, longer runs should compete at a lower priority for computing resources or be run only on demand or at longer intervals during the development process. The nightly test run must be good enough to detect almost all flaws added during development.

> **Key Point**
>
> It should be trivial to determine when a test run has succeeded or failed; a failed test should be trivial to reproduce.

Tests should be written such that output is generated only when an error is detected. A clean test is one without any output. At worst, the output should be less than a page long. Too often, testing systems generate voluminous output, making it difficult for developers to quickly ascertain the status of the test run. Test output that is difficult to quickly evaluate tends to be ineffectual and ignored.

When a test fails, the exact state of the software system and any inputs or process that must be used to reproduce the discovered error should be clearly displayed within the test output. If the developer is unable to efficiently reproduce a test failure, the test system will tend to be ignored. Reproducibility is the key to maximizing the rate at which the developer can remediate flaws discovered by the testing system and bring a reliable software product to market faster.

3.5.6 Development Process Efficiency

Although there is significant evidence that following a structured, comprehensive software quality management process security improves relative to the use of unstructured processes, these rigid methodologies often cause a loss in efficiency, delayed time to market, and frustration in the daily lives of software developers and managers. The key is to implement techniques that are compatible with the requirements of high-assurance quality standards yet designed and proven to improve efficiency of software development. The goal is to maximize the reliability to production cost ratio of software.

When software changes rapidly, the efficiency of the build process becomes a critical component of developer efficiency. Complex software projects are often characterized by complex build processes, in which the software not only takes a long time to recompile from scratch, but also may require many recompiles to exercise different production configurations. For example, a software system may have a "production" build, in which compiler optimizations are fully enabled and the software is configured for maximum speed and reliability; a "debug" build, in which the system has debugging information enabled so that developers can most easily debug the software; and a "checked" build, in which the system turns on additional sanity checks that may drastically reduce performance but increase the probability of finding unusual problems such as RAM hardware failures.

During development, it may not be practical for a developer to build all configurations to test a change. Therefore, an *autobuild* system should be used. When a change is committed to the configuration management system, dedicated build servers update their local checkouts of the software and rebuild all configurations. When a build fails, the autobuilder sends an automated e-mail to the component manager(s) of the affected component(s). An e-mail is also sent to the person in charge of the build system. The autobuild system ensures that erroneous changes causing build failures are immediately detected before they affect other developers.

Key Point

Use an autobuild system to quickly detect changes that break system builds.

With modern PCs and servers, developers are enjoying faster builds than ever before. However, as software complexity has grown in concert, build times remain an important factor relating to developer efficiency. Yet there are a couple of methods to reduce the effect of build times on developer productivity.

Key Point

Always ensure a developer has at least two development projects to work on at all times.

When a change is made and a long build is started, there is no excuse for a developer to be waiting for that build to complete. A developer should always have a secondary project to work on during inevitable breaks, such as those caused by waiting for builds, peer reviews, or the many other reasons why a foreground project is delayed. Ensuring that a developer has multiple projects to work on at all times is, ultimately, the responsibility of the developer's management. However, a developer naturally considers downtime waiting for a build to complete as a normal mode of operation in which the developer has no choice but to take a break. Therefore, it is important to teach the developer to proactively request more work in the case he finds himself blocked on all fronts.

Another technical solution to the waiting build problem is to reduce build times by bringing to bear the full power of corporate computer horsepower. A typical development site may employ many developers, working on the same or different software projects. There may be dozens, if not hundreds, of PCs on the desks of these developers. Sadly, these PCs typically spend the majority of their time idle.

Key Point

Employ distributed builds to maximize computer utilization and improve developer efficiency.

Some compiler vendors provide distributed build capability. At a minimum, management should take advantage of local parallelization of build operations on a single multi-core host computer. To truly scale, the parallel build system must be able to interrogate compute resources throughout a site, locate machines that are underutilized, and migrate the files required to accomplish distributed builds to these available resources. Ideally, a distributed build system should not require significant configuration. The only required piece of configuration information for the tool is the knowledge of which machines on the network should be candidates to share in the workload. Distributed builds greatly reduce build times, therefore shortening the build-edit-debug cycle that is so crucial to developers' productivity on a daily basis. Another beneficial side effect of distributed builds is the potential to reduce developer capital expenses by better utilizing compute resources throughout a site.

3.6 Independent Expert Validation

After the previous extended discussion on secure development process, we now return to the fifth rung in PHASE: independent expert validation.

Key Point

Intrinsic assurance evidence must be evaluated and confirmed by independent experts.

Consumers who read about a vendor's claim of "certifiable" should interpret such hyperbole as "not certified." For example, to achieve certification to the Evaluated Assurance Levels above five in the United States, a product not only must have all its design documentation, testing, and formal methods painstakingly evaluated by an accredited Common Criteria testing lab, but also must undergo penetration testing by the NSA's penetration testing experts who have complete access to the source code and essentially unlimited resources with which to craft attack vectors.

In the U.S. aircraft safety domain, the FAA must certify critical aircraft electronics (and their constituent software) against the RTCA/DO-178B standard. Earlier in this chapter, we described the five safety levels, A through E, with Level A being the highest, where a failure in the electronic system may be catastrophic. As mentioned earlier, DO-178B Level A specifies a rigorous assurance process, with many similarities to Common Criteria high-assurance requirements. In this case, the independent expert evaluators are the designated engineering representatives (DERs) of the FAA.

Regulatory bodies for safety certification of industrial control, automation, and automotive systems sometimes use the IEC 61508 standard internationally. IEC 61508 also is gaining acceptance in other industries. The railway application-specific interpretation of IEC 61508 is embodied in the CENELEC EN 50128, "Software for railway control and protection systems," and EN 50129, "Safety related electronic systems for signaling." These standards cover the entire software development life cycle and place a strong emphasis on analysis and mitigation of safety hazards, enforcement of a rigorous quality management system, and thorough verification and validation. TÜV, based in Germany, is the internationally recognized certification authority for the IEC 61508 and related safety standards. Exida is another firm that performs IEC 61508 and related certifications.

The automotive-specific interpretation of IEC 61508 is embodied in ISO 26262, whose adoption in the automotive industry is nascent.

Rail and other industries utilizing IEC 61508 go through rigorous development and testing requirements. Documentation includes traceability to critical safety requirements for the system being developed and its interface to the end product. The certification authority audits the entire system development process, including witnessing key tests and checking traceability of system attributes. Depending on the safety integrity level, the system architecture could be simple or increasingly complex. Suppliers that can start with an operating system that is already CENELEC certified can save money by not having to system test (and possibly even unit test, depending on the needed function) the operating system and related runtime components when the software itself is part of the safety case.

Software that fulfills all five PHASE principles (it's easy to see how the lack of any one of them can be fatal) can be trusted to manage and protect high-value assets used in security critical systems. PHASE does not attempt to rigorously define a "secure development process" or specifically how the principles of least privilege apply to a particular software component or system. Interpretations will necessarily vary depending on the situation and level of security required. Incorporation of these principles can also reduce development cost in the long run, by incorporating software development practices that focus on security. Furthermore, PHASE can be applied selectively and creatively, to improve both security and reliability without sacrificing reuse of legacy software.

High-assurance development processes, such as that encouraged by ISO/IEC quality standards as well as regulatory bodies that govern the use of software in safety and security systems, are generally believed to result in higher-reliability software. However, an organization that follows these processes to the letter is likely to get to market later than an organization that develops the same product without the overhead of following these quality standards. In this chapter, we have presented a set of guidance recommendations, part of a process that has been proven in use for 30 years at leading security-critical software development organizations, and have argued how these controls can actually increase developer productivity and reduce time to market through the use of methodologies that find software flaws faster and reduce developer dependencies on each other and on productivity-killing bureaucracy.

3.6.1 Common Criteria

The international standard for evaluating the security of IT systems is ISO/IEC 15408, more commonly known as the Common Criteria.

> **Key Point**
>
> With 26 participating countries, the Common Criteria standard represents an admirable achievement in obtaining global acceptance and is certainly the most comprehensive international security evaluation standardization effort ever conceived.

Under Common Criteria, products are evaluated against *Protection Profiles* that specify the product family's security functional requirements and assurance requirements. Functional requirements are the security policies or protections that a product claims to implement. Assurance requirements are the controls—such as configuration management and testing—that the developer must follow to ensure that the functional requirements are properly realized.

> **Key Point**
>
> In a Common Criteria protection profile, assurance requirements are collectively tagged with a security level (called the Evaluated Assurance Level, or EAL) that represents the overall confidence that stakeholders can have that the security functional requirements are actually met by a conforming product.

There are protection profiles for firewalls, antivirus applications, operating systems, mobile devices, and many other forms of technology. Protection profiles can be written by anyone, but the profiles themselves must be evaluated, ensuring that products are measured against a well-understood, valid, and accepted standard. In practice, a majority of protection profiles have been authored by government security organizations such as the NSA.

While the Common Criteria standard provides an excellent, proven technical framework for specifying and evaluating security for software, hardware, devices, and systems, the standard has had a disappointing political history. As with many well-intended tools, its potential utility does not always translate into productivity. The Common Criteria standard has suffered from poor policy decisions and inefficient evaluation processes. For example, a high-assurance protection profile, whose requirements were conceived and written in a few months, took six years for the U.S. government to evaluate due to interminable tweaking, arguing, and bureaucracy. The first product certification against this protection profile took four years, despite the fact that the product was field proven for a decade prior to the evaluation and no significant vulnerabilities were discovered throughout the evaluation. The inability to efficiently perform meaningful high-assurance certifications has relegated Common Criteria to low-level certifications that amount to massive, expensive paperwork exercises, with no confidence that certified products are secure against determined attackers with even a moderate level of sophistication. It is hoped government agencies can reform their approach to both what gets evaluated (avoiding expensive validations that provide no significant assurance to users) and how evaluations are performed, and derive what could be tremendous value from this important security approach and standard.

Nevertheless, let's focus on the important contribution of Common Criteria as a technical framework for independent expert validation.

Common Criteria assurance levels range from EAL1 to EAL7. The general meanings of the levels are shown in the following table:

EAL1	Functionally tested
EAL2	Structurally tested
EAL3	Methodically tested and checked
EAL4	Methodically designed, tested, and reviewed
EAL5	Semiformally designed and tested
EAL6	Semiformally verified design and tested
EAL7	Formally verified design and tested

General-purpose products, such as Windows, Linux, VMware, Oracle database servers, and Cisco routers, not specifically designed for high-assurance levels, have been certified at EAL4 or lower. The Common Criteria state that EAL4 "is the highest level at which it is likely to be economically feasible to retrofit an existing product line."[23] This is sensible because above EAL4, secure-by-design techniques with enhanced formality are required.

As one security expert has explained, "Security experts have been saying for years that the security of the Windows family of products is hopelessly inadequate. Now there is a rigorous government certification confirming this."[24] In addition, Windows and other products were evaluated against protection profiles that provide the following disclaimer: "not intended to be applicable to circumstances in which protection is required against determined attempts by hostile and well funded attackers to breach system security."[25]

Some protection profiles essentially ignore assurance level, seeking instead to define functional requirements to facilitate standards and increase interoperability. For example, the NSA has developed several EAL2 protection profiles that provide general guidance for the types of security controls that certain classes of products should have. One example is the *U.S. Government Protection Profile for General Purpose Operating Systems in a Networked Environment,* which describes requirements for user authentication, discretionary access control of operating system objects, cryptographic services, and audit services.[26] While the security controls are useful to understand, developers must not interpret the assurance portion of this specification as providing any meaningful level of confidence in those security functions.

3.6.2 Case Study: Operating System Protection Profiles

In addition to the aforementioned, the U.S. government has created several other operating system-related protection profiles. These operating system specifications have varying security levels, making them more or less appropriate for certain threat environments. The operating system profiles and their applicable threat environments are listed in the following table:

Name	Title	Security Level	Threat Environment
SKPP	**Separation Kernel in High Robustness Environments**	**EAL6+ / High Robustness**	**"management of classified and other high-valued information, whose confidentiality, integrity or releasability must be protected"[27]** **"presence of both sophisticated threat agents and high value resources"[28]**
CAPP	Controlled Access Protection Profile	EAL4+	"non-hostile and well-managed user community"[29] "inadvertent or casual attempts to breach the system security"[30] "not intended to be applicable to circumstances in which protection is required against determined attempts by hostile and well-funded attackers"[31]
CCOPP-OS	COTS Compartmentalized Operations Protection Profile—Operating Systems	EAL4	"not expected to adequately protect against sophisticated attacks"[58] "users are highly trusted not to attempt to maliciously subvert the system or to maliciously exploit the information stored thereon"[58]
LSPP	Labeled Security Protection Profile	EAL4+	"non-hostile and well-managed user community"[32] "inadvertent or casual attempts to breach the system security"[33] "not intended to be applicable to circumstances in which protection is required against determined attempts by hostile and well-funded attackers"[34]
SLOS	Single Level Operating Systems in Medium Robustness Environments	EAL4+	"suitable for use in unclassified environments"[35] Not appropriate for "organization's most sensitive/proprietary information" when exposed to "a publicly accessible network"[36] "likelihood of an attempted compromise is medium"[37] "motivation of the threat agents will be average"[38]

MLOS	Multilevel Operating Systems in Medium Robustness Environments	EAL4+	"suitable for use in unclassified environments"[39] Not appropriate for "organization's most sensitive/proprietary information" when exposed to "a publicly accessible network"[40] "likelihood of an attempted compromise is medium"[41] "motivation of the threat agents will be average"[42]

As can be seen from this table, only the SKPP is appropriate to protect high-value information exposed to the threat of sophisticated and determined attackers.

SKPP is the protection profile that was created by the NSA to specify security requirements for "high-robustness" embedded operating systems. High-robustness operating systems control computer systems that manage and protect high-value resources in the face of attacks by resourceful adversaries (see the following table). According to Department of Defense guidance, high robustness refers to "security services and mechanisms that provide the most stringent protection and rigorous security countermeasures."[43] An SKPP-compliant operating system must guarantee that a malicious application cannot harm (corrupt, deny service to, steal information from, etc.) any other application running on the computer.

Attacker threat level

		Low threat	**Medium threat**	**High threat**
Asset Value	**High Value**	Low	Medium	HIGH
	Medium Value	Low	Medium	Medium
	Low Value	Low	Low	Low

The requirements of the SKPP are far more stringent than any other operating system security standard. The resulting assurance, or confidence, that developers, users, and other stakeholders are therefore able to derive from an independent SKPP evaluation is extremely high and unprecedented in the world of computer security. SKPP requires an extremely rigorous development process, formal methods—which provide mathematical proof of the security—and penetration testing by the NSA's security experts who have complete access to the source code. Thus, SKPP assurance requirements represent an excellent example of a standard for independent expert validation that would provide high levels of security assurance to embedded systems developers and their customers. While SKPP-evaluated products are in widespread use, the SKPP itself is no longer being used for new product evaluations. This is due to the current inability of the U.S. government to perform these evaluations at a level of cost and schedule efficiency that can be justified in current market conditions. Nevertheless, security officials can still use the SKPP-based evidence in an accreditation that is performed outside the Common Criteria evaluation program.

Key Point

The following of applicable EAL6+ assurance guidelines, as exemplified in the SKPP, and obtaining independent evaluation of the associated assurance evidence by expert consultants would be an excellent approach for embedded systems developers who want to raise the bar on their product security.

We will explore these assurance requirements, comparing high- and low-assurance controls, in more detail in the following sections. Readers are also encouraged to read the SKPP, especially its assurance requirements.

Common Criteria development process assurance requirements, with numeric leveling comparison between SKPP and CAPP (the most widely applied profile in enterprise operating systems), are shown in the following table:

Requirement	Description	SKPP	CAPP	Notes
ACM_AUT	Configuration management automation	2	0	SKPP requires complete automation
ACM_CAP	Configuration management capabilities	5	3	SKPP requires advanced support
ACM_SCP	Configuration management scope	3	1	SKPP CM requires coverage of development tools
ADV_ARC	Architectural design	1	0	
ADV_FSP	Functional specification	4	1	SKPP requires *formal* specification
ADV_HLD	High-level design	4	2	SKPP requires semiformal high-level design
ADV_IMP	Implementation representation	3	0	SKPP requires rigorously defined transformation from representation to implementation
ADV_LLD	Low-level design	2	0	SKPP requires semiformal low-level design
ADV_RCR	Representation correspondence	3	1	SKPP requires *formal* correspondence

ALC_DVS	Development security	2	1	SKPP requires confidentiality and integrity measures during development
ALC_FLR	Flaw remediation	3	0	SKPP requires systematic flaw remediation
ALC_LCD	Life-cycle definition	2	0	SKPP requires a standardized life-cycle model
ALC_TAT	Tools and techniques	3	0	SKPP requires total system compliance with implementation standards
AMA_AMP	Assurance maintenance plan	1	0	
ATE_COV	Analysis of test coverage	3	2	SKPP requires complete test coverage of all functional requirements
ATE_DPT	Depth of testing	3	1	SKPP requires testing against high-level design, low-level design, and implementation
ATE_FUN	Functional testing	2	1	SKPP requires ordered functional testing
ATE_IND	Independent testing	3	2	SKPP requires independent testing of all requirements (CAPP only requires sampling)
AVA_VLA	Vulnerability assessment	4	1	SKPP requires *NSA penetration testing*; others do not

3.6.2.1 Configuration Management

While CAPP requires the use of a configuration management system and the ability to re-create the product from the configuration management system, the SKPP includes complete automation of the CM system, including an ability to automatically identify all aspects of the product that are affected by the modification of any single configuration item. SKPP also requires that ancillary items, not just product source code, be configuration managed. For example, any support files (e.g., scripts) used for the creation of production images must be kept under CM, and all tools used to generate the executable product must be covered under CM. This is very important due to the potential for tool changes (e.g., compiler) to cause unforeseen changes in the behavior of the product. A case study of this risk was presented earlier in this chapter.

SKPP requirements also cover CM-related plans and documentation, including demonstration of how CM-related security measures prevent unauthorized changes to the product and how acceptance procedures are followed for change reviews.

3.6.2.2 Functional Specification

Functional specification refers to the description of the interfaces and behaviors of the product. CAPP requires only an informal functional specification. SKPP goes much further, requiring that all details of the security functions be covered by the functional specification and that this specification be formal in style. This formality is a critical requirement leveling from EAL7. In the case of the Green Hills Software INTEGRITY-178B operating system that was successfully evaluated against SKPP, the behavior of the operating system was formally modeled in a functional language that lends itself well to automated theorem proving. This enabled the verification of security policies against the model. The formal definition of security policies and formal proofs of the policies against its model are not required in the other protection profiles.

3.6.2.3 Assurance Leveling

To better understand how the assurance levels apply to the assurance requirements families and their components, let's look at a sample Common Criteria 2.1 assurance component: ADV_HLD, "High-level Design." According to Common Criteria, "The high-level design requirements are intended to provide assurance that the TOE [target of evaluation] provides an architecture appropriate to implement the TOE security functional requirements." If a protection profile includes this assurance requirement (most do), then the rigor of the requirement increases with assurance level:

EAL1	No high-level design required
EAL2	ADV_HLD.1: Descriptive high-level design
EAL3	ADV_HLD.2: Security enforcing high-level design
EAL4	ADV_HLD.2: Security enforcing high-level design
EAL5	ADV_HLD.3: Semiformal high-level design
EAL6	ADV_HLD.4: Semiformal high-level explanation
EAL7	ADV_HLD.5: Formal high-level design

You can see from this example that the assurance requirements for a particular component do not necessarily increase with each corresponding increase in assurance level. Rather, requirements become more stringent when necessary to achieve the confidence goals of a new level.

The ADV_HLD component is further divided into elements that are the individual descriptive requirements. When requirements increase between assurance levels, elements are added to the requirements inherited from the preceding level. For example, an EAL5 protection profile will include the element ADV_HLD.3.9C:

> *"The high-level design shall describe the separation of the TOE into TSP [TOE Security Policy]-enforcing and other subsystems."*

For a firewall product, this might mean that the high-level design must point out the subsystems that enforce the traffic filtering rules. For an EAL6 protection profile, this requirement is augmented to ADV_HLD.4.10C:

> *"The high-level design shall justify that the identified means of achieving separation, including any protection mechanisms, are sufficient to ensure a clear and effective separation of TSP-enforcing from non-TSP-enforcing functions."*

In other words, evaluators will expect the vendor to provide high-level design evidence demonstrating that the traffic-filtering subsystems are protected from corruption, subversion, and so on, by other subsystems.

For an EAL7 protection profile, this requirement is further augmented to ADV_HLD.5.10C, which is the same as ADV_HLD.4.10C, except that the presentation, or justification, must be formal:

> *"A formal specification is written in a notation based upon well-established mathematical concepts... These mathematical concepts are used to define the syntax and semantics of the notation and the proof rules that support logical reasoning. The syntactic and semantic rules supporting a formal notation should define how to recognize constructs unambiguously and determine their meaning. There needs to be evidence that it is impossible to derive contradictions, and all rules supporting the notation need to be defined or referenced."*[44]

In the modern electronic world, unsubstantiated and misleading security claims are commonplace. Potential customers appreciate developers who back up their claims with independent assessments, whether formalized as in the Common Criteria example, or more informal yet rigorous. Some embedded security vendors provide independent consultation, evaluation, and written assessments for their customers; developers can then use the generated evidence to dramatically enhance their claims for product security. Similarly, using components that have already been independently validated is another effective approach to improving security and developers' claims thereof. This concept applies equally well to safety-critical systems. In the following case study, a more informal independent assessment is included in the development and deployment process for a high-assurance embedded product.

3.7 Case Study: HAWS—High-Assurance Web Server

This case study brings together all the PHASE concepts described in this chapter. In 2008, Green Hills Software developed a high-assurance web server (HAWS) to host a website immune to hackers. HAWS was deployed as an application on the high-assurance, security-certified INTEGRITY embedded real-time operating system technology. The web server's lead designer was Dan Hettena, senior operating systems, security, and virtualization architect at Green Hills Software.

Why did Green Hills decide to write a web server from scratch instead of attempting to lock down a commercial web server such as Apache? Apache is a tremendously powerful and successful open source web server, hosting a majority of the world's websites as of the time of this writing. The reason for not adopting Apache in this particular case is that the value of resources to be managed and protected by the web portal was considered extremely high, and the threat of sophisticated attackers wishing to subvert or deface the web portal was also considered extremely high. This combination of high-value resources exposed to a high-threat environment requires high assurance in security that cannot be feasibly retrofitted to software not originally developed to meet this standard.

The Apache code base consists of hundreds of thousands of lines of code developed using commercial general-purpose processes, not using a high-assurance security standard. Hundreds of software vulnerabilities have been found in Apache over the years, and the probability of severe, undiscovered vulnerabilities in any snapshot of the software is considered extremely high. One example of vulnerability from the U.S. CERT National Vulnerability Database is as follows:

> *CVE-2010-0425: "modules/arch/win32/mod_isapi.c in mod_isapi in the Apache HTTP Server does not ensure that processing is complete before calling isapi_unload...which allows remote attackers to execute arbitrary code via unspecified vectors related to a crafted request."*

The primary security policies of HAWS are to ensure that the web portal's content cannot be maliciously corrupted and that requests from legitimate clients not be denied service by other (malicious) clients. Physical security of the web server's computer, as well as upstream network quality of service, is beyond the scope of the HAWS design goals (although these issues were addressed at the system level).

In addition to these stringent integrity and availability policies, HAWS was designed to handle many thousands of concurrent connections with good response time performance.

It is important to note that Apache is a perfectly good choice for most systems (and in fact a port has been done to the same INTEGRITY operating system hosting HAWS), but a higher level of security was deemed necessary for this particular application. Let's run through the PHASE principles in turn, providing examples of how each principle was applied to the development of HAWS.

3.7.1 Minimal Implementation

In contrast to a typical commercial web server's hundreds of thousands of lines of code, the HAWS web services engine consists of approximately 300 lines of code. This feat was accomplished primarily by converting the sophisticated logic of parsing and responding to HTTP requests into a simple state machine that operates on a pre-loaded, read-only database of information corresponding to a particular instance of the hosted files serviced by HAWS. This carefully crafted piece of code single-handedly provides HTTP servicing compliant to HTML/1.1 (RFC 2616), the version of HTTP in common worldwide use. The initial website hosted on HAWS was professionally designed and includes sophisticated graphics (Adobe Flash) and JavaScript. The central state machine loop is shown here (code comments and macros elided). Readers are not expected to understand the code; it is provided to demonstrate an example of minimal implementation:

```
while (preempt_count > 0) {
  uint64_t inst = INST_TABLE(task->pc);
  uint32_t nextpc = task->pc + 1;
  preempt_count-;
  if (((inst >> 59) & 0xf) == 0xf && task->ibuf_off < task->ibuf_len &&
        task->ibuf[task->ibuf_off] == (uint8_t)(inst >> 48)) {
        task->ibuf_off++;
        task->inst_count++;
  } else {
       if ((inst >> 61) && task->ibuf_off == task->ibuf_len) {
           preempt_count -= 100;
           if (!refill_ibuf(task))
               return;
       }
       task->inst_count++;
       c1 = ((inst >> 61) & 0x1) ?
          task->ibuf[task->ibuf_off] : task->GPR((uint8_t)(inst >> 48));
       c2 = ((inst >> 61) & 0x1) ?
          (uint8_t)(inst >> 48) : ((inst >> 20) & 0xfffff);
       cond = (cop & 2) ? ((c1 == c2) ^ (cop&1)) : ((c1 < c2) | (cop&1));
       if ((inst >> 62) & (1 << cond))
           task->ibuf_off++;
       if (cond) {
           uint8_t xop8 = (inst >> 40) & 0xff;
           uint32_t xop20 = (inst >> 0) & 0xfffff;
           // Note: switch not used in order to optimize common cases
           if (opx == 0) {
               nextpc = xop20;
           } else if (opx == 1) {
               nextpc = task->GPR(xop8);
           } else if (opx == 2) {
               if (task->ibuf_off != 0)
                  task->ibuf[-(task->ibuf_off)] = xop8;
           } else if (opx == 3) {
                preempt_count -= 400;
                if (!continue_send(task, xop20))
                   return;
           } else if (opx == 4) {
```

```
                task->GPR(xop8) = xop20;
            } else if (opx == 5) {
                task->GPR(xop8) = (ticks - task->start_ticks) & 0xfffff;
            } else if (opx == 6) {
                free_nettask(task);
                return;
            }
        }
    }
    task->pc = nextpc;
}
```

It is left as an exercise for readers to browse the open source Apache code, comparing the preceding state-machine-driven implementation with a traditional input parsing loaded with gobs of special case logic. However, be warned: the Apache software dealing with HTTP-compliant input handling and response consists of thousands of lines of code.

By avoiding common parsing mechanisms, including large network read operations and string matching, HAWS avoids the kinds of runtime logic that is prone to vulnerability (e.g., buffer overflows).

Finally, the availability features of HAWS are implemented with simple policies instead of complicated heuristics. Nevertheless, these policies serve their purpose well; HAWS has proven impervious to sophisticated penetration testing and has not encountered unscheduled downtime or reported service unavailability since its deployment in 2008. The availability features are further described in the following sections.

3.7.2 Component Architecture

The HAWS web services engine runs as an isolated, memory-protected, user mode process. HAWS takes advantage of INTEGRITY's strict time and space partitioning policies to guarantee that the HAWS component is unable to affect other system components (the network device driver and TCP/IP stack) and vice versa. HAWS does not share address space with any other processes or plug-ins and does not dynamically incorporate any code (e.g., DLLs).

Each remote client connection has a strictly enforced maximum lifetime (temporal componentization), and HAWS rate limits new connection requests in order to prevent DoS attacks and provide quality-of-service assurance to valid connections.

3.7.3 Least Privilege

The HAWS service engine is provided no system resources other than the CPU time and memory it is allocated and the bidirectional stream of bytes corresponding to HTTP requests and responses. The service engine has no capability to access any file system or network device or to launch other programs. This allocation of resource capabilities is specified in the system security policy.

The aforementioned state machine operates on pre-processed data (the files served) that is mapped read-only directly into the service engine address space. Read-only access protects against inadvertent corruption of the served data and is a generally good design choice for security.

Other than HTTP, HAWS has no other network ports. In fact, the website content itself cannot be updated remotely (e.g., via secure FTP) because this was deemed an unnecessary risk. Instead, only a local authorized security administrator with physical access to the web server's computer may update the content.

3.7.4 Secure Development Process

HAWS was developed using a rigorous high-assurance process by Green Hills Software's real-time operating systems engineering group. The development process includes many of the tenets described in this chapter, including exhaustive testing, strict configuration management, peer review, automated enforcement of a proven coding standard, and formal design.

3.7.5 Independent Expert Validation

An independent, leading "white hat" hacking organization performed an analysis and penetration test of HAWS and one of its hosted websites over the course of an extended period of time. This organization has been providing unbiased, technically advanced, low-level security research since 1998. The security company was unable to find any vulnerability, launch any successful denial-of-service attacks, or deface the site.

While it is not intended to be a drop-in replacement for Apache or other full-featured commercial web servers, HAWS provides a practical, successful, deployed example of the application of PHASE concepts.

3.8 Model-Driven Design

Guy Broadfoot and Philippa Hopcroft author the following section on model-driven design (MDD).

Guy Broadfoot is a founder and CTO of Verum Software Technologies. A veteran of the software development industry, he combines deep academic expertise with solid technical experience in developing products and turning around failed software projects. Broadfoot has been designing and implementing software projects since 1965, and in recent years has developed a specialized interest in the application of formal methods. In 1986, Broadfoot founded Silverdata, a software company that developed products for marine navigation and hydrographic survey. He sold the product rights to private investors in 1998. He worked as a consultant to Mountside Software Engineering from 1998, where he was responsible for establishing a rigorous software project planning and financial control system. Broadfoot graduated with distinction from the University of Oxford in 2001 with a master's degree in software engineering.

3.8.1 Introduction to MDD

3.8.1.1 Trends Driving Adoption of MDD

This book has often stressed the trend toward increased software and systems complexity that breeds security vulnerabilities. As the majority of embedded software developers primarily use traditional programming languages, especially C and C++, this chapter has thus far focused on processes and techniques for improving reliability and thereby reducing security flaws in traditional code bases. However, another approach that has met with increasing success is the use of model-driven design (MDD).

Key Point

The premise of MDD is to raise the abstraction of software development from the low-level imperative programming language that is fraught with sharp edges and opportunities to shoot one's self in the foot to a higher-level modeling language that reduces the distance between design and implementation and by doing so reduces the flaws that lead to security and safety failures.

Modeling also lends itself better to formal proofs of specifications and security policies than does traditional programming languages. Indeed, a side benefit of using some MDD platforms (the ones that support formal methods and automatic code generation) is the ability to make formal arguments regarding the correspondence between specification, design, and implementation, a core challenge in all formal approaches. Therefore, the following discussion of MDD leans toward methods that lend themselves to formal analysis and therefore raise the assurance of quality, safety, and security.

This section introduces the subject of model-driven design as applied to embedded software development. It addresses the following questions:

What is model-driven design?
What benefits can be expected by applying it to embedded software development?
What types of modeling languages are currently in use?

What types of MDD platforms are there and how do they differ?
What should be considered when choosing the right MDD platform?
How does the use of an MDD platform comply with safety standards?

In discussions about MDD and MDD platforms, the term *verification* is often used and tends to mean different things to different people. According to the IEEE,[45] the term *verification* is given the following two definitions:

1. The process of evaluating a system or component to determine whether the products of a given development phase satisfy the conditions imposed at the start of that phase.
2. Formal proof of program correctness.

Most MDD vendors assume the first definition for the term *verification*; throughout this section, we do the same. In cases in which formal proof of program correctness is intended, we use the term *formal verification.*

This distinction is important: many MDD platforms offer facilities for verification, usually in the form of simulating execution on the development host before testing it on the target system. Test case generation features of one form or another often support this activity. However, few MDD platforms offer facilities for formal verification to provide proof of correctness of a software design, and for those that do, this is an important distinguishing capability. Where these facilities exist, they offer great benefits, especially in multi-threaded and multi-core designs.

We use the term *component* to mean some physical hardware or software element of a system that satisfies one or more interfaces. We regard a system as comprising a collection of components. The term *subsystem* is a modeling concept defining the grouping of related components and/or other subsystems. It is a modeling abstraction to assist in the stepwise partitioning of large and complex systems. The physical realization of a subsystem is always by means of the components that it comprises.

Software development methodologies have historically tended to concentrate on tools that assist the software developer with system composition, integration, and especially testing. Various component-based technologies, such as CORBA and Microsoft's COM, DCOM, and .NET have aimed at increasing software development productivity and quality by improving reuse through component abstraction in those domains where the technologies are applicable. Although increasingly sophisticated development tools, such as optimizing compilers, source language debuggers, test generators, and test execution platforms, have long been available to help software developers with implementation and testing, this conventional way of developing software is increasingly challenged by:

the relentless pressure to reduce time to market and introduce new products in ever-shorter cycles, frequently leading to hardware and software being developed in parallel and final hardware platforms not available during most of the time the software is under development;

the increasing complexity of embedded software, as essential product characteristics and end-user value are realized by software;
the introduction of multi-core technologies, creating an environment in which software design and coding errors are more likely and conventional testing techniques are less effective;
the continuing downward pressures on development costs;
the trend toward distributed development sites, often in different time zones.

All these challenges must be met while achieving ever-higher quality levels to meet rising user expectations and regulatory pressures.

Conventional software development is labor intensive and bounded by human error. According to a recent Venture Development Corporation survey,[46] software labor costs account for more than 50% of the total spent on embedded system development. Of this spending, according to Ebert and Jones,[47] typical industry experience finds up to 40% spent on testing and rework. It is hard to think of another engineering domain that habitually spends 40% of its development budget identifying and removing defects that shouldn't have been present in the first place.

Historically, organizations have attempted to address some of these issues by means of process improvement—that is, introducing practices, procedures, and standards governing every aspect of specification, design, implementation, and testing. While some of this rigor is required in any high-assurance development process, MDD can help to reduce that overhead, enabling higher assurance while improving developer efficiency, an important theme of this chapter. As discussed earlier, decreased efficiency due to development process overhead leads rapidly to diminishing return on investment and unsurprisingly to a loss of interest by senior management. Finally, beyond a relatively small threshold of complexity and size of a component, it is simply infeasible to achieve high-assurance security and safety with traditional software development techniques. MDD is a tool to address this scalability problem, enabling software developers to be more productive and deliver products more quickly and with significantly fewer defects.

3.8.1.2 What Is MDD?

As noted previously, model-driven design, or MDD, is an emerging paradigm that aims to improve the software development life cycle, particularly for large and complex software systems, by providing a higher level of abstraction for software design than is currently possible with conventional and widely used programming languages. Supported by an appropriate MDD platform, MDD claims to be that technological advance.

MDD is based on the systematic use of models as primary artifacts throughout the software engineering life cycle. MDD aims to elevate functional models to a central role in the specification, design, integration, and validation of software.

In conventional software development, the definitive statement of software function is a mixture of informal functional requirements documents and the program source code, and establishing the correspondence between the two is often difficult. By raising the abstraction level in which specifications and designs are expressed, MDD aims to replace the specification and source code approach with a functional model expressed in an unambiguous, precise modeling language that can be understood by both the domain experts who determine what the resulting system does and the software developers who implement it. Additionally, code generated automatically from the model assures correspondence between specification and implementation.

MDD uses models to represent a system's elements, the structural relationships between them, and their behavior and dynamic interactions. Modeling structural relationships supports design exploration and system partitioning; modeling behavior and interactions is essential for being able to verify designs by verifying models and for code generation. In practice, these last two points are the source of most of the benefits claimed for MDD.

For many embedded software developers, code generation in particular seems to be difficult to accept.

Key Point

Without code generation, much of the advantage of MDD is lost; in reality, deciding to adopt an MDD approach means accepting automatic code generation from models.

The quality of generated code and its suitability for a particular domain and use is an important consideration in choosing an MDD platform or even deciding to take the MDD route at all.

We distinguish between two types of models, namely those that are executable and those that are both executable and formally verifiable. We discuss this issue further in the following sections. Any MDD platforms offering models that are not executable have little added value and are therefore ignored.

3.8.1.3 The Potential Benefits of MDD

By raising the abstraction level in which specifications and designs are expressed and eliminating informal functional specifications, MDD has the potential to

1. Eliminate specification and design errors early in the development cycle where they are cheapest and easiest to rectify.
2. Increase the degree of automation that can be applied to the development process by means of automatic code generation.

3. Facilitate parallel hardware/software design by enabling models to be verified by simulating execution behavior on development hosts before the target system is available.
4. Reduce the required testing effort by applying automated formal verification techniques to the functional models in combination with simulating execution behavior instead of relying solely on testing the implemented program code.

Of course, eliminating errors is the primary impetus behind the discussion of MDD from a security perspective. However, often the most valued benefits of applying MDD to embedded software development are significant reductions in both cost and time to market needed to meet the required level of functionality and quality.

The use of a suitable modeling language contributes to reducing defects in the resulting product in two ways. First, it leads to increased precision and reduced ambiguity when specifying a system's required behavior. Second, it reduces the abstraction gap between the concepts and language of domain experts responsible for specifying the system's required behavior and the software developers who must build it. The result is a functional specification that clearly and unambiguously specifies the required functional behavior and is understandable to all product stakeholders, domain experts, and developers alike.

Automation is increased primarily through automatic code generation, reducing manual programming effort and thus saving cost and preventing defect injection during programming. Further automation is possible by using simulation and automated test case generation, both increasing test coverage and reducing testing costs.

The application of formal verification techniques in industry to establish that the functional behavior of the code generated from a model is as intended has been problematic. There are few MDD platforms that deliver fully automated verification for software, and at the time of this writing, only Verum's Analytical Software Design (ASD) platform can do this automatically, without specialized knowledge and for large-scale embedded software systems with extensive concurrent behavior.

Where such tools are available and applicable, automated verification carried out on functional models can reduce defects to unprecedented levels very quickly and provide significant savings in time and cost; this directly addresses the 40% of development effort currently spent testing and debugging software. We return to this subject later in this section.

3.8.2 Executable Models

Key Point

Models need to be executable; discount any MDD platform where this is not the case.

In addition to being rigorous and precise enough to allow for code generation and test case generation, executable models can be executed on the development system by means of simulation very early in the development life cycle.

Executable models provide the following significant benefits:

1. Provide rapid and early feedback on requirements and specifications and serve to validate that the "right" product is being built. This is what most MDD vendors mean when they use the term *verify*.
2. Allow functional testing to take place on the development host without requiring access to software running on the eventual target machine. This is particularly important when the target hardware and the software are developed in parallel.

Two well-known examples of such platforms are IBM's Rational Rhapsody products and Mentor Graphics BridgePoint, both of which use Unified Modeling Language (UML)-based modeling languages. These are discussed later in more detail.

Executable models are not necessarily formally verifiable, and most MDD platforms do not provide facilities for formally verifying the correctness of the software being designed. These platforms, of course, ensure that the models are well formed in the same sense that a compiler ensures a program conforms to the programming language syntax and grammar.

3.8.2.1 Formally Verifiable Executable Models

In addition to being executable, some MDD platforms produce models that are formally verifiable. This means that the MDD platform includes tools for automated formal verification based on mathematical principles.

Simulation, like testing, is a form of sampling. The sample is the set of test cases, or example scenarios, and the population is the set of all possible execution paths through the software. In practice, for anything other than very small developments, the sample size is statistically insignificant in comparison to the population size. As E. W. Dijkstra once famously observed, "program testing can be used to show the presence of bugs, but never to show their absence!"[48]

In contrast, formal verification uses mathematical techniques to compute and examine every possible trace of behavior in a design to establish whether or not a user-defined set of properties hold under every possible circumstance. If the property set is sufficiently large and complete in the sense of covering all aspects in the software's specification, then formal verification establishes whether or not a design is correct with respect to its specification. This is analogous to the formal verification methods available in some electronic design automation (EDA) platforms for verifying design intellectual property.

In practice, applying formal verification to software designs has proven to be difficult. There are two major challenges to overcome:

1. The ability to derive the property sets from the specifications and specify them in some formal notation—for example, temporal logic
2. Scalability

Deriving a complete set of properties to be verified such that all essential aspects of the software's specification are covered is a difficult and skilled job. Expressing the properties in a formal mathematical notation such as temporal logic also requires a skill set most embedded software developers will not have and results in specifications that are inaccessible to most project stakeholders. Therefore, it is generally impossible to review them together with domain experts and establish that they are a correct interpretation of the software's specification. Even for the few MDD platforms that provide some form of formal verification, its use is challenging and not widespread.

Key Point

The value of formal verification is that it symbolically examines the entire state space of a software design to determine whether or not the specified properties hold under all possible inputs.

In practice, this is rarely done; almost all useful software systems are far too complex to formally verify as a whole. Different MDD platforms use different approaches to these problems. Some MDD platforms restrict the class of designs that can be verified; for example, SCADE Suite (Esterel Technologies) deals with deterministic, synchronous designs. ASD (Verum Software Technologies BV) provides a different approach called *compositional verification* whereby an entire system is verified component by component in such a way that the proven properties still hold when the components are composed to form the complete system. This approach allows fully concurrent and asynchronous designs to be formally verified, both for compliance with the specified properties and also to show the absence of typical concurrent and asynchronous design errors such as deadlocks, live locks, and race conditions. Simulation and testing are particularly ineffective at eliminating such errors.

Key Point

Formal verification should not be seen as an alternative to verification based on simulation and testing; rather, it is complementary.

Formal verification with well-designed tools brings a number of significant benefits:

1. Enables early feedback on the correctness of component designs without requiring any test cases to be specified and before any code is generated. A component can be formally verified before any of the components it uses exist.
2. Gives early feedback on defects before code is generated.
3. Does not require simulation so no test sets need be prepared.
4. Can completely replace or significantly reduce the need for component-level (unit) testing.
5. Reduces integration and validation effort because software enters those phases with very low defect rates.

The value of these benefits increases rapidly as the scale and complexity of the software grows. For example, earlier in this chapter, we discussed the benefits of thorough code coverage testing. A byproduct of formal model verification is that complete multiple condition coverage is accomplished automatically.

3.8.3 Modeling Languages

Models must be expressed in a modeling language with a formally defined grammar and semantics capable of expressing both static structure and dynamic behavior at an abstract level removed from the programming domain. We can divide these languages into two groups:

> ***Vendor-specific languages:*** Languages developed and promoted by a specific vendor of an MDD platform such as MatLab and Simulink from MathWorks, Esterel from Esterel Technologies, and the ASD language used in Verum Software Technologies' ASD:Suite.
>
> ***Standardized languages:*** Languages that have been defined by industry groupings of interested users and MDD platform vendors and are most commonly based on the Unified Modeling Language (UML).

In practice, a significant difference between the vendor-specific modeling languages and the vendor-independent, UML-based languages is, of course, the degree of vendor lock-in that results from using them. The UML-based languages have the laudable goal of being vendor independent and, in principle, offering the possibility for models to be exchanged between platforms. Vendor-specific languages generally do not facilitate model exchange between platforms.

UML-based modeling languages are increasingly being adopted by MDD platforms and other tools from several major vendors and are standardized. Because of their increasing adoption by MDD platform vendors, UML-based modeling languages are discussed more fully in the following sections. Some vendor-specific languages are briefly discussed later together with the MDD platforms that support them.

3.8.3.1 The Unified Modeling Language

The Unified Modeling Language, or UML, is a general-purpose, graphically based modeling language arising out of the field of object-oriented programming and intended for describing software. The Object Management Group (OMG) standardizes UML, and its specifications are freely available for download on the public Internet.[49] Increasingly, MDD platforms use a modeling language derived from or strongly related to UML.

The UML standard prescribes the graphical representation of UML.

> **Key Point**
>
> Because UML is standardized, UML-based MDD platforms tend to present their models in similar ways, making it easier for users to understand and use MDD platforms from different vendors.

The UML standard also prescribes a machine-readable format in the form of XML files. The hope is that this will lead to models being interchangeable between MDD platform vendors.

UML version 2.0 defines 12 different diagram types together with their graphical elements. There are 5 structural diagrams types and 7 behavioral diagram types. The structural diagrams represent system elements and the static code-level structural relationships between them. These structural diagram types are

Class diagrams: Classes are design-time entities representing the basic concepts within a system and the relationships between them. Classes represent runtime entities called objects; each object is an instance of some class. Class diagrams are used to model the static relationships and functional partitioning of a system into its components.

Package diagrams: Packages are general-purpose containers for grouping related UML elements, including other packages. They are typically used to organize class diagrams, although they can group just about anything. The meaning of the grouping is model (and modeler) dependent. Packages play an important role in partitioning and organizing models of a large system because they provide a scoping mechanism for names; each package represents a distinct namespace.

Composite structure diagrams: When modeling large systems, one often repeats complex relationships such as models of real-world entities or repeatable design patterns; these are called composite structures and represent reusable pieces within a model.

Component diagrams: These are similar to class diagrams but instead of dealing with design concepts, they represent relationships between physical pieces of a system called components.

Deployment diagrams: These diagrams show the physical disposition of a complex system between different physical processing nodes within a real-world setting.

Behavioral diagrams capture the varieties of interaction and instantaneous states within a model as it executes over time, tracking how the system will act in a real-world environment, and observing the effects of an operation or event, including its results. The seven behavioral diagram types are:

Use-case diagrams: These diagrams are used to model user/system interactions. They define behavior, requirements, and constraints in the form of scripts or scenarios.

Interaction diagrams: This type is actually a family of diagrams focusing on interactions between modeled system elements. The family members are

- ***Sequence diagrams:*** These diagrams are by far the most common and the ones with which most developers are familiar. They show control flow as an ordered sequence of interactions between system elements, usually between instances of components.
- ***Communication diagrams:*** Communication diagrams tend to emphasize control dependencies between system elements rather than the sequences in which the interactions take place.
- ***Interaction overview diagrams:*** These are simplified interaction diagrams and something of a cross between a sequence diagram and an activity diagram. Their use is for summarizing overall control flow through the modeled system.
- ***Timing diagrams:*** These diagrams focus on specific timing behavior of interactions. They are used to represent the time-based flow of events in the system and for modeling real-time behavior.

State-chart diagrams: There are two flavors of these: one is used to model the evolving behavior of system elements, typically components, as finite state machines. The other is used to model the state of a protocol between various system elements.

Activity diagrams: These diagrams are used to represent interactions between concurrent processes. The state behavior of interacting system entities is represented in "swim lanes" and the interactions between them. These diagrams show how the system entities, usually components, run in parallel with each other and the synchronization points between them. The state behavior shown in the activity diagrams is usually an abstraction of the state behavior shown in the individual state-charts and represents the externally visible state behavior between the entities. Synchronization points are represented as "fork" and "join" operations.

UML has the advantage that it is widely, although by no means universally, known throughout the software development community. UML's graphical orientation makes it attractive to both users and MDD platform vendors. Also, there are many available resources such as books and training courses to help newcomers get started. For examples, readers are directed to the OMG website (www.omg.org).

UML has the disadvantage of being weak at describing functional behavior and tending to present the world in concepts from the object-oriented programming domain. The use of terms and concepts such as *classes, objects, aggregation,* and *inheritance* make it difficult for non-programmers to understand exactly what is being specified. This goes against one of the major advantages of an MDD approach—namely, that the use of more abstract models expressed in terms specific to the product being developed helps communication and common understanding between the different stakeholders in a product development environment, including customers, engineers from different disciplines, domain experts, and software developers. Software developers unaccustomed to programming in an object-oriented language such as C++ also tend to find the concepts difficult to grasp initially; most embedded software is still developed in C, and most embedded software developers still tend to come from that background.

The difficulty of precisely describing required functional behavior has resulted in tool vendors tending to define their own variants of UML to overcome this gap.

3.8.3.2 The System Modeling Language

UML provides an extension mechanism called *Profiles,* which enables UML to be customized and extended to form domain- and/or platform-specific modeling language variants seen as members of the UML family. One such UML variant is called the System Modeling Language, or SysML. Originally started as an open source project and now standardized by the OMG, SysML is a general-purpose modeling language for systems engineering. Also graphically based, SysML supports specification, analysis, design, verification, and validation of complex systems that include hardware and software.

As a modeling language for systems engineering purposes, SysML offers the following improvements over UML:

1. SysML is less software-centric. It supports the modeling of requirements, structure, behavior, and constraints to provide a robust description of a complex hardware/software system, its components, and its environment. Compared to UML, SysML does not rely so heavily on object-oriented programming concepts to represent ideas; this makes it easier for non-programmers and engineers from other disciplines to learn and understand.
2. SysML is a smaller language than UML. The total number of diagram types is reduced to nine, seven of which are reused from UML and two of which, requirements diagrams and parameter diagrams, are new to SysML to support requirements management, requirements tracing, and quantitative analysis.
3. SysML has more flexible and expressive semantics, enabling greater behavioral modeling. It also includes function-based, message-based, and state-based representations.
4. SysML provides more extensive and flexible support for tabular notations in the form of allocation tables. The allocation tables provide a support system for partitioning and allocating behavior, structure, and physical and performance constraints between system

elements. This is particularly useful during the architectural design phase when a system is partitioned into its subsystems and components.

As a consequence, MDD platform vendors are increasingly adopting SysML. A good introduction to SysML is the book *A Practical Guide to SysML.*[50]

3.8.3.3 Executable UML

Another UML variant defined by means of the UML Profile mechanism is called executable UML, or xtUML. xtUML is a graphical language extended by a non-graphical object action language for specifying object behavior at a high level of abstraction but with sufficient precision to enable model execution, simulation, and code generation.

An xtUML model makes use of the diagrams from UML. In some cases, it applies them to different concepts or has its own version of them. The core set of diagrams is

Domain charts: These are essentially UML package diagrams where the packages represent discrete areas of functionality, called domains, within a system. A domain chart identifies the domains and the relationships between them.

Class diagrams: These diagrams are used to partition domains into design-time entities representing the basic concepts within a domain and the relationships between them. The corresponding runtime entities are objects instantiated from classes.

State-chart diagrams: These diagrams are simplified versions of the UML state-chart diagrams. Within xtUML, a state machine described by an xtUML state-chart diagram defines the behavior of every class.

Action specification: This is not a diagram as such but rather an entry form via which object behavior can be specified using the xtUML object action language.

The remaining UML diagrams such as use-case, sequence, collaboration, and activity diagrams can also be reused in a straightforward way.

The xtUML object action language defines behavior in sufficient detail to enable execution. This language is an OMG-compliant object action language (OAL). It differs from a conventional programming language such as C++ or Java in that it describes behavior at a higher level, abstracting away specific programming languages and detailed decisions about how the software is organized. The object action language does not make coding decisions; it makes no statements about tasking structures, distribution, classes, or encapsulation. Instead, the object action language describes only data and behavior, enabling an xtUML model to be deployed in various software environments without change.

Conceptually, each object specified in an xtUML model executes concurrently and asynchronously with respect to all others. Each object is either executing a procedure or waiting for something to trigger its execution. Event sequences are defined per object; there is no concept of global or universal time. All required synchronization between objects must be explicitly modeled.

The job of mapping this conceptual structure onto a specific target platform is given to a model compiler. The model compiler not only generates executable code in the chosen programming language, but also provides the runtime mapping onto the target runtime environment in a semantics-preserving way. In principle, and depending on the availability of suitable model compilers, a model could be compiled into an implementation designed to execute on a distributed system using CORBA, on a small-footprint embedded system using C and targeted for "naked" silicon (without any operating system), or on a complex system using C++ or Java targeted for a multi-processor platform using a real-time operating system such as INTEGRITY or a general-purpose operating system such as Linux or embedded Windows.

xtUML is an embodiment of the Shlaer-Mellor method of software design on which the Mentor Graphics BridgePoint platform is based. An excellent introduction to xtUML can be found in *Executable UML, A Foundation for Model-Driven Architecture.*[51]

3.8.4 Types of MDD Platforms

The practical application of MDD requires the use of an MDD platform: a set of development tools and runtime frameworks that include the following:

1. Editing tools for creating and editing models.
2. Checking tools for checking models for completeness and ensuring they are well formed and meaningful.
3. Tools for verifying the designed behavior, during the specification and design project phases; for example, by simulating runtime execution, automated formal verification methods, or both.
4. Code generators for generating compiler-ready, good-quality, efficient source code corresponding to the functional behavior in the model. Most MDD platforms provide code generators for C and C++. Other languages such as Java and Ada are also available from some vendors.
5. Tools to improve testing efficiency; for example, automating test case generation, test execution, and results analysis.
6. Runtime frameworks that enable the generated code to run properly in a variety of different execution environments.

Embedded software performs a wide variety of functions in numerous engineering domains. Software ranges in scale and complexity from simple sensors, I/O interfaces, motion control, image processing, and supervisory machine control of individual machines at one end of the scale to supervisory control of an entire plant at the other.

> **Key Point**
>
> For MDD to be successfully applied, the MDD platform used must be appropriate not only to the scale and complexity of the system being developed but also to the engineering domains of the system.

A problem that arises in complex systems is that multiple engineering domains are encountered, and no single MDD platform covers them all.

To illustrate this point, in the following section, we use a modern high-speed digital pathology scanning system made by a leading manufacturer of healthcare and medical imaging devices as an example to show how different parts of a complex system require different approaches or platforms.

3.8.5 Case Study: A Digital Pathology Scanner

The purpose of the digital pathology scanner in this case study is to make high-definition, high-resolution color images of human tissue samples and store them in a standard picture archiving and communication system (PACS) for later review by clinical specialists (see Figure 3.6). The image quality must at least equal that of the best optical microscopes and must be suitable for subsequent computer-aided diagnostic examination. Tissue samples are prepared for examination and presented to the device on glass microscope slides; each slide can contain several samples. The slides are placed in racks, and the device has two robot arms with grippers for moving the slides between the slide racks and the scanning part of the device. During scanning, the slides are moved on a precision stage underneath three fixed line scan cameras and their associated optics and illumination sources. The stage moves in three axes; the two horizontal axes are used for moving the sample under the cameras during scanning, and the vertical axis is used to adjust the height of the samples below the cameras to ensure the best

Figure 3.6:
Digital pathology scanner.

focus for every spot on the image. When each image is complete, it is transmitted via a network connection to the PACS system for storage and later analysis. Because the resulting images are very large, the device can reach its throughput targets only by performing slide handling, scanning, image processing, and image transmission actions in parallel.

The embedded software controlling such a machine includes

1. Closed-loop motion-control software for controlling the robot arms and the precision stage; for example, this includes the software needed for stepper motor proportional-integral-derivative (PID) controllers.
2. Calibration software for robots, stage, and image acquisition.
3. Image processing software to enable feature extraction and continuous real-time focusing.
4. Communications software providing remote access to the image servers and implementing multi-layered protocols defined by the PACS standard.
5. Supervisory machine control software to tie everything together and supervise the concurrent behavior of robots, image scanning, image processing, data transmission, and execution of operator commands.
6. A modern graphical user interface (GUI) for the operator.

The software architecture of the sample system is divided into three major layers as follows:

1. Motion Control and Algorithms
2. Supervisory Machine Control
3. Graphical User Interface (GUI)

The Motion Control and Algorithms layer comprises the software implementing the closed-loop control algorithms for the motion controllers together with the image processing and auto-focus algorithms and some hardware abstraction interface code enabling the access to hardware interfaces and signals.

The Supervisory Machine Control layer is a core part of the machine. Built on the Motion Control and Algorithms layer, the software progressively realizes the necessary abstractions leading away from the implementation and programming domains toward the end-user and application domain and implements a complete but "headless" system. All functionality is present, including everything needed to support the GUI to the operator, but excluding the GUI itself.

In the sample system, the GUI layer was implemented in a separate process space for robustness and to facilitate its development by subcontracting it to a third party with specialized human interface design skills. It is implemented using Adobe Flash and coupled to the Supervisory Machine Control layer via an off-the-shelf CORBA Object Request Broker (ORB).

As is typical with complex systems, various technologies are needed to realize the complete machine, and the software is a mix of custom-built components and off-the-shelf components supplied by multiple suppliers.

The chosen MDD approach must be able to support this variety. There is no single MDD platform suitable for all aspects of all types of embedded software design and implementation and capable of supporting the necessary variety of technologies and mixtures of custom-made, off-the-shelf, and legacy software. The benefits likely to be realized from MDD in practice depend on the match between the chosen MDD platform and the nature of the software being developed. Be prepared to mix and match multiple platforms for a complex embedded system development. The mix of modeling languages and other development environments used to build the digital pathology scanner is shown in Figure 3.7. Each component and the choice of modeling system are described further in the following sections.

Furthermore, different MDD platforms require very different skill sets to use them, and these skills might not be present in your development teams, requiring extensive training or the use of expensive outside specialists. Be aware that some MDD platform vendors have the supply of

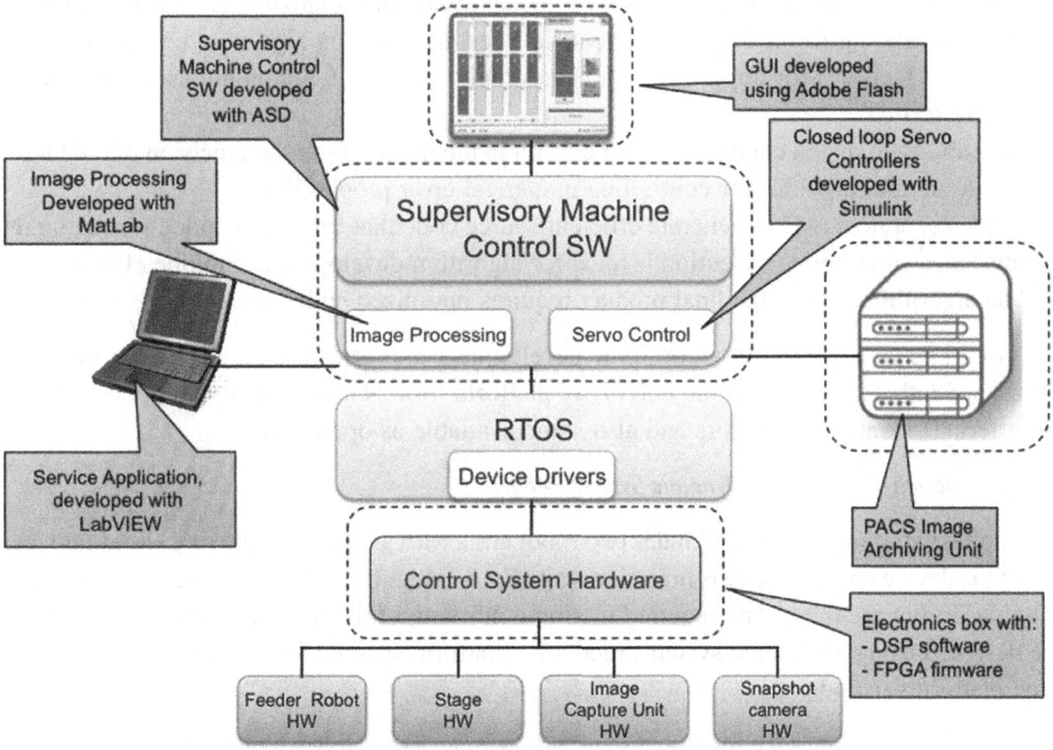

Figure 3.7:
Digital pathology scanner subsystems and their associated development/modeling environments.

these consulting skills as an essential part of their business model, and some are service companies disguised as product companies.

3.8.5.1 Algorithm Development

The digital pathology scanner accepts multiple tissue samples per slide, and each sample must be scanned and stored separately in its own high-resolution image. Before a high-resolution scan starts, individual tissue samples must be located as regions of interest on the slide and their boundaries identified. This is done by taking a low-resolution image of the entire slide and applying image-processing algorithms to identify the regions containing the samples.

Development of these image-processing algorithms is performed using a modeling environment designed for algorithm development. Typically, such platforms provide

1. A (usually) text-based, fourth-generation modeling language designed to model mathematical processes and algorithms as easily as possible. For example, in the MATLAB language from The MathWorks Inc., matrices are first-order objects, and an extensive set of mathematical functions and operators works directly on them. An image is represented as a matrix of pixel values containing the image data.
2. Interpretive model execution combined with easy ways of visualizing numerical and image data in different forms to provide immediate feedback during algorithm development.
3. Multiple facilities for importing data values and a variety of commonly used data and image formats.
4. The ability to match computational precision in the models with the precision of the target hardware and facilities for controlling numerical error propagation.
5. Code generators able to generate efficient source code that can be compiled for the target machine. Interpretive execution is ideal for algorithm development; runtime efficiency of the algorithms within the final product requires optimized code generation.

Two well-known examples of algorithmic development platforms are the MATLAB platform from The MathWorks Inc. and the LabVIEW platform from National Instruments Corp. There are offerings from other vendors and also some available as open source.

3.8.5.2 Motion Control and Dynamic System Modeling

The digital pathology scanner includes two robot arms with grippers for moving slides into and out of the device and a three-axis precision stage for moving the slides under the fixed-line scan cameras during scanning. The required motion-control algorithms for the closed-loop control are designed using a dynamic systems modeling platform specific for this purpose. Typically, such platforms include

1. Graphical user interface for visual modeling with an extensible set of predefined modeling blocks. Models are typically developed in the form of dataflow block diagrams using an interactive graphical editor.

2. Domain-specific modeling aimed at control engineers developing motion-control or other closed-loop control applications.
3. Interpretive model execution combined with easy ways of visualizing numerical and image data in different forms to provide immediate feedback during algorithm development.
4. A library of driver software for interfacing with common instruments and I/O devices.
5. Code generators able to generate efficient source code. This is used both for efficient high-speed simulation during algorithm development and to generate a version deployable on the target hardware in the final product.

Two well-known examples of such platforms are Simulink from The MathWorks Inc. and LabVIEW from National Instruments Corp.

3.8.5.3 Supervisory Machine Control Layer Software

The supervisory machine control software is typically the place where the more general-purpose MDD platforms such as the IBM Rational Rhapsody, Mentor Graphics BridgePoint, and Verum Software Technologies ASD would be applied. They are used to partition the software into its various subsystems, components, and component instances; define component interfaces; and specify the functional behavior and detailed design of the components.

The supervisory machine control software in the sample system consists of 350,000 executable C++ statements. Compared with complex medical imaging systems, manufacturing machines, and telecommunications equipment, this is not an enormous application. According to VDC,[52] the average amount of software in complex embedded systems now exceeds one million executable C language statements.

The pathology scanner was designed to meet strict throughput targets; indeed, this is one of the unique advantages of the product over its competition. To achieve this performance, the scanner was designed from scratch to overlap and execute concurrently as many different phases of its operation as possible. The design of the supervisory machine control software therefore has the following characteristics:

Event driven: The system must handle all incoming events in every possible sequence in which they can occur.

Reactive: The system must always react to incoming events, including those from the machine operator. There can be no "dead" periods when incoming events are ignored or lost.

Asynchronous and concurrent: To meet system throughput targets, the supervisory machine control software must schedule as many processing phases and mechanical movements in parallel as possible.

All these characteristics present formidable challenges for both designers and testing-based verification.

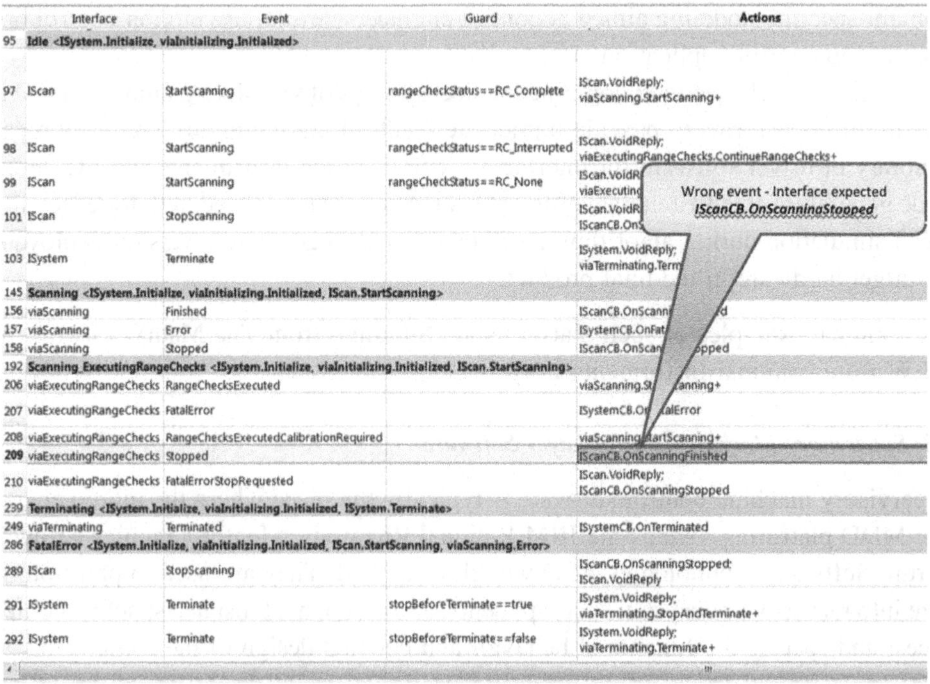
Interface
Event
Guard
Actions
95 Idle <ISystem.Initialize, viaInitializing.Initialized>
97 IScan StartScanning rangeCheckStatus==RC_Complete IScan.VoidReply; viaScanning.StartScanning+
98 IScan StartScanning rangeCheckStatus==RC_Interrupted IScan.VoidReply; viaExecutingRangeChecks.ContinueRangeChecks+
99 IScan StartScanning rangeCheckStatus==RC_None
101 IScan StopScanning
103 ISystem Terminate ISystem.VoidReply;
145 Scanning <ISystem.Initialize, viaInitializing.Initialized, IScan.StartScanning>
156 viaScanning Finished
157 viaScanning Error
158 viaScanning Stopped
192 Scanning_ExecutingRangeChecks <ISystem.Initialize, viaInitializing.Initialized, IScan.StartScanning>
206 viaExecutingRangeChecks RangeChecksExecuted
207 viaExecutingRangeChecks FatalError
208 viaExecutingRangeChecks RangeChecksExecutedCalibrationRequired
209 viaExecutingRangeChecks Stopped IScanCB.OnScanningFinished
210 viaExecutingRangeChecks FatalErrorStopRequested IScan.VoidReply; IScanCB.OnScanningStopped
239 Terminating <ISystem.Initialize, viaInitializing.Initialized, ISystem.Terminate>
249 viaTerminating Terminated ISystemCB.OnTerminated
286 FatalError <ISystem.Initialize, viaInitializing.Initialized, IScan.StartScanning, viaScanning.Error>
289 IScan StopScanning IScanCB.OnScanningStopped; IScan.VoidReply
291 ISystem Terminate stopBeforeTerminate==true ISystem.VoidReply; viaTerminating.StopAndTerminate+
292 ISystem Terminate stopBeforeTerminate==false ISystem.VoidReply; viaTerminating.Terminate+
Wrong event - Interface expected
IScanCB.OnScanningStopped

Figure 3.8:
ASD sequence-based specification.

Modern trends in chip design are providing increased processing capacity by means of multi-core solutions rather than single-core solutions with increased clock speeds. Multi-core platforms lead to true concurrency with multiple threads executing instruction streams simultaneously. The sample system is built on such a multi-core platform and was designed from scratch to take maximum advantage of it.

Concurrent designs are the most difficult to make and the most difficult to verify by simulation and testing. In practice, the number of interleaved execution paths of individual threads and tasks is extremely large, and no economically feasible amount of testing can make much of a statement as to the software correctness or the percentage of execution paths (as opposed to executable statements) actually tested. Differences in runtime execution performance between a simulated platform and the target platform may alter the set of possible interleaved execution paths. Errors in interleaved execution paths that occur only on the target system may therefore go undetected on the simulated platform. Similar effects occur when adding test instrumentation to the software under test; errors disappear when debug statements are added in an attempt to find them.

This is exactly the situation in which formal verification pays dividends in comparison with simulation or testing-based verification approaches alone.

The supervisory machine control software of the pathology scanner was fully developed using the ASD platform from Verum Software Technologies. This platform was designed from scratch to incorporate formal verification of industrial-scale systems. ASD models are both executable and formally verifiable, and this was the principal reason for choosing ASD. Formal verification does not suffer from any of the simulation and testing shortcomings described previously; all possible interleaved execution paths are computed and evaluated to verify a software design.

Figure 3.8 shows the sequence-based specification (SBS), as shown in the ASD development environment, of a portion of the supervisor software: the glass path controller, the part of the scanner that controls the glass slide handling into, out of, and within the machine. As shown in Figure 3.8, the verification step failed because the interface model (not shown) required the design to send an *OnScanningStopped* notification event at this point in its execution. The designer accidentally specified an *OnScanningFinished* event instead. The formal verification detected this non-compliance between the required behavior specified in the interface model and the designed behavior specified in the design model.

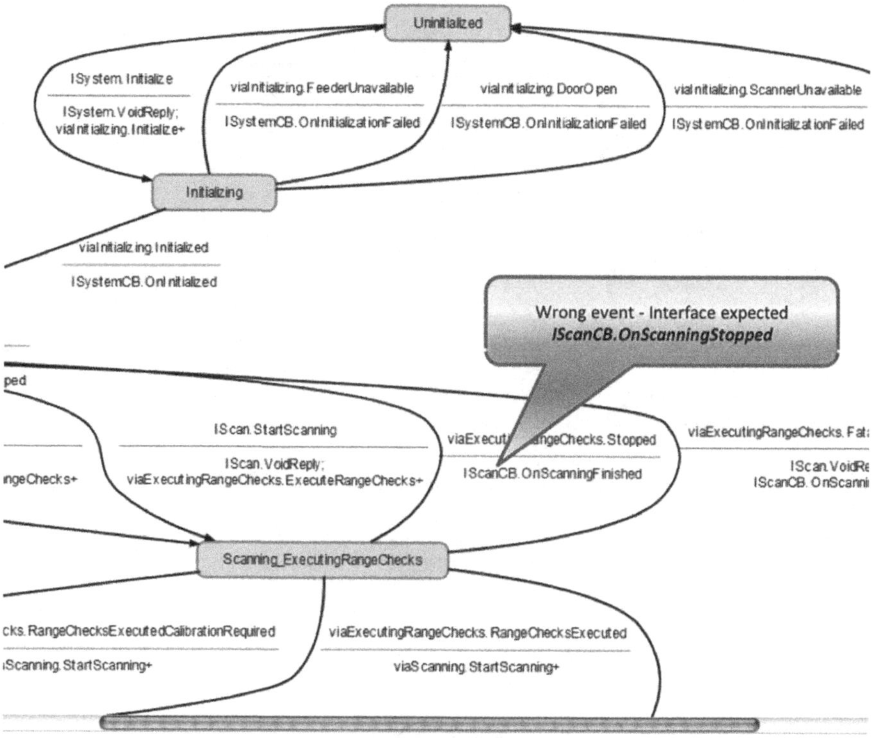

Figure 3.9:
State flow diagram (partial) for glass path controller.

The SBS shows the way that an ASD user specifies the design logic in terms of rules that map a sequence of responses to every input trigger.

The state flow diagram for the glass path controller is shown in Figure 3.9. Also shown is the incorrect *OnScanningFinished* event.

This state flow diagram is automatically generated by ASD from the SBS input by the user. The state flow is a convenient visual display for understanding state transitions caused by events. The arc representing the illegal transition is clearly displayed near the center of Figure 3.9.

When formal analysis detects an error, ASD displays the shortest sequence of events that lead to the failure. The sequence diagram elucidating the illegal *OnScanningFinished* event is shown in Figure 3.10.

To get a sense of the code-generation capability of ASD, let's look at the following code fragment, which is automatically generated C code for the corrected model, corresponding to the *Stopped* transition method for the *ExecutingRangeChecks* state. This event causes

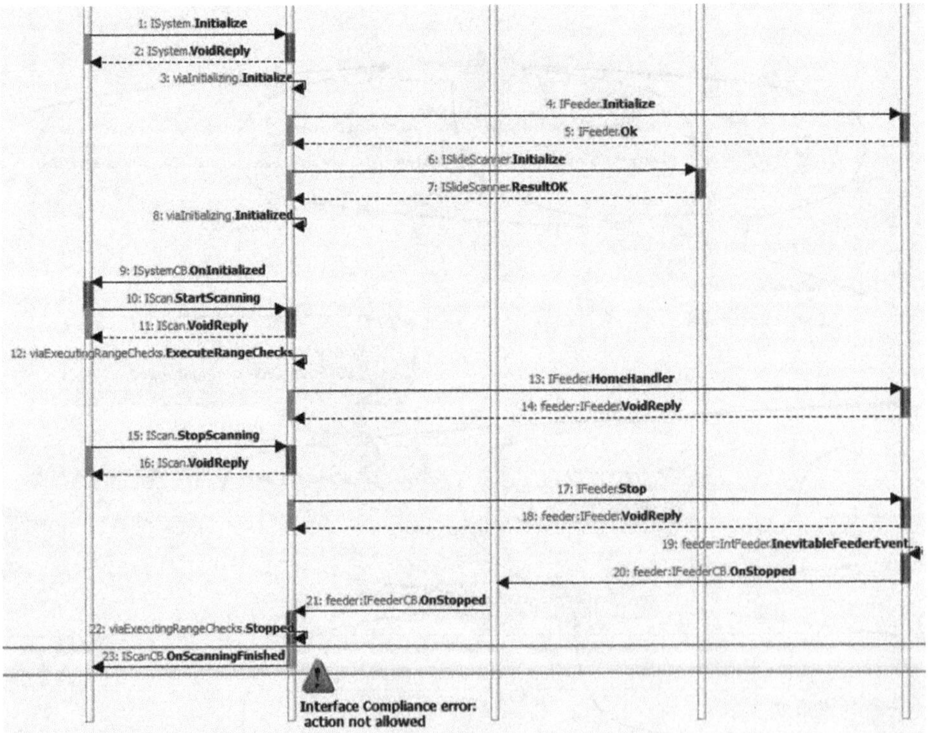

Figure 3.10:
Sequence diagram showing how design implementation does not match the interface specification.

transition to the *Idle* state. It is clear from this code that no other state transition is legal from the *ExecutingRangeChecks* state.

```
static void viaExecutingRangeChecks_Stopped(GlassPathController* self)
{
 ASDDBG_ENTRYTRACE();
 switch (self->impl.asd_data.machine){
  case GLASSPATHCONTROLLER_MACHINE:
   switch (self->impl.asd_data.state){
    case GLASSPATHCONTROLLER_SCANNING_EXECUTINGRANGECHECKS_STATE: {
     /* Rulecase line number: 209
      * The client interrupted the range check by a StopScanning() request.*/
     self->impl.asd_data.state = GLASSPATHCONTROLLER_IDLE_STATE;
     IScanCB_OnScanningStopped(self->IScanCB_intf);
     self->impl.asd_data.predicate.rangeCheckStatus =
        GLASSPATHCONTROLLER_RC_Interrupted_PRED;
    }
    break;
    default:
     asd_illegal(ASD_ILLEGAL_MSG);
    break;
   }
   break;
  default:
   asd_illegal(ASD_ILLEGAL_MSG);
   break;
 }
 ASDDBG_EXITTRACE();
}
```

3.8.6 Selecting an MDD Platform

3.8.6.1 Examples of Commercial MDD Platforms

3.8.6.1.1 Analytical Software Design Platform

The vendor of Analytical Software Design (ASD) describes the product as a Software Design Automation (SDA) platform rather than an MDD platform. The reason is that ASD was conceived with practical formal verification at its core. It is the ability to formally verify industrial-scale software designs and to make this available to all software developers without requiring specialist knowledge of formal verification techniques that the vendor claims is its distinguishing feature and considered in a different product category from MDD platforms that lack this ability. The term *SDA* was derived from EDA platforms used for hardware design on the grounds that ASD makes extensive use of formal verification techniques in a manner analogous to EDA platforms' use of formal verification techniques for hardware logic designs.

ASD is a component-based technology for constructing correct and complete industrial-scale systems from formally verified components. ASD uses a non-graphical tabular form of the sequence-based specification language in which a system response is defined for every possible sequence of input events.[53] The ASD:Suite provides the following:

1. A Model Editor used for constructing models of components.
2. A Model Compare tool for comparing different versions of models and showing their differences.
3. Fully automated formal verification of designs against specifications.
4. Code generation in MISRA C, C++, and Java.
5. Support for a wide variety of runtime platforms, from "naked" silicon through complex multi-processor and multi-core systems with a real-time or general-purpose operating system.

The ASD:Suite itself has a client/server architecture; the Model Editor and Model Compare tool are installed on the user's desktop while the model verifiers and code generators are provided as a hosted service. Automated formal verification is extremely resource intensive, and the hosted service enables users to access this capability without investing in powerful hardware systems. Another advantage of the hosted service is development security: because the code generators are secured in Verum's data center, developers need not worry about protecting against the kinds of internal Trojan horse attacks described earlier in this chapter.

ASD recognizes the fundamental distinction between an interface and an implementation and uses two distinct types of models: an interface model and a design model.

An ASD interface model captures not only the syntax of a component's interface, such as the operations and methods, parameter data types, and so on, but also its externally visible behavior in the form of a sequence-based specification. This externally visible behavior is what an interface-consuming component can assume to be true of every compliant implementation while remaining free of all internal behavior and implementation details. ASD interface models can be made completely independent of the target language.

In contrast, an ASD design model is a specification of the internal behavior that defines the implementation. All the behavioral source code is generated from the design model; the interface model is used to generate source code representing the external interface of the component. When the target language is C++, for example, the interface models are used to generate the abstract class declarations, and the design models are used to generate the class implementations. ASD design models are also independent of the target programming language. All supported languages can be generated from any ASD model.

Earlier, we discussed the difficulties of applying formal verification to software designs. In particular, formal verification requires specialized skills lacking in most development organizations, and modern system designs are often too complex for the application of formal techniques.

ASD addresses these issues in two ways: first, by including the externally visible behavior as part of a component's interface specification, and second, by using a technique called compositional verification in which systems are verified component by component in such a way as to ensure that all properties verified at the component level hold when the components form the complete system.

An ASD component design model references a number of interface models. One of these, the *implemented interface model*, is the specification of all the externally visible functional behavior the design must implement. The others, *used interface models*, are the interface models of the other components being used by the design. Together, these referenced interface models form the functional specification against which the design will be formally verified for compliance.

The usability and economic issue of deriving property sets needed for design verification is solved because ASD users do not need to specify them by hand; these properties are generated automatically by the ASD platform from the set of referenced interface models. ASD users require no specialized knowledge to be able to formally verify software designs.

The scalability issue of applying formal verification to complete systems comprising newly modeled designs, legacy code, and off-the-shelf software components is also handled. New designs modeled with ASD have both an ASD interface model and an ASD design model, as mentioned previously. Legacy code or off-the-shelf components will not have an ASD design model; instead, the user must analyze the components' conventional specifications and then construct ASD interface models representing those components.

ASD's verification is fully automated. A component is verified by submitting for verification its ASD design model together with its set of referenced ASD interface models. The ASD platform translates the ASD models into mathematical models in a process algebra called Communicating Sequential Processes (CSP).[54] The resulting CSP models are exhaustively examined by a finite-state model-checking engine running on the vendor's server farm.

ASD currently supports two execution models: namely, the standard execution model (SEM) and the single-threaded execution model (STM). In both execution models, every ASD component instance behaves as a Hoare Monitor and automatically provides the necessary mutual exclusion and synchronization.[55] In the standard execution model, all components execute concurrently and asynchronously. However, ASD recognizes that most interactions between components in practical designs are synchronous in nature, ultimately being implemented on the target environment by synchronous procedure calls. Instead of requiring the user to model the synchronization explicitly for this most common case, ASD does this automatically. The user has only to model explicitly those synchronizations necessary to coordinate asynchronous interactions between concurrently executing components.

The single-threaded execution model, in contrast, assumes only one thread of control through an entire stack of single-threaded components. Furthermore, this has to be guaranteed by the environment, and the generated code includes no mechanisms for ensuring mutual exclusion at runtime. This model was designed for simple, deeply embedded applications executing on target platforms with highly constrained resources. The model is also widely used in the telecommunications domain when many thousands or millions of component instances are required to support active calls through a switching node.

Components of both executable models can be mixed in a single subsystem; this is particularly useful when interfacing with GUI frameworks. Typically, modern GUI frameworks are themselves single threaded and driven by a so-called message pump: they require that actions synchronously invoked by the framework complete their processing on the calling thread. Being able to mix components with different execution models enables, for example, a single threaded set of components to interface directly with a GUI framework while invoking standard model components for carrying out the actions requested via the GUI either synchronously or asynchronously, as called for by the design.

3.8.6.1.2 IBM Rational Rhapsody

Rhapsody is a UML/SysML-based visual modeling platform aimed at embedded software developers. It is available in Eclipse and Microsoft Windows versions. Rhapsody is aimed at the complete development life cycle, from requirements through to deployment. Rhapsody is available in a number of editions, each being targeted at a specific community of users. Depending on the edition chosen, Rhapsody's features include

1. Integrated requirements and modeling environment using industry-standard SysML or UML diagrams.
2. Full life-cycle traceability and analysis from requirements to design. Requirements can be imported from external sources such as DOORS, IBM Rational Requisite Pro, Microsoft Word, and Excel. Requirements can also be managed directly from within Rhapsody and are fully traceable to use cases and software components. It provides
   - Static model checking analysis to improve design consistency.
   - Support for graphical difference and merging capabilities.
   - Visualization of architecture and design with industry-standard UML.
   - Generation of C, C++, or Java code frames.
   - Full behavioral code generation for C, C++, Java, and Ada.

Documentation can be generated from the models, reducing the compliance effort required for safety critical software.

Rhapsody models are executable, and verification is by means of simulation and testing. Rhapsody does not provide formal verification of design correctness.

3.8.6.1.3 Mentor Graphics BridgePoint

Mentor Graphics BridgePoint is another UML-based visual modeling platform aimed at embedded software developers with particular emphasis on real-time systems. BridgePoint is based on the Shlaer-Mellor method and uses executable UML (xtUML) as its modeling language. BridgePoint is aimed at all development phases, from architecture through to deployment.

In addition to the core xtUML language elements mentioned earlier (domain charts, class diagrams, state machine diagrams, and action specifications), BridgePoint also supports use-case diagrams, sequence diagrams, communication diagrams, and activity diagrams. Features include

1. Static model checking analysis, which helps design consistency during model construction. As model elements change, action specifications are automatically updated as necessary to keep the models consistent.
2. The ability to visualize architecture and design with industry-standard UML.
3. Full behavioral code generation for C and C++.
4. Model compilers for translating xtUML to C or C++.

BridgePoint models are platform-independent models (PIMs) that can be directly executed in xtUML without requiring code generation or translation from xtUML. Direct execution supports verification during the design phase.

3.8.6.1.4 SCADE Suite

SCADE (Safety Critical Application Development Environment) is both a language and a tool set specifically aimed at safety-critical applications. SCADE resulted from a joint effort by Airbus, Schneider Electric, and Verilog (a former French startup company, not the hardware description language of the same name). SCADE is based on the theory of synchronous languages for deterministic real-time applications and in particular on the Lustre and Esterel languages.[56,57]

SCADE implements a synchronous, cycle-based execution model, the common sampling/actuating model from control engineering, by executing a continuous loop alternating between reading input sensors and executing application actions. On each cycle, input sensors are read, and cycle outputs are computed and fed back to the environment. SCADE is used to develop the application logic within this continuous control loop.

The SCADE Suite from Esterel Technologies is a member of the product family addressing all phases of software development for control software. Its features include

1. Visual modeling of both data flow and control flow using a combination of data flow block diagrams and state machine diagrams.

2. Visual modeling of GUI interfaces.
3. C code generators producing small-footprint code compliance with a number of safety standards, including DO-178B, IEC 61508, EN 50128, and IEC 60880.
4. Ada code generators.
5. Customizable adapters to enable the generated code to run on any target environment, from naked silicon to those with an onboard RTOS. INTEGRITY from Green Hills Software, VxWorks from Wind River, and PikeOS from Sysgo are among those directly supported by the suite.
6. Verification based on simulation.
7. Formal verification of functional behavior and timing properties.

Other members of the SCADE family include SCADE System, for the model-driven design of critical system architectures, and SCADE Display, providing a visual modeling environment for developing critical displays and graphical human/machine interfaces for safety-critical and mission-critical systems.

The SCADE timing verifier statically examines the compiled and linked code of a subsystem or a complete system. Users specify information about the target CPU such as clock rate, cache behavior, and pipeline structure. In addition, the tool requires that the upper bound of the number of iterations of every loop is known and specified. This loop information can be extracted automatically from SCADE Suite-generated code but has to be specified manually for legacy code or third-party code modules.

The approach taken by SCADE Suite is possible because of the sequential, deterministic behavior of the systems it produces. Developers must investigate whether or not this approach scales sufficiently and is otherwise feasible for a particular application.

Because of its safety and mission-critical embedded systems application focus, the vendor has prequalified the platform and, in particular, its code generators for compliance with a number of safety standards, including IEC 61508 and its derivative standards ISO 26262, EN/ISO 13849, and IEC 60324.

3.8.6.2 MDD Platform Choice Is Strategic

Adopting an MDD approach to software development is a decision with many consequences, not all of which are apparent. Some of the most important considerations are discussed in the following sections.

In the EDA world, the choice of EDA platform has turned out in practice to be a strategic decision. It is unusual for users to switch vendors and platforms because transition costs are very high. Costs include a large investment in retraining users and porting existing intellectual property into the new modeling environment.

While some MDD platforms claim otherwise, the same reality applies to MDD. With the passing of time, much of an organization's software intellectual property will exist in the form of models specific to one MDD platform and thus one vendor.

Key Point

While portable models that can be seamlessly imported with full semantics preservation between vendors' platforms are desirable, this ability does not yet exist in practice; every MDD platform has many unstated assumptions as to the precise meaning and interpretation of its models, and these assumptions are built into the code generators, simulators, test case generators, formal verifiers, and supporting runtime platforms.

Even for MDD platforms implementing the UML and/or SysML standards, the problem remains. All UML-based languages are designed for extensibility, with many semantic variation points. Every MDD platform vendor is likely to have made vendor-specific semantic extensions.

3.8.6.3 Skills

Software engineers will need training in the use of an MDD platform, and the amount of training varies between platforms. While an important consideration, training is not the most important when considering the skills required.

As noted earlier, model-driven design is fundamentally about working at higher levels of abstraction than is usual or possible with conventional software development practices. Consequently, model-driven software development values higher-level software architecture and design skills above programming skills. Many experienced and highly skilled programmers find it difficult to develop software architectures and designs at the level of abstraction required for successful modeling. The larger and more complex the system being designed, the more serious this problem becomes.

It is therefore important to assess the level of these skills within the existing organization as part of the decision process. Bridging the gap between the skill set required and those skills available will require training followed by mentoring and coaching. MDD is sometimes adopted by organizations as an attempt to overcome a deficiency in design skill, but this impetus rarely translates into success. Instead of overcoming the skills deficit, it merely serves to highlight them.

3.8.6.4 Impact on Existing Work Flows

Experience has shown MDD to be compatible with incremental and agile methods of development.

Key Point

Compared to conventional methods, MDD enables verification to start earlier, as soon as models are completed; verification is not dependent on handwritten code or test cases. To benefit from this opportunity, however, an organization must establish requirements much earlier in the development life cycle and to a greater degree of precision than is usually the case with conventional software development.

Some MDD platforms, such as Rhapsody, BridgePoint, and ASD, are intended for the development of complete, complex software systems and implement an "end-to-end" approach for all phases of the software development life cycle, from requirements to maintenance after deployment. A settled architecture that cleanly partitions the system into coherent components is a prerequisite for making successful functional models. Many MDD platforms require this process to be followed, and, especially for organizations at a lower maturity level (e.g., CMMI level two or below), the impact of this process can be severe.

When considering MDD as a way of working and selecting an MDD platform, developers must understand what the MDD platforms under consideration require from an organization and how their use will change the way the organization currently operates.

3.8.6.5 Interfacing with Legacy Code and Off-the-Shelf Software

Most complex systems contain a combination of software components that are newly developed, reused from legacy code, and supplied by third parties. Legacy code and third-party components are often not developed using an MDD platform, and no models of those components exist. In addition, as in the pathology scanner case study, there will be a layer of handwritten software implementing hardware abstraction interfaces, device drivers, and the underlying operating system.

Key Point

When MDD-designed and -generated code is mixed with handwritten code, there are two principal issues to consider: first, how can the MDD models be verified when their correct runtime execution depends on behavior in unmodeled handwritten code? Second, how will the model-generated code interface with the handwritten code at runtime?

The first issue is a big one; much of the benefit from MDD comes from being able to verify the design models without needing the target platform (thus reducing the 40% of development money spent on test and integration). For this economy to be realized, the externally visible functional behavior of the unmodeled code must be captured and specified in some way, whether the verification is achieved informally by simulation or by means of formal verification.

MDD platforms address this problem differently. Some platforms allow unmodeled components to be invoked directly from the model during verification. Another common approach is to provide a language in which the functional behavior of the unmodeled code can be specified at a high level of abstraction. For example, the ASD platform addresses this problem by requiring ASD interface models representing the functional behavior of the unmodeled code.

Regardless of approach taken, the problem of ensuring the correspondence of these models with the handwritten code remains with the user. Verification based on the models is valid only to the extent that the legacy code or third-party components behave according to their models.

The second issue concerns the conventions assumed by the MDD platform when generating code and the assumptions the platform makes about memory models and runtime execution process and thread structure. In many cases, considerable effort can be expended handwriting code wrappers to transfer control and data between generated and handwritten code.

3.8.6.6 Code Generation

Code generation has yet to gain widespread acceptance in the embedded software world. However, as mentioned earlier, much of the value of MDD is lost unless code is generated directly from the models. The MDD platform vendor takes responsibility for ensuring that the generated code executes exactly as modeled. If code is handwritten using the verified models as a specification, or the generated code is modified, then all bets are off. Having verified the models, the developer will have to re-verify the handwritten or modified code by conventional testing, nullifying the advantages of MDD.

Key Point

When an organization chooses an MDD platform, the efficiency, portability, and compliance characteristics of the generated code are of vital importance, especially in the embedded domain.

Developers and managers must examine sample code generated from realistic models to be satisfied with at least the following:

Memory footprint: Is the generated code small enough to fit in the target system? Is the memory footprint properly distributed between ROM and RAM memory (if applicable)? In the case of severely resource-constrained systems, these issues can be showstoppers.

Memory allocation model: Some target environments, such as those compliant with the MISRA standards discussed earlier in this chapter, require all resources to be allocated before execution starts and prohibit dynamic memory allocation; consequently, recursion and the language-specific equivalents of *malloc*() and *free*() are prohibited. Does the MDD platform provide a code generation option for complying with these language subsets (if applicable)?

Merging of handwritten code/legacy code: Some code generators use proprietary calling conventions in the generated code. To what extent are these conventions compatible with the system's legacy code? And if incompatible, what is the solution to make the generated code compatible? If wrappers are needed, are these completely handwritten (creating extra effort and maintenance as well as potential fault injection risk), or does the code generator provide some assistance with this process?

Threading and runtime execution model: Most MDD platforms have particular execution and threading models that cannot be modified by the developer. For example, some platforms have a strict message-passing model to describe interactions between components and assume complete concurrency, with one thread being allocated to each component. The platform supplies a means by which these generated threads are mapped onto actual operating system threads at runtime. There are domains in which this threading model simply does not work — for example, in the telecommunications domain where very large numbers of component instances exist and there may simply be insufficient system resources to support this number of threads. Designers need to understand the execution models of the candidate MDD platforms and determine if they are sufficient for the target environment.

3.8.6.7 Runtime Execution Speed

Unless the target is deeply embedded or so performance constrained that developers must regularly count cycles, execution speed is one potential problem that most generated code does not have. Experience has shown that in all but the most performance-constrained target environments, automatically generated code performance is adequate.

For performance-critical applications, some platforms offer facilities for verifying timing properties. For example, the SCADE Suite has features for computing and verifying worst-case execution time for sequential applications by statically examining the compiled and linked executable of a subsystem or a complete system.

3.8.6.8 Runtime System Requirements

Different applications have varying target platform requirements that must be supported by the chosen MDD platform. For example:

1. If the target system is a deeply embedded "naked" silicon (without an operating system) with limited resources, will the generated code execute in the target environment "out of the box," or will a simple task scheduler and interrupt service routines need to be developed by hand?

2. If the target system has an operating system, does the generated code support it? If not, how can the generated code be ported to the target environment? Is there an operating system abstraction layer that needs to be ported, or must the generated code be modified? If so, is there a development kit provided by the platform vendor for this with clear specifications and compliance tests?
3. If the application is a hard real-time application (i.e., one in which timing constraints are correctness properties), is the generated code and runtime execution model sufficiently deterministic with respect to timing behavior? Does the system allow interrupts to be serviced quickly enough?

3.8.7 Using MDD in Safety- and Security-Critical Systems

The use of MDD platforms in safety-, security-, and mission-critical domains such as aerospace, military applications, rail transportation, and automotive is gradually increasing. The standards governing software in such domains increasingly recognize that unit testing can be reduced if an MDD platform is used that provides sufficient verification facilities on the model itself.

The detailed requirements governing the use of tools such as MDD platforms differ in detail from one standard to the next. Broadly speaking, the standards follow the same general approach:

- The standards distinguish between the operational embedded software shipped as part of the product from the tools used to make that software. Certification usually applies to the operational software, not to the tools. When tools are covered in a certification, their requirements are quite different and, in general, not as rigorous.
- The various standards for safety- and security-critical software—for example, IEC 61508, EN 50128, ISO 26262, and ISO 15408 (Common Criteria)—often require the tool users to undertake an assessment of the tool to categorize it according to whether or not the tool itself can introduce errors into the operational embedded software and to perform an assessment of the tool against the applicable criteria in the standard to qualify the tool for safe and secure use.
- All users of software development tools must qualify them to establish that they comply with the applicable standards. The responsibility for this falls on the tool user and not on the tool vendor. At the minimum, the user is required to ensure that a hazardous operation analysis is performed and a Safe Use Guide is written, specifying, for example, the checks that users must perform on the resulting output (e.g., generated code) and tool features to be avoided because they are considered hazardous. Also, known errors in the version of the platform being used must be identified together with practices for avoiding them.
- The user must establish that the MDD platform vendor's software development practices and organization comply with the applicable standards. For example, EN 50182 requires

that the MDD platform manufacturer have a named person external to the development team who is responsible for ensuring each release conforms to requirements and remains qualified. This person must report to the company's most senior management and must have the authority to stop a release. In some European Union jurisdictions, this person has a legal responsibility.

When considering MDD platforms for use in a safety-, security-, or mission-critical domain, an organization should consider at least the following:

1. Is the MDD platform already in use within a domain governed by the same or similar standards and regulations as the organization's and/or project's domain?
2. Has the platform vendor already conducted a qualification assessment supervised by a recognized external certification organization? Undertaking the assessment necessary to qualify an MDD platform for safe use is a time-consuming and expensive process; prequalification by the MDD vendor significantly reduces the costs of qualification to the user. Prequalification also provides evidence that the expensive and time-consuming qualification assessment is likely to be successful and shows vendor commitment to, and knowledge of, the development organization's safety- or security-critical domain.
3. What form of verification does the MDD platform offer? Especially in the safety- and security-critical areas, formal verification is particularly valuable.
4. To what extent can the MDD platform's verification be used to eliminate or reduce the amount of conventional unit testing? Can the MDD platform also simplify and reduce integration testing and result in savings with respect to end-to-end validation?
5. Does the generated code comply with the applicable standards and regulations? For example, in the automotive domain, generated code must be compliant with the MISRA C standard.
6. Does the platform assist in producing the embedded software documentation required by the applicable standards?

Because of its potential development cost and time efficiencies and ability to reduce the occurrence of software design flaws, model-driven design is an important software development technique worth considering for embedded systems that have stringent safety, security, and/or reliability requirements. Deciding on the most appropriate MDD platform for a particular component and for the overall system requires that engineers and managers understand the core technical advantages and disadvantages of available solutions. Consult your embedded software vendors for further advice and evaluation.

3.9 Key Points

1. Software assurance refers to the level of confidence that the software end user and other relevant stakeholders (e.g., certifiers) have that the security policies and functions claimed by that software are actually fulfilled.

2. PHASE—Principles of High-Assurance Software Engineering—prescribes a set of five principles to be used in the creation of ultra-reliable software and systems: minimal implementation, component architecture, least privilege, secure development process, and independent expert validation.
3. An important software robustness principle is to compose large software systems from small components, each of which is easily maintained by, ideally, a single engineer who understands every single line of code.
4. An important corollary to the component architecture principle is that safety and/or security-enforcing functionality should be placed into separate components so that critical operations are protected from compromise by non-critical portions of the system.
5. Componentization provides many benefits, including improved testability, auditability, data isolation, and damage limitation.
6. Ensure that no single software partition is larger than a single developer can fully comprehend.
7. Ensure all developers know who the component managers are.
8. If possible, use an operating system that employs true application partitioning.
9. Contrary to popular belief, designers should strive for a one-to-one ratio between threads and processes.
10. Components must be given access to only those resources (e.g., communication pathways, I/O devices, system services, information) that are absolutely required.
11. For critical safety- and security-enforcing components, the software development process must meet a much higher level of assurance than is used for general-purpose components.
12. An extremely important aspect of maintaining secure software over the long term is to utilize an effective change management regimen.
13. Use asynchronous code reviews with e-mail correspondence or carefully controlled live meetings.
14. Use the configuration management system to automate enforcement of peer reviews for every modification to critical code.
15. Apply a level of process rigor, including code reviews and other controls, that is commensurate with the criticality level of the component.
16. By making security a part of peer reviews, management will create a culture of security focus throughout the development team.
17. Developers and users require assured bit provenance: confidence that every single binary bit of production software originates from its corresponding known-good version of source code.
18. Develop and deploy a coding standard that governs software development of all critical components.
19. Maximize the use of automated verification of the coding standard; minimize the use of manually verified coding rules.

20. Prohibit compiler warnings.
21. Take advantage of the compiler's strictest language settings for security and reliability.
22. The goal of Embedded C++ is to provide embedded systems developers who come from a C language background with a programming language upgrade that brings the major object-oriented benefits of C++ without some of its risky baggage.
23. Use automated tools to enforce a complexity metric maximum and ensure that this maximum is meaningful (such as a McCabe value of 20).
24. The best coverage for software flaw detection via static analysis requires multiple tools from multiple vendors to be used in concert.
25. For organizations new to static analyzers, MISRA, and other automated code-quality enforcement tools discussed in this chapter, managers and their development teams need to prepare themselves for what may be a painful initial deployment.
26. Management must allow for exceptions to coding standard rules for specific portions of critical software in which the risk and cost of retrofit clearly exceed the benefit.
27. It is absolutely critical that management decide the course, invest in the initial retrofit cost and schedule, and then stay the course until a high-quality coding standard is firmly entrenched into the development system and indoctrinated throughout the engineering organization.
28. Management should carefully limit the number of coding standard controls that must be manually enforced; if that number is too large, peer reviews will either become overly inefficient or enforcement of those rules will become lax and ultimately irrelevant.
29. A secure development process should employ dynamic code analysis in addition to static code analysis.
30. Whenever possible, a static analysis tool should be used to check for the largest potential runtime stack memory requirements for a program or for all threads in a multi-threaded program.
31. A comprehensive test regimen, including functional, regression, performance, and coverage testing, is well known to be one of the best mechanisms to assure that software is reliable and secure.
32. We strongly recommend the use of MC/DC coverage testing for the most critical components of an embedded system—for example, the operating system kernel, network security protocols, and cryptographic components.
33. A lack of complete test coverage almost always points to a lack of other important validation (e.g., functional testing), design flaws, or simply latent code that has unknown and potentially security-relevant impact.
34. The testing system should be running 24x7.
35. The testing system should run on actively developed products as well as currently shipping products.

36. The testing system should be able to effectively test a software project in less than one night.
37. It should be trivial to determine when a test run has succeeded or failed; a failed test should be trivial to reproduce.
38. Use an autobuild system to quickly detect changes that break system builds.
39. Always ensure a developer has at least two development projects to work on at all times.
40. Employ distributed builds to maximize computer utilization and improve developer efficiency.
41. Intrinsic assurance evidence must be evaluated and confirmed by independent experts.
42. With 26 participating countries, the Common Criteria standard represents an admirable achievement in obtaining global acceptance and certainly the most comprehensive international security evaluation standardization effort ever conceived.
43. In a Common Criteria protection profile, assurance requirements are collectively tagged with a security level (called the Evaluated Assurance Level, or EAL) that represents the overall confidence that stakeholders can have that the security functional requirements are actually met by a conforming product.
44. Following applicable EAL6+ assurance guidelines, as exemplified in the SKPP, and obtaining independent evaluation of the associated assurance evidence by expert consultants would be an excellent approach for embedded systems developers who want to raise the bar on their product security.
45. Independent expert validation is a critical requirement for any embedded systems organization that wishes to improve product security and make justifiable claims about that security.
46. The premise of MDD is to raise the abstraction of software development from the low-level imperative programming language that is fraught with sharp edges and opportunities to shoot one's self in the foot to a higher-level modeling language that reduces the distance between design and implementation and by doing so reduces the flaws that lead to security and safety failures.
47. Without code generation, much of the advantage of MDD is lost; in reality, deciding to adopt an MDD approach means accepting automatic code generation from models.
48. Models need to be executable; discount any MDD platform where this is not the case.
49. The value of formal verification is that it symbolically examines the entire state space of a software design to determine whether or not the specified properties hold under all possible inputs.
50. Formal verification should not be seen as an alternative to verification based on simulation and testing; rather, it is complementary.
51. Because UML is standardized, UML-based MDD platforms tend to present their models in very similar ways, making it easier for users to understand and use MDD platforms from different vendors.

52. For MDD to be successfully applied, the MDD platform used must be appropriate not only to the scale and complexity of the system being developed, but also to the engineering domains of the system.
53. While portable models that can be seamlessly imported with full semantics preservation between vendors' platforms are desirable, this ability does not yet exist in practice; every MDD platform has many unstated assumptions as to the precise meaning and interpretation of its models, and these assumptions are built into the code generators, simulators, test case generators, formal verifiers, and supporting runtime platforms.
54. Compared to conventional methods, MDD enables verification to start earlier, as soon as models are completed; verification is not dependent on handwritten code or test cases. To benefit from this opportunity, however, an organization must establish requirements much earlier in the development life cycle and to a greater degree of precision than is usually the case with conventional software development.
55. When MDD-designed and -generated code is mixed with handwritten code, there are two principal issues to consider: first, how can the MDD models be verified when their correct runtime execution depends on behavior in unmodeled handwritten code? Second, how will the model-generated code interface with the handwritten code at runtime?
56. When an organization chooses an MDD platform, the efficiency, portability, and compliance characteristics of the generated code are of vital importance, especially in the embedded domain.

3.10 Bibliography and Notes

1. ILTIS—Railway Traffic Control. Siemens Switzerland Transportation Systems. http://www.aonix.com/pdf/CaseStudy_ILTIS.pdf.
2. http://en.wikipedia.org/wiki/Comparison_of_programming_languages.
3. DO-178B. *Software Considerations in Airborne Systems and Equipment Certification.* RTCA, Inc.; 1992.
4. MISRA-C:2004. *Guidelines for the Use of the C Language in Critical Systems.* The Motor Industry Software Reliability Association; October 2004.
5. Working with Import Libraries and Export Files. Microsoft Development Network. http://msdn.microsoft.com/library/default.asp?url=/library/en-us/vccore/html/_core_working_with_import_libraries_and_export_files.asp.
6. MISRA C++:2008. *Guidelines for the Use of the C++ Language in Critical Systems.* The Motor Industry Software Reliability Association; June 2008.
7. *The Embedded C++ Specification, Version WP-AM-003.* The Embedded C++ Technical Committee; October 1999.
8. Haden M. *Embedded C++ Slashes Code Size and Boosts Execution.* http://www.ghs-rtos.com/wp/ec++article2.html; 1998.
9. Watson AH, McCabe TJ. *NIST Special Publication 500-235: Structured Testing: A Testing Methodology Using the Cyclomatic Complexity Metric*; September 1996.
10. DiBona C, Ockman S, Stone M, editors. *Voices from the Open Source Revolution.* Sebastol, CA: O'Reilly; 1999.
11. "CDRH Software Forensics Lab: Applying Rocket Science to Device Analysis"; Chloe Taft, "Medical Devices Technology, " *The Gray Sheet* (October, 2007).

12. Boogerd C, Moonen L. *Assessing the Value of Coding Standards: An Empirical Study.* IEEE International Conference on Software Maintenance; October 2008. 277–86.
13. Barr M. *Embedded C Coding Standard.* Netrino; 2009.
14. www.splint.org.
15. ISO/IEC 9899-1999 (December 1999).
16. C—The C1X Charter. Document WG14N1250 (June 2007).
17. ISO/IEC 14882:2011. Programming Language—C++ (August 2011).
18. National Institute of Standards and Technology National Vulnerability Database. http://web.nvd.nist.gov/view/vuln/detail?vulnId=CVE-2010-2836.
19. *General Principles of Software Validation. Final Guidance for Industry and FDA Staff.* U.S. Food and Drug Administration, Center for Devices and Radiological Health; January 11, 2002.
20. Document RTCA/DO-178B. *Software Considerations in Airborne Systems and Equipment Certification.* RTCA; December 1992.
21. Dupuy A, Leveson N. *An Empirical Evaluation of the MC/DC Coverage Criterion on the HETE-2 Satellite Software.* Proceedings of the Digital Aviation Systems Conference (DASC); October 2000.
22. Sergiy VA, Bowen JP. From MC/DC to RC/DC: Formalization and Analysis of Control-Flow Testing Criteria. *Formal Aspects of Computing* 2002;**18**(1).
23. *Common Criteria for Information Technology Security Evaluation, Part 3: Security Assurance Requirements, Version 2.1.* Section 6.2.4; August 1999.
24. Shapiro JS. Understanding the Windows EAL4 Evaluation. *Computer* February 2003;**36**(2):103–5.
25. Labeled Security Protection Profile, Version 1.b. Section 1.2, p. 9. http://www.niap-ccevs.org/cc-scheme/pp/PP_OS_LS_V1.b.pdf
26. U.S. Government Protection Profile for General Purpose Operating Systems in a Networked Environment. Version 1.0; August 30, 2010.
27. U.S. Government Protection Profile for Separation Kernels in Environment Requiring High Robustness, Version 1.03. Section 1.2, paragraph 9, p. 10. http://www.niap-ccevs.org/cc-scheme/pp/pp_skpp_hr_v1.03.pdf.
28. Ibid., Section 2.8, paragraph 102, p. 47.
29. Controlled Access Protection Profile, Version 1.d. Section 1.2, p. 9. http://www.niap-ccevs.org/cc-scheme/pp/pp_os_ca_v1.d.pdf
30. Ibid.
31. Ibid.
32. Labeled Security Protection Profile, Version 1.b. Section 1.2, p. 9. http://www.niap-ccevs.org/cc-scheme/pp/PP_OS_LS_V1.b.pdf
33. Ibid.
34. Ibid.
35. U.S. Government Protection Profile for Single-level Operating Systems in Medium Robustness Environments, Version 1.91. Section 1.2, paragraph 11, p. 10. http://www.niap-ccevs.org/cc-scheme/pp/PP_OS_SL_MR2.0_V1.91.pdf
36. Ibid.
37. Ibid., Section 3.1, paragraph 64, p. 28.
38. Ibid.
39. U.S. Government Protection Profile for Multilevel Operating Systems in Medium Robustness Environments, Version 1.91. Section 1.2, paragraph 11, p. 10. http://www.niap-ccevs.org/cc-scheme/pp/pp_os_ml_mr2.0_v1.91.pdf
40. Ibid.
41. Ibid., Section 3.1, paragraph 66, p. 30.
42. Ibid.
43. *Department of Defense Instruction 8500.2, Information Assurance (IA) Implementation*; February 6, 2003.

44. *Common Criteria for Information Technology Security Evaluation, Part 3: Security Assurance Requirements, Version 2.1.* Section 10, paragraph 308; August 1999.
45. IEEE Std 610.12-1990. *IEEE Standard Glossary of Software Engineering Terminology*; September 1990.
46. VDC Research. *2011 Embedded Engineer Survey*; 2011.
47. Ebert C, Jones C. Embedded Software: Facts, Figures, and Future. *Computer* April 2009;**42**(4):42–52.
48. Dijkstra EW. *Notes on Structured Programming*. Technological University Eindhoven, Department of Mathematics. The Netherlands; August 1969.
49. Object Management Group. *OMG Unified Modeling Language (OMG UML), Infrastructure, Version 2.4.1*; August 2011.
50. Friedenthal S, Moore A, Steiner R. *A Practical Guide to SysML: The Systems Modeling Language*. Burlington, MA: Elsevier; 2008.
51. Mellor SJ, Balcer MJ. *Executable UML, A Foundation for Model-Driven Architecture*. Reading, MA: Addison-Wesley; 2002.
52. VDC Research. *2011 Embedded Engineer Survey*; 2011.
53. Prowell SJ, Poore JH. Specification, Foundations of Sequence-Based Software. *IEEE Transactions on Software Engineering*; May 2003:417–25.
54. Roscoe AW. *Understanding Concurrent Systems*. London: Springer Verlag London Ltd.; 2010.
55. Hoare CA. Monitors: An Operating System Structuring Concept. *Communications of the ACM*; October 1974::547–8.
56. Halbwachs N, Caspi P, Raymond P, Pilaud D. The Synchronous Dataflow Programming Language Lustre. *Proceedings of the IEEE* September 1991;**79**(9):1305–20.
57. Berry G, Gonthier G. The Esterel Synchronous Programming Language, Design, Semantics and Implementation. *Science of Computer Programming*; February 1992:87–152.
58. COTS Compartmentalized Operations Protection Profile - Operating Systems (CCOPP-OS), Version 2.0; June 19, 2008.

CHAPTER 4

Embedded Cryptography

Chapter Outline

Embedded Systems Security. DOI: 10.1016/B978-0-12-386886-2.00004-7

4.1 Introduction

This chapter introduces and describes the fundamental concepts of modern digital cryptography. Many textbooks describe the history of cryptography, starting in Roman times with the invention and use of Caesar's cipher, consisting of a single alphabetic substitution based on a secret number of cyclic shifts in the alphabet, to the poly-alphabetic Enigma cipher used by the Germans during World War II. All these ciphers used mechanical rotors and wheels to substitute characters in some random manner. Most of the ciphers, like Caesar's cipher, were either fundamentally flawed or relatively easily broken, like the Enigma. The history of these ciphers is so fascinating and extensive that we could not possibly do it justice covering it here. For readers interested in a more thorough background, we recommend the following books that cover the fascinating evolution of cryptography.[1–3] Readers who already have a good understanding of basic cryptography may wish to skip ahead to the case studies later in this chapter.

In this chapter, we begin with fundamental principles of cryptography, including the description of a cipher that is unconditionally secure: the one-time pad (OTP). And using the one-time pad as a foundation, we grow the discourse across the core modern-day symmetric key ciphers: keystream and codebook. For each, we discuss the applicable modes of operation, cryptographic synchronization, and key management aspects required for a secure cryptosystem.

We then cover public key cryptography and its extensive use in modern cryptography for dynamically establishing secret keys between two communicating entities.

We follow the fundamental descriptions with practical examples of digital cryptosystems, pointing out threats and vulnerabilities and mitigation techniques to counter these threats.

While this chapter provides a good overview of both symmetric (stream and block ciphers) and asymmetric cryptography (public key) principles and methods of operation, we do not attempt to describe in detail the cryptographic algorithm specifications. Detailed specifications for unclassified algorithms are available in the public domain, readily accessible on the Internet. For symmetric cryptography, a suggested reading is the U.S. National Institute of Standards and Technology (NIST) Federal Information Processing Standards Publication (FIPS PUBS) Advanced Encryption Standard (AES):

- NIST FIPS PUB 197, "Advanced Encryption Standard, 2001," http://csrc.nist.gov/publications/fips/fips197/fips-197.pdf

For asymmetric cryptography, the following algorithms are used extensively today:

- ***RSA:*** See PKCS #1: "RSA Cryptography Standard, Version 2.1," ftp://ftp.rsasecurity.com/pub/pkcs/pkcs-1/pkcs-1v2-1.pdf
- ***Diffie-Hellman:*** See PKCS #3: "Diffie-Hellman Key Agreement Standard," ftp://ftp.rsasecurity.com/pub/pkcs/doc/pkcs-3.doc
- ***Elliptic Curve (ECC):*** See NIST FIPS PUB 186-3, "Digital Signature Standard, 2009," http://csrc.nist.gov/publications/fips/fips186-3/fips_186-3.pdf. Also, see NIST Special Publication 800-56A, "Recommendation for Pair-Wise Key Establishment Schemes Using Discrete Logarithm Cryptography" (March, 2007).

This chapter's endnotes include additional sources for further reading.

Finally, we also discuss a topic that is often misunderstood or neglected by embedded systems designers: how to design and incorporate secure cryptographic elements into embedded systems that have applicable confidentiality or authentication requirements. We illustrate this using several case studies in the later part of this chapter.

4.2 U.S. Government Cryptographic Guidance

There are two major sources of cryptographic guidance and standards originating from the U.S. government: NIST (mentioned previously) and the NSA. In addition to the FIPS PUBS, NIST also writes "special publications," such as SP 800-57, which provides recommendations for key management and is referred to several times in this chapter and in Chapter 5 on data protection protocols. Whereas NIST collects good commercial practices from the worldwide cryptographic community and distills these practices into important standards and recommendations, the NSA is entrusted with the protection of the U.S. government's most sensitive (classified) information.

Key Point

It behooves embedded systems developers to understand NSA cryptographic guidance, even though NSA certification is not required for most embedded systems.

FIPS 140-2 and NSA cryptographic certification and additional background on NSA cryptographic strategy are covered later in this chapter.

4.2.1 NSA Suite B

To complement the existing national policy on the use of AES to protect national Security Information (CNSSP-15),[4] the NSA announced Suite B cryptography at the 2005 RSA conference.

Key Point

NSA Suite B includes a suite of unclassified cryptographic algorithms providing a variety of cryptographic services required in modern cryptography, including encryption, hashing, digital signatures, and key exchanges.

The entire suite of cryptographic algorithms is intended to protect both classified (Top Secret and Secret) and unclassified national security systems and information. The NSA's Suite B website contains references to the preferred algorithms.[5] In addition, Internet Engineering Task Force (IETF) standards, in the form of RFCs, document Suite B guidance with respect to the use of cryptography in many popular network security protocols, such as IPsec. The role of Suite B in recent NSA certification and approval strategy is discussed later in this chapter.

While Suite B guidance is mentioned throughout this chapter as we discuss the applicable classes of cryptographic functions, Table 4.1, which lists algorithm choice and key lengths for secret and top-secret data protection, may serve as a handy quick reference.

Table 4.1: NSA suite B guidance for algorithm and key lengths

Function	Algorithm	Secret Key Size	Top Secret Key Size
Symmetric Key	AES	128	256
Public Key Signature	ECDSA	256	384
Key Agreement	ECDH	256	384
Hash	SHA-2	256	384

4.3 The One-Time Pad

Throughout recorded history, cryptography has been used to protect information. The one-time pad system, first introduced in 1917 by Gilbert Vernam, who was awarded a patent on the invention,[6] is an extremely simple encryption scheme that can be implemented, literally, with a couple of note pads. Also of great importance is the formal theoretical proof of the security of the OTP by Claude Shannon in 1945 (declassified and published in 1949).[7] In particular, Shannon proved the *perfect secrecy* of the OTP: an adversary, having intercepted encrypted communications (ciphertext) and possessing unlimited resources in which to attempt to recover the original message (plaintext) from the encrypted message, is provably unable to distinguish the ciphertext from random data. This is also referred to as unconditionally secure; OTP provides no information about the original message.

Key Point

Because of its proven, unconditional, and simple security, the one-time pad is often considered the basis of all good encryption systems.

Since cryptography almost invariably involves trade-offs and compromises, the one-time pad remains a refreshing beacon of purity in the cryptographic world.

Emulation and digital implementation of the unbreakable one-time pad system can be best described by first developing a digital binary equivalent of the one-time pad, as shown in Figure 4.1.

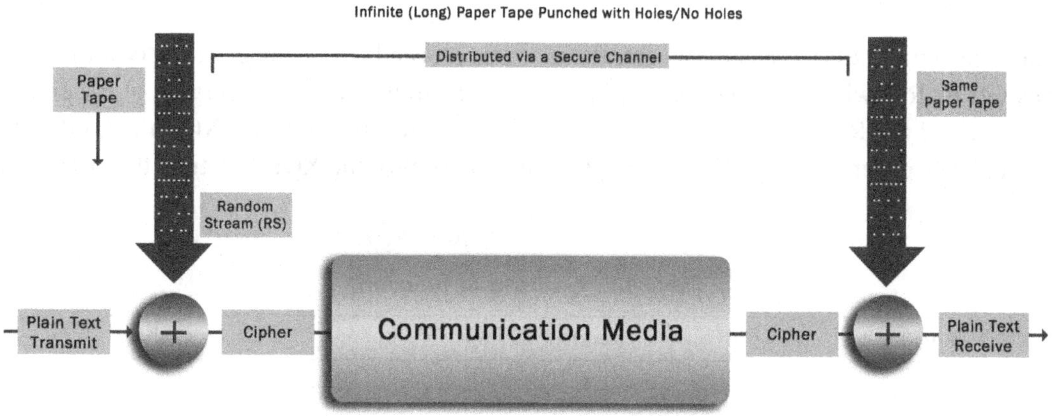

Figure 4.1:
Binary equivalent of one-time pad.

Two parties desire to communicate and protect the communication from eavesdroppers. One of the parties, prior to secure communication, develops a very long paper tape punched with a hole for logic or binary one and no hole for logic or binary zero. The sequence of paper tape punches is randomly determined by the toss of a coin, with heads denoting logic one, and tails denoting logic zero. Each coin toss event has a 50% probability of being heads or tails, and thus the paper tape contains a truly random sequence. After this long paper tape is punched as described, one and only one duplicate copy is made of the paper tape. We now have two identical paper tapes containing the same random stream of ones and zeros. One of the paper tapes is securely couriered (a discussion of potential courier mechanisms is left out of this discussion and independent of the core encryption concept we are describing) to the other end of the communication link, and then we establish the following secure communication protocol.

At the transmitting end, the message to be encrypted is interpreted as an equivalent string of ones and zeros. For example, the message "hello" may be converted to a binary sequence of six bytes (48 bits total) where each byte contains the digital ASCII representation of the corresponding string character (followed by a terminating zero). We call this string the plain text transmit (PTT). The PTT is now operated on by the exclusive - or (XOR) function with the random stream from the paper tape (Table 4.2).

The output of the XOR operated on the entire PTT is called the cipher text transmit (CTT). For every bit of PTT processed, the paper tape is moved forward to the next hole/no hole position. The used portion of the paper tape, as it is moved forward, is permanently discarded (it must never be reused). We follow this process of applying a bit of the paper tape's random stream to a bit of the PTT until the entire PTT string is enciphered or encrypted. We then send the enciphered message across the communication channel. For this example, assume it is a point-to-point wire, like a fiber cable, connected from the transmitter of the message to the receiver of the message.

Once the entire message has been sent, the receiver proceeds to decipher or decrypt the message by following the same identical process as the transmitter: for every bit of the cipher text received (CTR, as the CTT becomes the CTR at the receive end), we XOR the CTR with the duplicate paper tape's random stream. By the properties of the XOR function, the process is

Table 4.2: Exclusive-or function

PTT	Random Stream	Cipher
0	0	0
0	1	1
1	0	1
1	1	0

Table 4.3: One-time pad sequence: plain text receive (PTR) matches plain text transmit (PTT) when random streams are equal

Transmitter				Receiver	
PTT	RS	CTT	CTR	RS	PTR
1	0	1	1	0	1
0	1	1	1	1	0
1	0	1	1	0	1
0	1	1	1	1	0
1	1	0	0	1	1
1	1	0	0	1	1
1	1	0	0	1	1
0	1	1	1	1	0
0	1	1	1	1	0
0	0	0	0	0	0
1	0	1	1	0	1
1	0	1	1	0	1
0	1	1	1	1	0
0	0	0	0	0	0
0	0	0	0	0	0
0	1	1	1	1	0
1	1	0	0	1	1

reversible, and thus we obtain the plain text receive (PTR), which matches the PTT. As an example, assume that the PTT and random stream (RS) are as follows:

```
PTT: 10101110001100001
RS:  01011111100010011
```

The XOR operations are shown in Table 4.3.

As shown in the table, PTT matches PTR, and we thus successfully encrypted and decrypted a message containing a string of ones and zeros. The message could be very long (as long as the very long paper tape). For every subsequent PTT message, we follow the same process, discarding the used tape as we pull it forward, until the paper tape is exhausted. For subsequent messages, we restart the whole process, generating and distributing a new random paper tape.

Intuitively, we can state that this method of enciphering a message is unconditionally secure. We assume that sender and receiver are able to keep their respective plaintexts physically secure; thus, an eavesdropper monitoring the communication channel can only obtain the cipher text (CTT or CTR). Since an eavesdropper does not have the random stream encoded on the paper tape, attempting to guess any bit of the plaintext from the cipher text results in a 50% chance that he obtains the incorrect plaintext bit. Thus, the attacker is unable to distinguish the encrypted message from random data (see Figure 4.2). We can thus claim that the system is unconditionally secure.

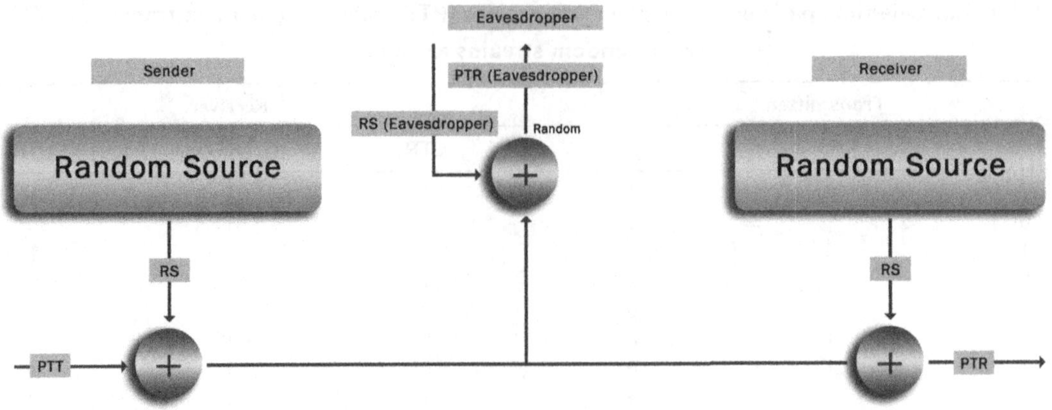

Figure 4.2:
An eavesdropper without access to the original one-time pad random source cannot recover plaintext.

With current technology, the one-time pad system is very simple to implement. Instead of a long paper tape, we can employ a pair of duplicate memory devices, such as USB thumb drives, to store the random sequence and deliver to both ends of a communications channel. With a thumb drive (available since 2010), up to 32 billion bytes of messages can be securely communicated using this method.

While this system is simple and easily implementable, it is not efficient since it requires a random stream as long as the message itself and introduces the logistical challenge of securely distributing the media to all potential communicators. Furthermore, a compromise of the random stream encoded on the flash media results in a compromise of all messages between any pair of nodes ever sent using it, as shown in Figure 4.3.

Furthermore, if one node is compromised in a shared paper tape system, the shared paper tape must be revoked from use, requiring regeneration and redistribution of a new random sequence to all remaining uncompromised nodes. It is easy to see how physical distribution and protection of shared paper tapes quickly become untenable. In general, it is a bad idea to share a single long-term random stream between a large set of nodes.

If each unique pair of communicators uses a per-pair unique paper tape, then compromise of a single paper tape affects only a single communications channel between one pair of nodes. All other pairs of nodes and their communications are unaffected.

How can we retain our desired property of unconditional security while avoiding the long stream distribution problem? One way would be to securely distribute a short random stream and use this short sequence to locally and deterministically generate a long, shared random stream. This would improve efficiency, and we would need to protect only the short random

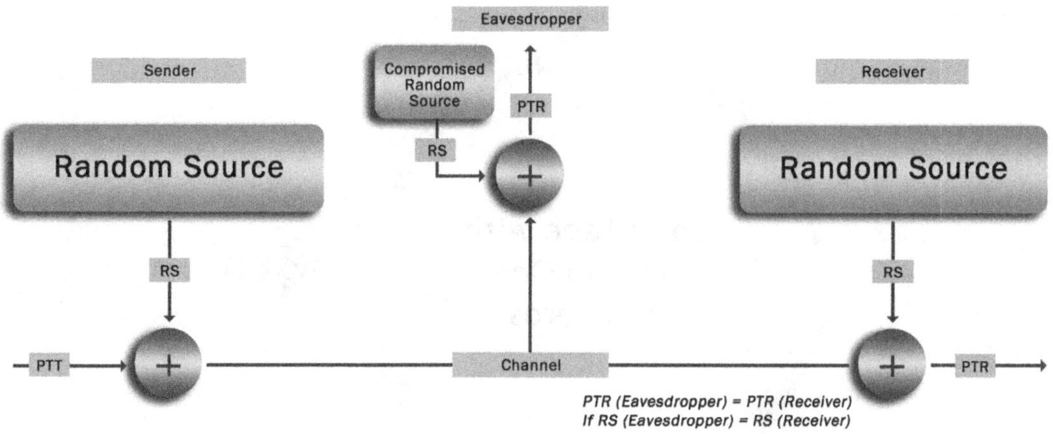

Figure 4.3:
A compromised random source allows an eavesdropper to recover all messages.

stream as it is distributed to all nodes. Moreover, if we can implement such a system efficiently in either hardware or software, we would be guaranteed that the system is secure.

To describe the concept of using a short stream to generate a long one, assume that the paper tape shown in Figure 4.1 is not very long but tied end to end. This looped tape still contains a random stream, but the random stream will repeat itself after the loop has completed one cycle. We would like to implement this finite loop in such a manner that both the transmitter and receiver generate the same long random stream using a short random stream that was previously distributed securely, as shown in Figure 4.4. The input sequence is called the *key,* and the resulting long stream is now called a random *keystream.*

Only the key is securely distributed to the receiver, and because both endpoints use the same looping mechanism to generate the random keystream, we are assured that both endpoints have the same random keystream. Note that before encrypting any messages, we must synchronize the long keystreams at both ends; in other words, even if both ends have the same key, if one end starts the keystream at a different point in the long loop, they will not be able to communicate since they are not using the same sequence of random bits.

In practice, duplicating the short sequence to create the long loop is a faulty approach: the repeating sequence would be easily discernible by an attacker, defeating the required randomness property. Instead, we must use a mechanism that takes a short random key and generates a long stream that is also random.

We can digitally implement this keystream derived by a relatively short key. The generation of this keystream is driven by a cryptographic algorithm and can be implemented in hardware, software, or a combination, depending on the data rate of the encryption required and the speed of the computer that runs the software. For voice applications, the required data rate may

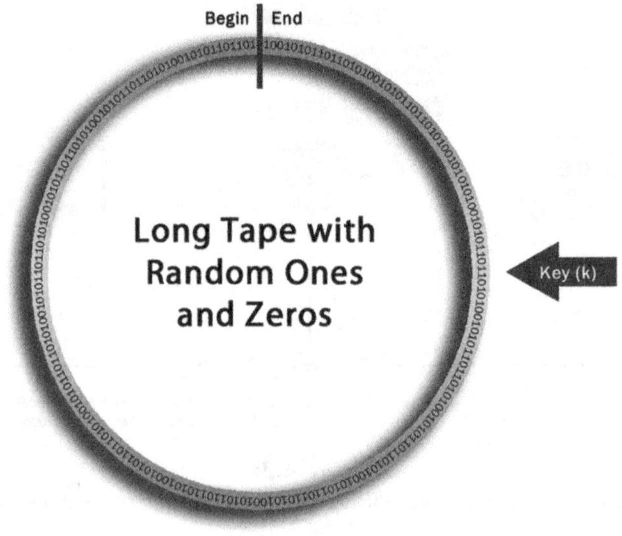

Figure 4.4:
Long random keystream.

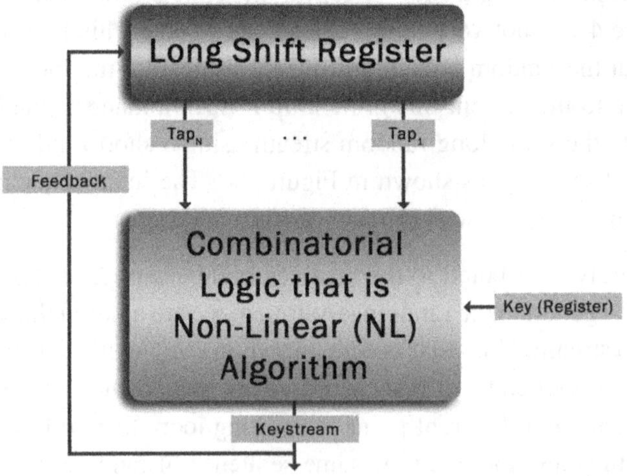

Figure 4.5:
Simplified keystream generator.

be no higher than 64 kilobytes per second, while video encryption may require gigabits per second. A potential implementation of a keystream is shown in Figure 4.5. Although this implementation is infeasible for a secure cryptosystem, we use it to highlight the concept.

The keystream is implemented using a long shift register (80 bits or greater to give us a reasonably long loop), whose taps (bit output stages of the shift register) are combined with

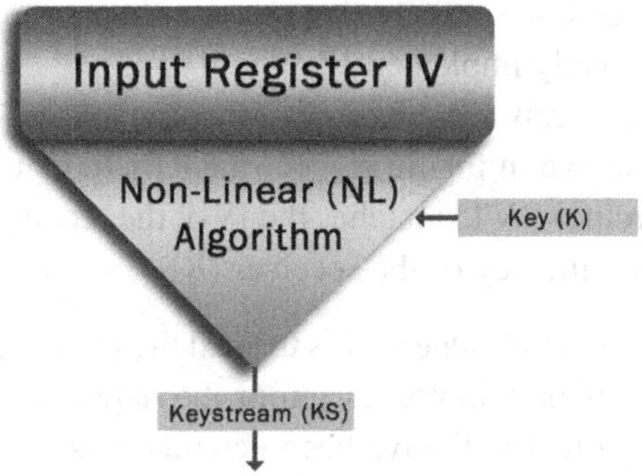

Figure 4.6:
Symbolic representation of keystream generator.

non-linear combinatorial elements and the key. This combinatorial block thoroughly mixes the outputs of the shift register with the key and is called the *cryptographic algorithm* of the keystream. We wire the keystream output to the first stage of the shift register, and if we continually clock this configuration, we obtain a digital long loop tape that is connected end to end, generating a random keystream whose cycle repeats itself after 2 to the N (2^N) bits, where N is the number of bits in the shift register. The shift register is initialized with a random sequence of N bits, and we denote this initial state as the initialization vector (IV). The initialization vector is the starting point of our digital tape. We symbolically represent this keystream generator as shown in Figure 4.6.

We have now developed a cryptosystem as shown in Figure 4.7.

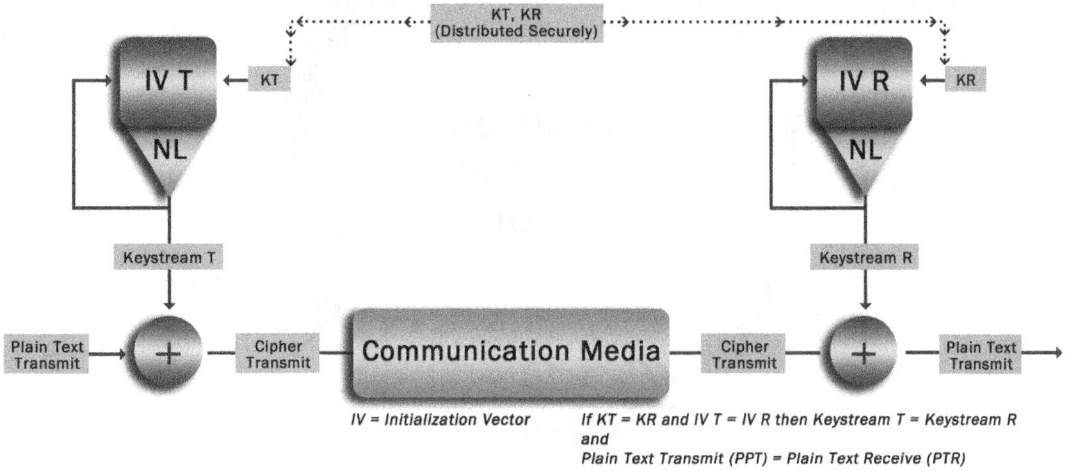

Figure 4.7:
Keystream-based cryptosystem.

Figure 4.7 is identical to Figure 4.1, except that the one-time pad has been changed to a keystream generator, identically implemented at both ends and driven by a common key that was previously distributed securely via courier or other means and loaded into the keystream generator at both ends. As shown in Figure 4.7, for the PTT to match the PTR, we must make sure that the IV of the transmitter (IVT) matches the IV of the receiver (IVR) and that the key of the transmitter (KT) matches the key of the receiver (KR).

How do we synchronize IVT to IVR? One way is to send the IV in the clear from the transmitter to the receiver. The receiver is now in possession of the required initialization sequence and knows that subsequent cipher text will have been generated using the start of the keystream generated by this IV. An eavesdropper receiving the IV will also be able to synchronize her evil key generator to the same starting point as the legitimate transmitter and receiver. However, the eavesdropper's PTR will be random since she lacks possession of the key and therefore cannot produce an equivalent keystream as the transmitter and receiver. If the key is compromised and the eavesdropper knows the cryptographic algorithm (as is often the case), the eavesdropper will be able to decrypt all the messages by using the same IV synchronization.

For each message, the transmitter generates a new IV to avoid reuse of a previous keystream (see Figure 4.8).

What happens when a new key is introduced to our key generator? A new random keystream is generated, and the new one does not match the old, as shown in Figure 4.9.

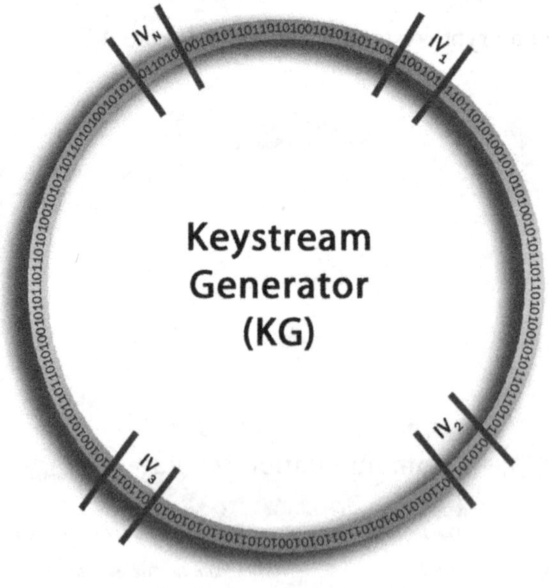

Figure 4.8:
New IV generated for each cryptoperiod.

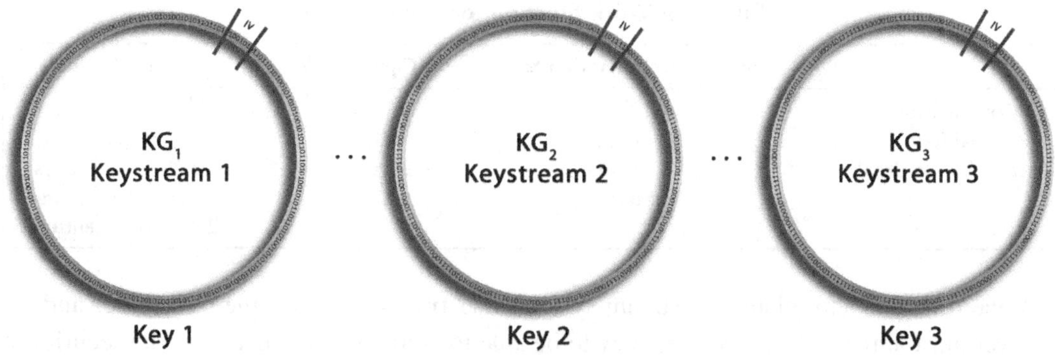

Figure 4.9:
Different keystreams for different keys.

By periodically creating a new key, we have the equivalent of an arbitrarily long digital tape with an arbitrarily long random keystream. This enables the cryptosystem to handle message traffic for the lifetime of the equipment without ever reusing the same random keystream.

If a key is compromised, all the ciphertext associated with that key is compromised, but not the ciphertext encrypted with any other key.

> **Key Point**
>
> The lifetime of a key's use within a cryptosystem is called the key's cryptoperiod.

Suppose an eavesdropper stores all ciphertext and subsequently mounts an attack by exhaustively trying all keys to decrypt the content. Assuming he succeeds, he will be able to decrypt only the messages generated during the cracked key's cryptoperiod (see Figure 4.10). Any messages received with another key will require a new exhaustive attack that is at best extremely time consuming and at worst infeasible for keys of sufficient length.

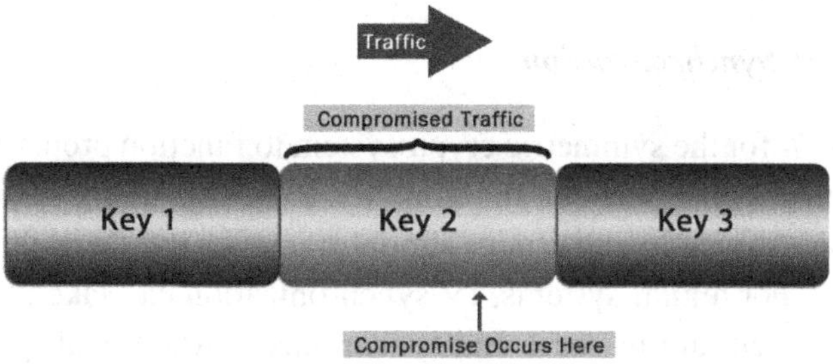

Figure 4.10:
Compromise of a session key results in compromise of traffic only within the compromised key's cryptoperiod.

Table 4.4: Work function for some ciphers

	Standard Combination Safe	Cipher Lock	120-Bit Variable
Combinations	10^6	60	2^{120}=1.33 × 10^{36}
Trial Rate	10/minute	1/sec	1/μsec
Mean Time to Break	833 hours 34 days	30 seconds	0.72 × 10^{25} days 2.1 × 10^{22} years 2.1 × 10^{20} centuries

The one-time pad and related keystream systems use *symmetric* keys: the transmitter and receiver must agree and use the same key to be able to communicate securely. The security of the system rests in the key, which must remain secret. Once a key is divulged, copied, or compromised, anyone can decrypt or encrypt messages even if the crypto algorithm is proprietary and not public.

Key Point

For the purposes of determining cryptographic strength, proprietary or secret algorithms are always assumed broken (fully understood and reproducible by an attacker) through either reverse engineering or other means.

Moreover, if a portion of the key is compromised, the system is more vulnerable to an exhaustive attack by reducing the size of the key space that must be searched. For example, while an 80-bit key would normally require the attacker to try 2^80 different keys, the attacker needs only 2^40 searches if one-half of the key bits have been compromised.

Table 4.4 presents some examples of the time required to unlock or break a combinatorial lock or key relative to its bits of freedom. The length of time required to break a particular key is called its *work function*. As shown in the table, an exhaustive attack for a 120-bit key takes a really long time (RLT). If that key is changed frequently, as in every message, the exhaustive attack becomes an exercise in futility, even with unlimited computing power.

4.3.1 Cryptographic Synchronization

As shown in Figure 4.7, for the symmetric cryptosystem to function properly, the IV used at the transmit and receive endpoints must be the same for a corresponding sequence of ciphertext generated with the IV. Typically, the sender simply transmits the IV in the clear prior to the ciphertext for that IV. For digital systems, IV synchronization can take place only if both transmit and receive clocks are in phase and synchronized. Additional synchronization techniques, including clock recovery from the data stream, must be employed when the transmitter and receiver are physically separated. For the purpose of this discussion, we assume that transmit and receive clocks are already synchronized and only the IV requires synchronization. In practice, clock synchronization is usually handled by hardware transparently to the system or software developer.

As previously described, for long keystreams, the IV register is normally 80 bits long or greater. These 80 bits must be received perfectly with no errors to achieve cryptographic synchronization (i.e., IVT = IVR). If even one of the received IV bits is in error, there is a 50% probability that each transmit keystream bit will mismatch with the corresponding receiver keystream bit. Cryptographic synchronization will not be achieved, and the resulting plaintext received will be random data. Moreover, the communication endpoints will not know that cryptographic synchronization has failed and will continue to encrypt and decrypt until the plaintext processor (e.g., software or human) determines that the data being received is incorrect and issues a cryptographic re-synchronization command.

Bit errors occur naturally for some communication media, and the system designer must take this fact into account when designing the communications protocol to ensure that the IV is perfectly received even in the presence of communication errors. Some communication channels, such as radio frequency (RF) channels, may have as much as a 10% bit error rate. The system designer must use a robust IV error-encoding scheme, which may include error detection, error correction, or both, depending on the quality of the channel. Error encoding/decoding can be performed in hardware, software, or a combination. Furthermore, upper-level protocols, such as IPsec, can be used to provide error detection and integrity protection of transmitted IVs.

For the purposes of the following discussion, we assume the IV is perfectly received into the receiver key generator. Once the IV is loaded, encryption/decryption can take place even in the presence of channel errors. As shown in Figure 4.7, once the IVT matches to the IVR, the keystream generated at both ends is not affected by any transmission errors in the ciphertext. An error in the ciphertext will simply result in a corresponding bit error in the received plaintext; assuming clock synchronization is robust, cryptographic synchronization is maintained. We can state that this scheme exhibits *single-bit error extension*, sufficient for many cryptographic applications. For example, voice communications can tolerate a few plaintext errors, not affecting the intelligibility due to the high redundancy in speech.

However, for data applications, lack of integrity in even a single PTR bit can adversely affect the message content while still appearing to the receiver as a legitimate message. An eavesdropper monitoring the channel may at certain times actively force an error in the channel such that the message received is altered. For example, let's consider a bank transaction message: "transfer 10,000 dollars to this account." An eavesdropper may alter the content of the message to "transfer 100,000 dollars to this account" (see Figure 4.11), not a desirable outcome for the bank but highly satisfying to the eavesdropper if it happens to be his account. Even if the attacker has not broken the encryption and can modify only ciphertext, if the attacker sees patterns in messages and knows where the dollar amount is located, he can make

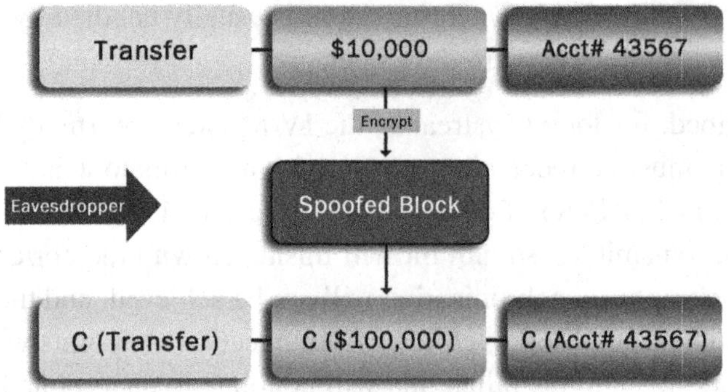

Figure 4.11:
Eavesdropper spoofing attack.

educated guesses about how to insert malicious ciphertext. This type of an attack is called a *spoofed message attack*, a form of *man-in-the-middle* attack. Cryptography can provide countermeasures for message spoofing and many other attacks by providing specialized cryptographic modes of operation.

4.4 Cryptographic Modes

Key Point

Cryptographic modes enable the application of encryption to specific use cases that have varying requirements:

- Some modes enable simultaneous origin authentication and data protection.
- Some modes have been created simply for improved execution time efficiency.
- Some modes propagate bit errors and thus thwart spoofing attacks.
- Other modes sacrifice anti-spoofing in exchange for bit error tolerance.

4.4.1 Output Feedback

As shown in Figure 4.7, the keystream cipher internally feeds the keystream output back into the IV register. This internal feedback is called *Output Feedback (OFB)* or *Key Auto Key (KAK)*. This cryptographic mode is error tolerant in the channel but vulnerable to spoofing.

One or more bits of feedback may be used depending on the implementation. For example, the key generation function may be structured to XOR each byte of PTT with one byte of keystream and one byte of feedback. While output is generated in byte (or larger) blocks, the communications channel often requires serialization of the stream. Parallel-to-serial conversion is usually handled transparently by the physical and/or data link interface. OFB mode implemented with N bits of feedback is shown in Figure 4.12. OFB mode guarantees

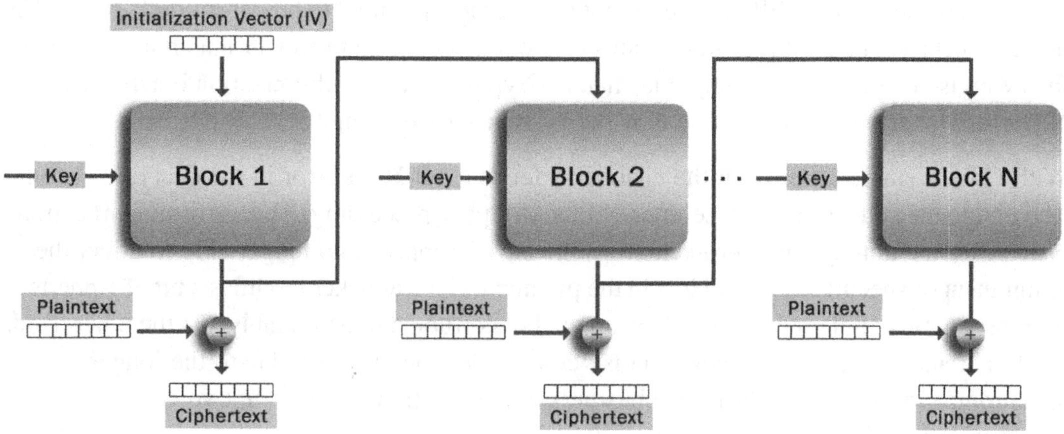

Figure 4.12:
Output feedback (OFB) mode.

message confidentiality but does not provide good message integrity; an attacker who is able to manipulate the channel can effect the modification of any individual bit in the plaintext.

4.4.2 Cipher Feedback

A variation of OFB is *Cipher Feedback (CFB)* or *Cipher Text Auto Key (CTAK)* mode, as shown in Figure 4.13:

Unlike OFB, CFB mode feeds the cipher output into the IV register instead of the keystream. Similar to OFB, cipher feedback can be one or N bits. Because the initial N bit output of the

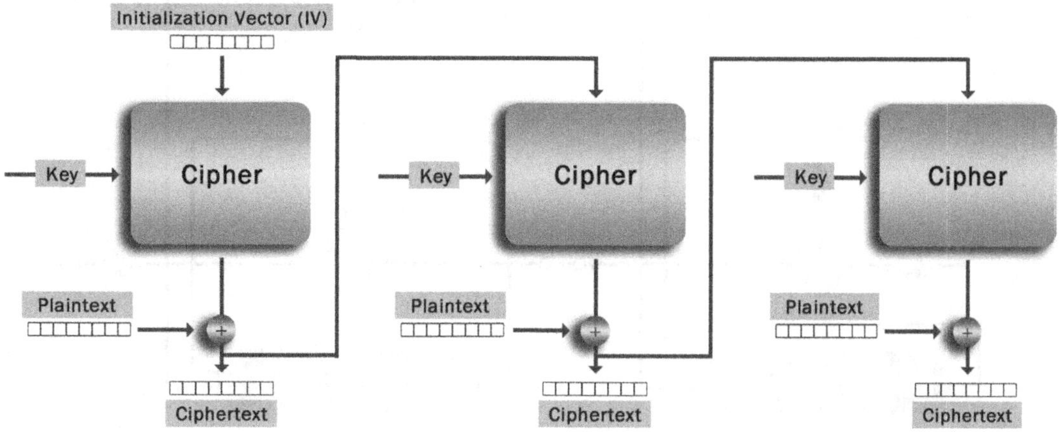

Figure 4.13:
Cipher feedback (CFB) mode.

cipher is used as the IV, CFB mode does not require an explicit IV to be transmitted. Thus, CFB is self-synchronizing, with the keystreams of the transmitter and receiver matching as soon as the IV register values at both endpoints match. Cryptographic synchronization is achieved after N bits of cipher processing, where N is the IV register's bit length.

With CFB, a single-bit error in the channel extends to an N-bit error in the decrypted result. The endpoints self-resynchronize after N bits of cipher processing. This error magnification provides some anti-spoofing protection, as the eavesdropper is no longer able to effect the assignment of specific individual bits in the plaintext. If the attacker modifies a bit of ciphertext in transit, N bits of plaintext will be corrupted, but corrupted unpredictably. On the other hand, CFB exhibits degraded efficiency in noisy communication channels due to the longer synchronization gaps. This lack of robustness may be sufficient to disrupt voice communication.

4.4.3 OFB with CFB Protection

This hybrid mode augments the efficiency benefits of OFB with data integrity protection. One implementation method uses an OFB mode key generator as the main encryptor and a smaller CFB key generator for anti-spoofing protection (see Figure 4.14).

Five bits of error extension are provided in this example. While this thwarts simple spoofing attacks, a larger CFB key generator (e.g., 32 bits or more of feedback) provides improved protection.

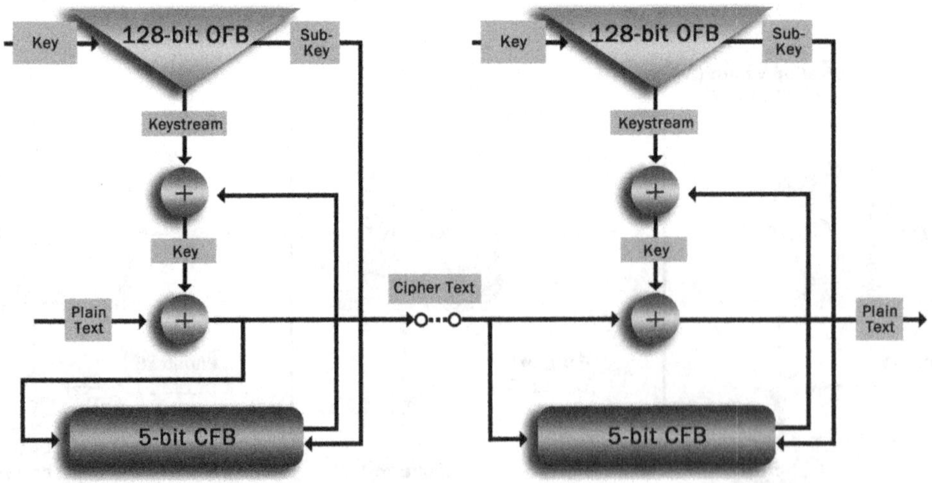

Figure 4.14:
OFB with anti-spoofing CFB.

4.4.4 Traffic Flow Security

Key Point

A desirable trait of secure communications algorithms and protocols, traffic flow security (TFS) aims to prevent an eavesdropper from deducing private information from patterns in communication rather than the communicated encrypted data itself.

For example, because the Secure Shell (SSH) protocol transmits a message for each keystroke during an interactive session, researchers have used the time between keystrokes as input to statistical analysis techniques to recover passwords much faster than an exhaustive search.[8] Removing structural or timing patterns in communication thwarts these kinds of attacks. TFS is one form of the more general concept of *transmission security*, which aims to protect against attacks that fall outside the scope of cryptanalytic attacks on the encrypted data. Another example of transmission security is the use of anti-jamming techniques on radio frequency (RF) communications. We mention TFS here because it is pertinent to the choice of encryption mode. In particular, the use of initialization vectors in the clear within the communications stream may allow an eavesdropper to deduce the starts and ends of messages or sessions. Keep this in mind as we discuss more modes in the following sections.

4.4.5 Counter Mode

Counter Mode, sometimes called *Long Cycle Mode (LCM)*, is shown in Figure 4.15. LCM has similar security properties to OFB; the main difference is IV generation: in OFB, the IV is random, but LCM's IV is generated using an initial value that is incremented after every cryptoperiod (for the session in which the same key is used).

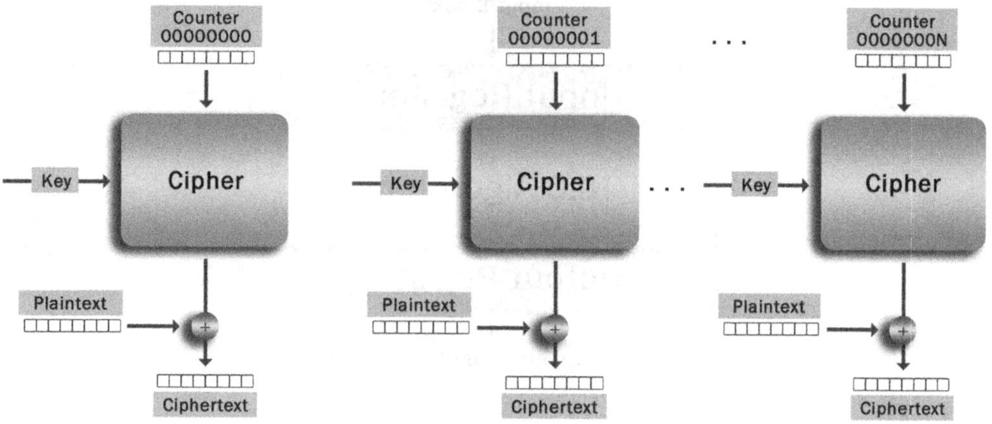

Figure 4.15:
Long cycle (counter) mode.

Similar to OFB, LCM has no error extension and is thus vulnerable to message spoofing. The IV counter can be a true *maximum length sequence generator*, providing the added benefit of a very long non-repeating keystream. In LCM, the IV counter can be based on a real-time clock. When a real-time clock is used, cryptographic synchronization can be based on time-of-day (TOD), a date value that never repeats. If the TOD is synchronized at both endpoints, IV transmission is unnecessary. Therefore, counter modes using TOD are useful in traffic flow security, as described earlier.

Another benefit of this approach is applicability to *broadcast receivers* such as digital set-top boxes. Many such systems lack full-duplex, two-way communication. Instead, a single transmitter broadcasts information to a large number of one-way receivers. The use of TOD-based counter mode allows a receiver to enter the communication stream at an arbitrary time without requiring an explicit cryptographic synchronization. For example, the TOD counter may increment once per day. When a receiver is powered up, it can quickly check its internal real-time clock and derive the current TOD counter, providing instant synchronization to encrypted broadcast content.

4.5 Block Ciphers

The IV register is used as the starting point of a keystream cipher. Suppose that instead of using the IV register as a starting point, we use it as an input register for plaintext, as shown in Figure 4.16, and then apply the non-linear cryptographic algorithm to the input register.

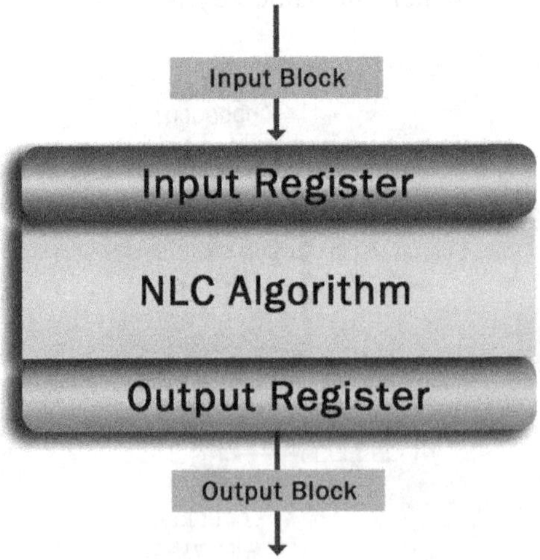

Figure 4.16:
Generalized block cipher.

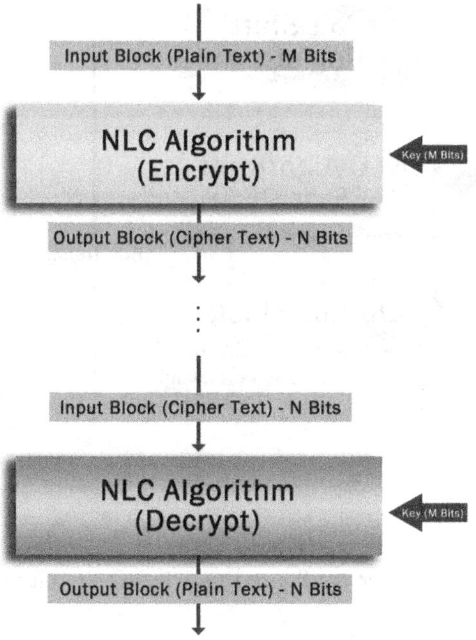

Figure 4.17:
Encryption and decryption using N-bit block cipher.

This configuration results in encryption of blocks of plaintext where a block can be as long as 128 bits for modern algorithms such as AES. Block ciphers function similarly to keystream ciphers but have different properties. Block ciphers are useful for data-at-rest encrypted storage, but they can also be used to implement a keystream-like bit-by-bit encrypted communications mechanism. For block ciphers, the encryption and decryption process is shown in Figure 4.17.

A block cipher transforms an input block (a string of input bits of fixed length) into an output block that is a string of output bits of the same fixed length. The cipher is a function of a key and some non-linear transformation. The block cipher (cryptographic algorithm) may use the same non-linear function used for keystream ciphers. The cipher thoroughly mixes the bits of the input block such that every bit of the output block depends jointly on every bit of the input block and every bit of the key. This mixing of the input block with the key provides protection against cryptanalysis by an adversary.

Key Point

Block ciphers must be designed such that knowledge of large amounts of matched input and output blocks is insufficient to determine the key other than by exhaustive search of all possible keys for the applicable key length.

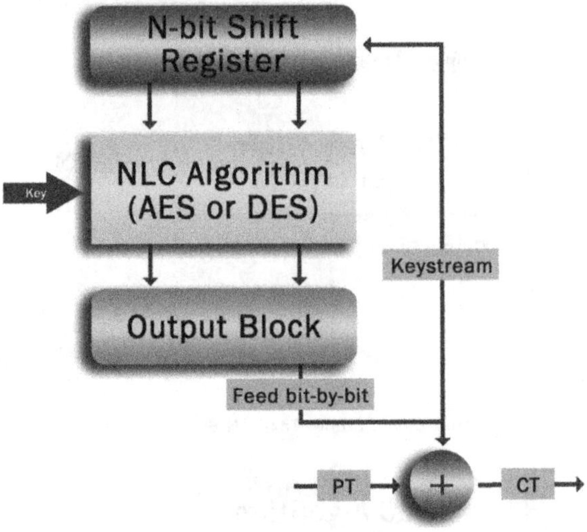

Figure 4.18:
Block cipher configured as a keystream cipher.

The mixing must make sure that small changes to the input block or small changes to the key result in major changes in the output block: the goal is 50% variation of the output bits when the cipher is run with an input block or key that has a single different bit.

Most modern algorithms, such as the Data Encryption Standard (DES) or Advanced Encryption Standard (AES), are designed as block ciphers but can be configured as keystream ciphers in all modes previously discussed (OFB, CFB, LCM). Figure 4.18 depicts a block cipher configured as a keystream cipher.

Block ciphers are sometimes termed *electronic codebook* (ECB); the names are used interchangeably. ECB ciphers always generate the same output block given a particular input block. Thus, large streams of plaintext can easily reveal patterns in the plaintext. This also makes ECB vulnerable to reordered ciphertext attacks: reordering (or injecting) ciphertext blocks into the communications channel causes the exact same reordering of the decrypted plaintext, potentially changing the message to benefit the attacker without the receiver being aware of the tamper. All ECB ciphers must have a reversible deciphering process.

Key Point

AES is the preferred block cipher in NSA Suite B.

Two hundred fifty-six-bit keys are used to protect top-secret information, and 128-bit or 256-bit keys are used to protect secret information. This key length guidance is applicable to all encryption and authenticated encryption modes. Current NIST guidance states that AES with 128-bit keys will be sufficient for general-purpose products until the year 2030.

4.5.1 Additional Cryptographic Block Cipher Modes

ECB mode is useful for encryption of small data or keys but is not used for bulk information protection due to the reordered block vulnerability. The *Cipher Block Chaining (CBC)* mode, shown in Figure 4.19, addresses this reordering attack.

In CBC mode, each input block (except for the initial block) is XOR'd to the previous output block. This feedback loop prevents the reordering attack since modification of a ciphertext block will necessarily corrupt all subsequent decrypted blocks.

CBC mode can be used only with block ciphers. A single-bit channel error results in corruption of the current plaintext block and a corresponding error in the plaintext bit of the next block. An active eavesdropper, attempting to modify the ciphertext, will cause significant loss of recovered plaintext. Thus, CBC mode prevents message spoofing. Because of this property, CBC mode is sometimes used in *message authentication* schemes. In such a scheme, a secret key is used to generate a Message Authentication Code (MAC) from the plaintext. If the receiver is able to compute the same MAC, then she is assured that the message arrived intact (data integrity) and must have come from the holder of the same secret key (origin authenticity). One such scheme is the *CBC Message Authentication Code (CBC-MAC)* algorithm. In CBC-MAC, the last cipher block of a CBC message encryption is input into the CBC cipher as an additional input block, enciphered, and the result appended to the ciphertext output. This result block is the MAC. At the receiver, a MAC is generated using the same key and verified to match bit for bit with the received MAC. Because a ciphertext block depends on all preceding plaintext blocks, a matching final block (authentication code) is possible if and only if the complete message arrived intact. MACs are covered in more depth later in this chapter.

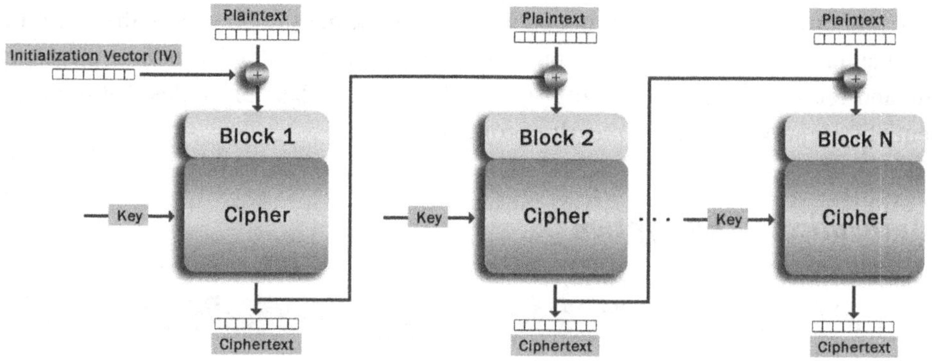

Figure 4.19:
Cipher block chaining (CBC) encryption mode.

As can be seen from Figure 4.19, CBC mode uses an initialization vector (IV) as input to the cryptographic function. The security of CBC mode requires that the IV not be predictable (i.e., a simple counter will not do), and thus randomly generated numbers are often used for this purpose. The IV need not be kept secret and is usually transmitted directly to the receiver that uses the same IV for the symmetric decryption process. Random number generation is discussed later in this chapter.

> **Key Point**
>
> For key encryption (otherwise known as key wrapping), variable-width ECB is the preferred encryption mode.

We have already seen that for ECB, a single-bit difference in input block or key results in corruption of all associated plaintext (with at least 50% variation in output block). While ECB is usually employed with 64- or 128-bit blocks, extension to a larger block size (to handle a key that is larger than a single block) is affected through the use of a *variable-width ECB*, essentially a wrapper around a fixed-width ECB that exhibits the same security properties as the fixed-width ECB. The NIST AES key wrap algorithm uses a variable-width ECB.[9] For example, let's consider a 256-bit input that we wish to wrap using a 128-bit key encryption key and the AES-ECB algorithm. The naïve approach would apply AES-ECB twice, once for each block of the input in sequence. However, if given another 256-bit input that matches the previous input, except with the two blocks swapped, then the 256-bit result will also match the previous result except for the same block swap. Thus, the block-by-block encryption reveals patterns in the input data. In contrast, the NIST key wrap specification ensures that a 256-bit input follows the same property as a single-block ECB: any single-bit change in the input will yield a completely different 256-bit result, as if the AES block size was actually 256 bits.

There are numerous other encryption modes in use today, but the preceding discussion covers the fundamental modes and how they protect against common attacks such as message spoofing and reordering. The following section discusses a class of modes that combine confidentiality (encryption) and integrity protection.

4.6 Authenticated Encryption

> **Key Point**
>
> Authenticated encryption can be more computationally efficient than executing independent encryption and authentication algorithms.

There are numerous authenticated encryption algorithms; in this section we discuss two of the more popular modes: CCM and GCM.

4.6.1 CCM

CCM mode (Counter with CBC-MAC), used in the ZigBee wireless standard, is a combination of counter mode for encryption and CBC-MAC for authentication. CCM is also mandated as in 802.11i (also known as WPA2), the modern authentication scheme for Wi-Fi. The 802.11i standard is discussed in Chapter 5. In CCM, counter mode is used to transform the entire plaintext into ciphertext. Then the CBC-MAC is used to compute the digest over the entire ciphertext. CCM is defined in RFC 3610,[10] and compatibly is defined in NIST Special Publication 800-38C.[11]

4.6.2 Galois Counter Mode

Key Point

Galois Counter Mode (GCM) is the Suite B-recommended authenticated encryption cipher.

AES-GCM is referenced in several Suite B guidance documents, including RFC 4869 (IPsec) and RFC 5430 (TLS). AES-GCM is desirable because it can be implemented efficiently in hardware. Like CBC, GCM requires an initialization vector. However, unlike CBC, the GCM IV need not be unpredictable, although it must be unique for each invocation with a given key.

On the downside, performance-optimized software implementations of GCM often require a large amount of RAM in the form of dynamically computed tables. For resource-constrained embedded devices that lack AES-GCM hardware offload, the footprint overhead may be prohibitive. Otherwise, the use of AES-GCM for authenticated encryption is a very good choice, both for its ease of use and for its preferred treatment by the U.S. government. On modern Intel Architecture (IA) chipsets, the Intel AES-NI instruction set can be used to implement optimized AES-GCM without any additional cryptographic hardware.[12]

4.7 Public Key Cryptography

Martin Hellman and Whitfield Diffie first published a public key cryptosystem in 1976. Several contemporaries of Hellman and Diffie also lay claim to the invention of public key cryptography in the 1970s. While the concepts introduced were revolutionary and described a set of security services that symmetric cryptography could not provide, public key cryptography was not extensively used until the early 2000s when the Internet age truly exploded. The growth of the Internet and its associated e-commerce transactions now required additional digital security services, enabled by public key cryptography. These security services include integrity and authentication of large multimedia files, digital signatures, and non-repudiation for all types of commercial and legal transactions.

Many educational sources attribute the creation of public key cryptography to rectifying an inherent practical weakness in symmetric key cryptography: its key distribution problem, also sometimes called the *bootstrapping problem*. Tor any two endpoints in a large network to communicate using symmetric cryptography, the network operators must first distribute a shared symmetric key between the two parties. This distribution could be done as a private in-person meeting or via some trustworthy courier service. The in-person proofing provides assurance that only authorized communicators are using the key. In a worldwide network, this kind of distribution is clearly impractical. With public key cryptography, the symmetric key is replaced with an asymmetric key pair in which only one of the keys, the private key, must be kept secret by one of the communicating parties; the other party uses the paired key, the public key. Because this second key need not be kept secret, the distribution problem is supposedly solved: one party can create the key pair and send the public half in the clear over the network to any other party that needs to communicate with it. The holder of the private key is the only party than can decrypt information sent by holders of the paired public key.

Key Point

Unfortunately, in most realistic uses of public key cryptography, the key distribution problem is no simpler than the symmetric key-sharing problem due to the need to distribute and perform a trusted installation of a common authority certificate.

The issue is that the ability to successfully encrypt data on one end and decrypt it successfully on another end is not sufficient by itself. At least one of the endpoints usually requires proof that a message is coming from a specific, "known-good" party. For example, before a shopper will send credit card information over the World Wide Web to a merchant, he wants assurance that he is indeed sending this private data to the merchant and not a malicious entity impersonating the merchant. The merchant sends its public key to be used for the encryption, but how does this shopper know that the merchant (and only the merchant) actually holds the public key's private other half? A man-in-the-middle eavesdropping on the network can substitute the merchant's public key with her own public key, and then the credit card information is easily stolen by the intruder. Thus, in most applications, the public key must be authenticated prior to use. The common mechanism for public key authentication, which we describe later, requires an attestation of the public key by an authority that both endpoints mutually trust. In such a system, the purchaser must hold a priori a copy of the trusted authority's public key which is used to validate the merchant's public key. In other words, the problem of physically distributing symmetric keys has been replaced with the problem of physically distributing the authority's public key! The bootstrapping problem remains

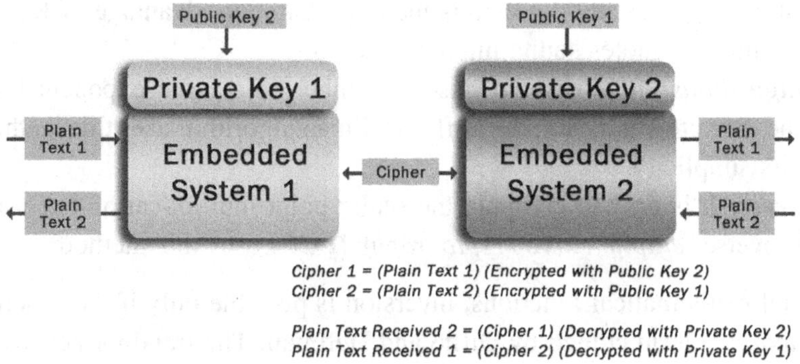

Figure 4.20:
Public key encryption/decryption.

significant. Nevertheless, public key cryptography, as we soon discuss, is critically important in many protection schemes due to its asymmetric nature.

To reiterate, in public key cryptography, as shown in Figure 4.20, the sender and receiver of a message have two distinct, but paired, keys: the encryption key, which is made public; and the decryption key, which is known only by the receiver. We term the encryption key the public key (PK); and its paired key, the private receiver key (PRK).

Any transmitter wishing to send a message to a particular receiver enciphers the message with the PK and sends the ciphertext over an insecure channel; only the intended receiver, who has exclusive knowledge of the PRK, can decipher the message.

The mathematics of public key cryptography is such that knowledge of either the PK or PRK by an adversary is insufficient to determine the other paired key. In addition, public key cryptography allows use of the private decryption key to encipher a message and for the public encryption key to decipher it. This property is very useful for origin authentication: anyone holding PK can decrypt (verify) the message, and successful decryption proves that the message originated from the holder of the unique, private PRK.

The mathematics of public key cryptography requires knowledge of basic number theory, modular arithmetic, and elliptic curve arithmetic. While we do not cover this math in detail in this book, it is very important to understand public key concepts and how they are derived and implemented. These concepts are described in the following sections.

Public key cryptography is based on special mathematical functions that are relatively easy to calculate but very difficult to invert. Thus, these special functions are called *one-way functions*. Some examples of these special mathematical functions are

- ***Factoring the product of two large prime numbers:*** It is easy to compute the product of two prime numbers, but there is no known method to factor the primes other than by trying all possible combinations to arrive at the primes themselves; the RSA algorithm (named after Rivest-Shamir-Adleman) uses this method. A major advantage of RSA is its simplicity, which promotes sound implementation.

- ***Discrete algorithms:*** It is easy to raise a number to a large exponent but difficult to calculate the discrete logarithm; the Diffie-Hellman algorithm uses this method and is also elegant in its simplicity.
- ***Elliptic curves:*** It is easy to compute the scalar point multiplication but very difficult to obtain the inverse; *elliptic curve cryptography (ECC)* uses this method.

For these special mathematical functions, inversion is possible only if the function contains a *trapdoor function*, a term coined by Diffie and Hellman. The trapdoor acts as the key to unlock what would otherwise be a one-way, non-reversible transformation. These trap doors must be well constructed and formulated when the public key pair (PK and PRK) is developed. For example, it is easy to multiply two very large prime numbers (each made up of 500 to 2,000 or more binary digits), but factoring the product of these two prime numbers is computationally infeasible without a trapdoor.

4.7.1 RSA

Let's consider P and Q, two 500-digit prime numbers, and N = P * Q. N is called the *modulus* in the *RSA* public key algorithm. Given publicly disclosed N and secret P and Q, it would take approximately five billion years to factor P and Q on a computer that can execute two billion instructions per second.

However, if a better method to factor RSA moduli is discovered, then this time could be drastically reduced. Of course, the steady increases in computer speeds have consistently increased the size of keys that can be recovered with brute-force factorization. For example, as of this writing, RSA-768 (RSA requiring factoring of a 768-bit modulus to break) is no longer secure.[13] While factoring 768-bit moduli may seem impressive in the 2010s most likely you, the reader, are having a little chuckle at this feat if you are reading in the 2020s. The rate of increase in modulus size factorization between 1991 and 2010 is shown in Figure 4.21. Most systems in use in 2012 employ 1,024 digits or greater RSA keys. However, cryptographers believe that a 1,024-bit modulus is already nearing (or at) its end of life as a safe choice. NIST recommends the use of 2,048-bit moduli and anticipates this choice will remain safe at least until the year 2030.[14]

4.7.2 Equivalent Key Strength

Key Point

The cryptography community traditionally estimates the cryptographic strength corresponding to key lengths of public key and other algorithms as a measure of bits of strength in a symmetric key algorithm, such as AES.

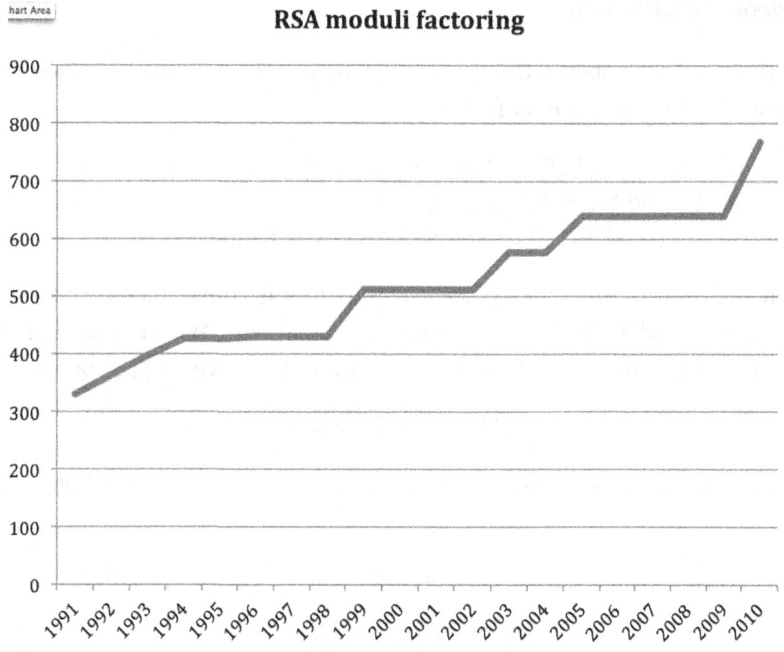

Figure 4.21:
RSA modulus factorization over time.

For example, a 3,072-bit RSA key is estimated to have equivalent strength to 128-bit AES.

> **Key Point**
>
> NIST special publication 800-57 provides estimates on the length of keys for particular algorithms based on the length of time the data needs to be protected.

These quanta anticipate that compute power and/or improved cracking techniques will render the previous generation key lengths inadequate. NIST equivalent key strength for common symmetric and public key cryptosystems is summarized in Table 4.5.

Table 4.5: NIST equivalent key strength

Bits of Security	Symmetric	RSA/DH	ECC
128	AES-128	3072	256
192	AES-192	7680	384
256	AES-256	15360	512

4.7.3 Trapdoor Construction

We illustrate the trapdoor construction process using the RSA algorithm. Two important modular arithmetic properties are as follows:

$$\begin{aligned}&\text{Given primes P, Q and N} = \text{P}^*\text{Q and W} = (\text{P}-1)^*(\text{Q}-1):\\&\text{X}^\wedge\text{Y}(\text{mod N}) = \text{X}^\wedge(\text{Y}(\text{mod W}))\\&((\text{X}^\wedge\text{Y}(\text{mod N}))^\wedge\text{Z})(\text{mod N}) = \text{X}^\wedge(\text{Y}^*\text{Z})(\text{mod N})\end{aligned}$$

For encryption and decryption to take place, the trap door is constructed by choosing a random integer, E, between 3 and W that has no common factors with W. We now find an integer, D, that is the inverse of E modulo W. In modulo notation, this inverse is expressed as follows:

$$(\text{D}^*\text{E})(\text{mod W}) = 1$$

Integers E and N are public; W is discarded once E and D are computed; all other quantities are secret.

Given these parameters, the RSA encryption operation is defined as follows:

$$\text{CTT} = \text{PTT}^\wedge\text{E}(\text{mod N})$$

The decryption operation is as follows:

$$\text{PTR} = \text{CTR}^\wedge\text{D}(\text{mod N}),\ \text{where CTR} = \text{CTT}$$

The modular arithmetic and choice of E and D prove that PTR==PTT as follows:

$$\begin{aligned}\text{PTR} &= \text{CTR}^\wedge\text{D}(\text{mod N})\\&= ((\text{PTT}^\wedge\text{E}(\text{mod N}))^\wedge\text{D})(\text{mod N})\\&= \text{PTT}^\wedge(\text{E}^*\text{D})(\text{mod N})\\&= \text{PTT}^\wedge((\text{E}^*\text{D})(\text{mod W}))\\&= \text{PTT}^\wedge 1\\&= \text{PTT}\end{aligned}$$

Given the selected E, D is the trapdoor that allows the decrypted cipher to match the plaintext.

The following example demonstrates the mathematics with simple numbers:

$$\text{Let P} = 3, \text{Q} = 11, \text{N} = \text{P}^*\text{Q} = 33, \text{W} = (\text{P}-1)^*(\text{Q}-1) = 2^*10 = 20$$

Randomly choose a number E that is relatively prime (no common factors) to W = 20. W's factors are 1, 2, 4, 5, 10, and 20. Therefore, choices for E are 3, 7, 9, 11, 13, 17, and 19. We randomly select E = 7. In modulo arithmetic, we compute D such that (E*D) (mod W) = 1. Thus, D = 3. Let PTT=2.

$$\text{Encryption} : \text{PTT}^\wedge\text{E}(\text{mod N}) = 2^\wedge 7(\text{mod } 33) = 128(\text{mod } 33) = 29(\text{CTT})$$
$$\text{Decryption} : \text{CTR}^\wedge\text{D}(\text{mod N}) = 29^\wedge 3(\text{mod } 33) = 23,389(\text{mod } 33) = 2(\text{PTT})$$

Because in practice, very large integers, often called *bignums* by cryptography practitioners, are used for the public and private portions of a key, public key cryptography is a computationally intensive process. For example, for RSA-2048, the modulus N used during the encryption is a 2,048-bit bignum. In general, public key algorithms are at least 1,000 times slower than symmetric algorithms using keys of relatively equal strength.

Key Point

Because of its computational inefficiency relative to symmetric key, public key cryptography is generally not used for encryption and decryption, particularly for the bulk processing of data, such as in a VPN session or full-disk encryption device. Public key cryptography, however, is preferred for digital signatures (providing origin authentication) as well as for the secure distribution of ephemeral symmetric keys.

These ephemeral keys are called session keys and are created through a public key exchange, or *key agreement*, between two parties. The parties then use the secret session key to protect the confidentiality of traffic sent during the communications session.

4.8 Key Agreement

As shown in Figure 4.22, properly applied cryptography provides a scheme to develop a secret key shared between two cooperating entities.

Initially, the two parties, whom we call Alice and Bob, do not share any common cryptographic secret. Both Alice and Bob generate an ephemeral public key pair and send the public portion to each other. Both Alice and Bob, using their own key pair and the newly received public key half, develop two distinct shared secrets as well as a combined common secret. The algorithm used to develop the shared secret is called a *key agreement algorithm*. This secret is used to

Figure 4.22:
Shared secret key creation.

derive a one-time session key, kept private to only Alice and Bob. The session key is used as the symmetric key for keystream or codebook encryption.

The *Diffie-Hellman algorithm* is the most common key agreement algorithm. To illustrate the key agreement concept using Diffie-Hellman, let's assume that Alice creates a public key pair in which E1 is the public key, and D1 is the private key. Similarly, Bob creates E2 and D2. Public Diffie-Hellman parameters G and P are known in advance to Bob and Alice (for example, an overall key agreement protocol may include transmitting G and P between Alice and Bob). E1 and E2 are computed as follows:

$$\text{Alice}: \text{E1} = \text{G}^\wedge\text{D1}(\text{mod P}); \text{Alice sends E1 to Bob}$$
$$\text{Bob}: \text{E2} = \text{G}^\wedge\text{D2}(\text{mod P}); \text{Bob sends E2 to Alice}$$

Both Alice and Bob then compute the shared secret as follows:

$$\begin{aligned}
\text{Alice} : S1 &= \text{E2}^\wedge\text{D1}(\text{mod P}) \\
&= ((\text{G}^\wedge\text{D2}(\text{mod P}))^\wedge\text{D1})(\text{mod P}) \\
&= (\text{G}^\wedge(\text{D2}^*\text{D1}))(\text{mod P}) \\
\text{Bob} : \text{S2} &= \text{E1}^\wedge\text{D2}(\text{mod P}) \\
&= ((\text{G}^\wedge\text{D1}(\text{mod P}))^\wedge\text{D2})(\text{mod P}) \\
&= (\text{G}^\wedge(\text{D1}^*\text{D2}))(\text{mod P}) \\
\text{Thus, S1} &= \text{S2}
\end{aligned}$$

The resultant shared secret S==S1==S2 was calculated independently within Alice's and Bob's respective computing devices and kept secret. Because S depends on the private portions of Alice's and Bob's ephemeral key pairs, no other entity is able to compute the same shared secret. The S1 and S2 results are shortened in a specified manner and used as the secret session key for symmetric cryptography.

Key Point

Diffie-Hellman is an important algorithm because it enables perfect forward secrecy to be achieved by network security protocols: session keys cannot be compromised even if other session keys or one of the long-term private keys used for origin authentication is compromised.

Because Diffie-Hellman does not itself provide for any origin authentication, it is considered an anonymous agreement algorithm.

Key Point

Digital signature functionality must be added to a protocol that employs Diffie-Hellman to be able to use its key agreement for secure peer-to-peer communications.

Such is the case with IKE and TLS, common network security protocols discussed in Chapter 5.

4.8.1 Man-in-the-Middle Attack on Diffie-Hellman

The classic Diffie-Hellman key agreement scheme suffers from *man-in-the-middle* attack vulnerabilities. One such attack is shown in Figure 4.23.

Figure 4.23:
Man-in-the-middle attack on Diffie-Hellman.

Since Alice's and Bob's ephemeral public keys are made public, an adversary, Mallory, who is an active eavesdropper, can intercept them and introduce his own public key for distribution to Alice and Bob. Alice computes a shared session key, AMK, that Mallory is able to compute, and Bob computes a different shared key, BMK, that Mallory too can compute. Traffic originating from Alice is decrypted by Mallory using AMK and re-encrypted with BMK and sent to Bob. Similarly, traffic from Bob is decrypted using BMK and re-encrypted using AMK. As a result, Mallory can now read all traffic between Alice and Bob without their knowledge of the deception.

To prevent this attack, we must ensure that the Diffie-Hellman ephemeral public keys are actually generated by Alice and Bob and not by an adversary. We say that the public keys must be *authenticated.*

4.9 Public Key Authentication

There are many modern methods of public key authentication, including

- ***Publishing all public keys to a trusted key server:*** The key server's database is essentially a whitelist of pre-authenticated clients. All clients are able to communicate with the key server on a secure (encrypted and authenticated) channel using pre-exchanged public keys. If Alice wants to send Bob an authenticated message, then Alice signs the message and sends it to the key server. Bob asks the key server to validate the message. The key server performs the authentication using Alice's public key and then informs Bob over the secure channel between Bob and the server. With this approach, each endpoint needs to have a pre-established, authenticated public key only for the key server, not for all other endpoints. The downside is that all messages must pass through the trusted key server. Another downside is risk to availability: if the server goes down, no messages can be authenticated.

- ***Sending public keys using a pre-established, authenticated secure channel:*** Instead of a whitelist, the key server initially possesses no public keys. Each client must register its public key dynamically using the secure connection it has with the key server.
- ***Embedding the public key in a digital envelope signed by a mutually trusted third party:*** The public key is concatenated with identity/attribute data of a user and the result cryptographically signed by the third party using the third party's secret signature key; this signed envelope is known as a *public key certificate.* The certificate binds this user identity to its public key. Thus, signature verification using this public key provides assurance that the signed data originated from the associated identity. Each client must have pre-installed the public key, called the *root certificate*, of the trusted third-party signature authority, called the *certification authority (CA).*
- ***Having a mutually trusted entity (web of trust) vouch for the public keys:*** Instead of a single certificate authority for a large set of clients, authorization is decentralized. Any endpoint can *vouch* for a public key by signing it, and multiple endpoints can vouch for the same key. A policy decision is made locally regarding whether to trust a public key, for example, based on the number of signatures. Pretty Good Privacy (PGP) is a sample cryptosystem that uses the web of trust concept.

The most common approach used for trusted distribution of public keys is to use certificates signed by a trusted certification authority (CA); use of a CA usually requires a *public key infrastructure* (PKI) used to manage the life cycle of certificates across all endpoints that share the CA.

The United States National Security Agency (NSA) operates such a PKI for securing classified information. NSA generates the public key pair for all entities and issues the private key and associated public key certificate to its registered users through a secure distribution process. Other sovereign governments employ similar centralized key generation.

Commercial certificate authorities such as VeriSign use a different approach to PKI: usually the public signatures of commercial CAs are embedded in browsers or other local storage of an endpoint device. Endpoints generate their own public key pair and have a CA sign the endpoint-generated certificate containing the endpoint's CA-verified identity and public key. The endpoint registration/signing process may use e-mail or some other communications method between endpoint and CA. The CA never sees the endpoints' secret keys.

4.9.1 Certificate Types

X.509 is the most prevalent standardized format for public key certificates. RFC 5280 defines the latest standard, X.509v3.[15] For embedded systems that must securely communicate with other systems not directly controlled by the embedded vendor, X.509 certificates and a third-party CA will be commonly used.

NSA Suite B guidance for the use of public key certificates is documented in RFC 5759.[16]

> **Key Point**
>
> NSA Suite B specifies the use of X.509 v3 certificates using elliptic curve cryptography (ECC)-based digital signatures.

4.9.1.1 Custom Certificate Approaches for Embedded Systems

For embedded systems whose peers can be more directly controlled (need not interoperate with arbitrary endpoints), other certificate formats can be used. In particular, some highly memory-constrained embedded systems may find the storage requirements for a minimum set of X.509 certificates, each of which may be on the order of 1KB in size, to be prohibitive.

Some resource-constrained embedded systems have adopted *implicit certificates*, in which the public key is reconstructed from the certificate. The implicit certificate is the same size as the public key. In the case of elliptic curve cryptography, this results in a certificate that is far smaller than a traditional X.509 certificate.

A well-known elliptic curve implicit certificate system is Elliptic Curve Qu-Vanstone (ECQV).[17]

The user of an implicit certificate computes the public key from the public key's associated identity and the certificate authority's public key (which is typically pre-installed) and then uses it in some public key operation (e.g., Elliptic Curve Digital Signature Algorithm, or ECDSA). The operation succeeds only if the public key is valid; hence, we call this approach implicitly certified.

At the time of this writing, the Certicom division of RIM has an unexpired patent on ECQV, making its use a barrier to most embedded systems developers.

> **Key Point**
>
> In some cases, the embedded systems designer may benefit from a cryptographic vendor who can provide custom, resource-optimized certificates.

For example, an optimized certificate may include a statically defined format of binary identification information, public key, and signature. Such a format will also be significantly smaller than a typical X.509 certificate. Embedded systems designers should consult their systems software/cryptography supplier to understand the choices available.

4.10 Elliptic Curve Cryptography

Despite the security and ubiquity of RSA, NSA Suite B specifies elliptic curve cryptography (ECC) for public key functionality. The reason is efficiency. According to NIST SP 800-57, a 224-bit ECC key used for digital signatures is equivalent in strength to RSA-2048.

Key Point

While the performance difference between RSA and elliptic curve algorithms with their respective contemporary key sizes is not dramatic, ECC provides better security strength efficiency per bit and therefore is deemed by NSA and most cryptographers as the superior long-term choice.

For example, after the year 2030, the NIST recommendation for ECC key length is 256 bits, just 14% higher than the pre-2030 guidance. But the RSA recommendation (3072) is 50% higher than the pre-2030 guidance.

ECC employs mathematics that is more complicated than RSA, leading to more complex implementations, which is another reason why some still prefer RSA. Another reason for the lack of rapid ECC adoption is the fear of RIM's ECC patents (obtained via acquisition of Certicom).

Key Point

While some implementation options for ECC are covered by patents, all major cryptographic toolkit vendors (as well as the open source OpenSSL developers) have created ECC implementations that follow public specifications and are believed to be free from patent risk.

Yet another reason for the lack of ECC adoption is legacy inertia: many deployed computer systems and certification authorities (even within the U.S. Department of Defense (DoD), whose cryptographic policies are controlled by the NSA) are set up to use RSA keys and RSA-based digital certificates.

Key Point

For embedded systems that operate on networks using legacy RSA-based PKI, ECC cannot be practically used until these infrastructures are upgraded to use ECC.

The conversion from RSA to ECC within the U.S. DoD is ongoing.

4.10.1 Elliptic Curve Digital Signatures

The digital signature algorithm for ECC is called Elliptic Curve Digital Signature Algorithm (ECDSA) and is the lone digital signature algorithm in NSA Suite B.

When making a decision regarding whether to use ECC or RSA, embedded systems designers need to consider the length of in-field deployment and whether NSA-approval may be required (in which case Suite B is a safe bet).

4.10.2 Elliptic Curve Anonymous Key Agreement

Like RSA, Diffie-Hellman is not in Suite B. Rather, Suite B specifies Elliptic Curve Diffie-Hellman (ECDH), which computes the ephemeral public/private key pairs based on elliptic curves instead of exponentials. Like RSA and ECDSA, the Suite B preference for ECDH lies in its relative bit efficiency. Interestingly, a different algorithm, ECMQV, used to be in Suite B instead of ECDH. While ECMQV is known to be more performance efficient than ECDH, the NSA was unable to satisfactorily resolve patent encumbrances of ECMQV. Although there is no technical requirement to use them together, generally speaking, if an embedded design chooses RSA and requires key agreement, DH is typically used. And if a design uses ECDSA, then ECDH would be used.

For all elliptic curve algorithms, NSA Suite B guidance stipulates specific curves that also imply key lengths. For top-secret information, 384-bit keys are required. For secret and below, 256-bit or 384-bit keys can be used. To meet NIST recommendations, ECDSA and ECDH require key lengths of at least 224 bits until 2030 and 256 bits thereafter (until the next guidance revision).

4.11 Cryptographic Hashes

Public keys are very large (usually thousands of bits), and their secure physical distribution and manual installation are cumbersome. A common practice is to electronically distribute the public key but physically distribute a fingerprint of the public key. This fingerprint is called a *hash*, as shown in Figure 4.24:

The hash is used to verify the authenticity and integrity of the public key.

A hash algorithm converts an arbitrary stream of bits into a fixed-length message, the hash value. Cryptographic hashes should be true one-way functions; in other words, given an arbitrary hash value h, it should be computationally infeasible to determine any message whose hash value matches h. Cryptographic hashes should also be resistant to collisions; in other words, given a message, m, and its hash value, h, it should be computationally infeasible to find another message, different from m, that also has a hash value of h. A related property to

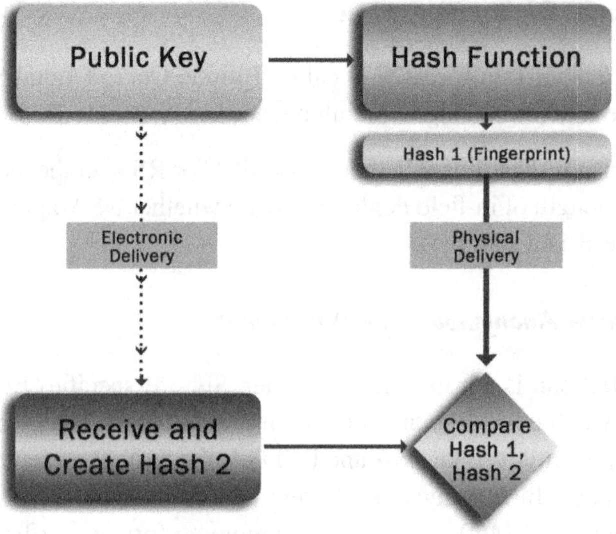

Figure 4.24:
Using a hash to reduce the size of delivered public keys.

collision resistance is randomness; the hash function should act as a random function while still being efficient to compute. These cryptographic hash properties ensure that attackers cannot modify a hash without being detected by anyone possessing the original message and also ensure that the attacker cannot derive any information about the original message given just the hash.

Cryptographic hashes, by themselves, are most often used for data integrity protection and tamper or fault detection. For example, the authors of a security-critical software product should generate cryptographic hashes of the software images that make up the product and then securely distribute these hashes to the software's end users. The end user can generate the hash upon software delivery and compare the hash to the known-good value to ensure that the software has not been modified prior to receipt. The same hash verification can be performed at boot time in a fielded product to ensure that the contents of memory being used to store the software have not been corrupted.

4.11.1 Secure Hash Algorithm

The Secure Hash Algorithm (SHA)-1 has been a historically popular hash algorithm that generates a 160-bit digest. SHA-1 has suffered from collision weakness discoveries. In February 2005, researchers reported narrowing the collision search space to 2^{69} computations, well below the brute-force strength of 2^{80} computations.[18] In August of the same year, an improvement to the attack reduced the timing complexity to 2^{63}.[19] Since then, several researchers have claimed to reduce the complexity further, in the area of $2^{\wedge 57}$ computations. Because of its fruitful cryptanalytic properties, SHA-1 should be relegated only to basic data-integrity-checking responsibilities.

SHA-1's current replacement is SHA-2, a family of hash algorithms with a selection of result bit lengths that are used to further qualify the algorithm name: SHA-224, SHA-256, SHA-384, and SHA-512. Note that SHA-1 produces a 160-bit result. NIST is in the final stages of selecting SHA-3 through a public algorithm competition. The winner should be proclaimed in 2012. SHA-1 and SHA-2 algorithm families are specified in FIPS 180-2.[20]

NSA Suite B stipulates the use of SHA-384 for top-secret information and the use of SHA-256 or SHA-384 for secret and below. In contrast, NIST guidance states that SHA-224 is sufficient for general-purpose products until the year 2030.

4.11.2 MMO

While SHA variants are the only NIST- and FIPS-approved hash algorithms, they are by no means the only popular ones. For example, the Matyas-Meyer-Oseas algorithm, typically used with AES (AES-MMO), is an interesting algorithm used in ZigBee. The nice thing about MMO is that it executes quickly and with negligible memory overhead when the embedded processor already has an offload engine for the underlying encryption algorithm. The following sample C code implementation shows the simplicity of MMO:

```
// This is the Matyas-Meyer-Oseas hash algorithm applied to the AES block
// cipher. The algorithm is standardized in ISO/IEC 10118-2.
// This implementation uses AES-ECB-128 in order to match the keysize
// with the AES block size (avoiding a transformation of intermediate blocks
// to a different key size for AES). The initialization vector must also
// be of length equal to the AES block size. Finally, this implementation
// requires that the input plaintext be of length equal to a multiple of
// the AES block size (no padding is performed).
// The output tag is the contents of the final output block, of size equal to
// the AES block size.
void AesMmo(const uint32_t *iv, const uint32_t *pt, uint32_t len,
        uint32_t *tag)
{
    uint32_t i       = 0;
    blk *key         = (blk *)tag;
    const blk *ptb   = (blk *)pt;
    assert((len % AES_BLOCK_SIZE) == 0);
    for (*key = *(blk *)iv; i < len / AES_BLOCK_SIZE; i++) {
        AesEcbEncrypt(&ptb[i], key);
        XORBLK(key, &ptb[i], key); // 128-bit XOR (ptb ^ key)
    }
}
```

4.12 Message Authentication Codes

Message Authentication Codes, or MACs, provide both integrity protection and origin authentication, and hence are associated with a private key, much like a symmetric encryption algorithm. A MAC takes as input this private key and a message, and similar to a hash function, generates a digest (the authentication code). The recipient of a message and its authentication code executes the same algorithm with the same pre-shared private key. If the codes are equal, this implies not only that the message arrived intact, but also that the message must have been hashed using the same private key.

In a network security protocol such as IPsec, a MAC may be used to provide integrity protection and origin authentication of messages sent between two peers that have previously established a shared private key.

While there are numerous MAC algorithms (CBC-MAC is described earlier in this chapter), the *keyed-hash* algorithm, HMAC, is the most common.

While HMAC using an SHA-1 hash, referred to as HMAC-SHA-1, does not suffer from the collision weaknesses discovered in SHA-1, the SHA-2 variants of HMAC (e.g., HMAC-SHA-256) are required by Suite B and are the best choice for most modern embedded designs. HMAC is a simple and elegant algorithm, defined (equivalently) in RFC 2104 and FIPS 198.[21,22]

4.13 Random Number Generation

Many cryptographic algorithms require the use of random numbers. For example, RSA and ECC keys must be generated with random numbers to ensure uniqueness. When lots of bits are required, pseudo-random number generator (PRNG) algorithms can be used, as long as they are seeded with a true random number generator (TRNG), usually derived from a natural entropy source in hardware. Just to keep the acronym soup flowing, another term used by FIPS for PRNG is deterministic random bit generator (DRBG).

> **Key Point**
>
> Of critical import to embedded systems designers is the requirement that they use an embedded processor with a true random number generator (TRNG).

Without such a feature in the processor, designers will find it challenging to locate sources of entropy for a TRNG and to prove that the approach is valid. Most embedded processors with security engines of any sort will have a TRNG. However, buyer beware: some hardware security engines that are documented to generate "random numbers" may actually be outputting pseudo-random numbers using deterministic algorithms. It is necessary to validate that random numbers are truly random. One way to do this, other than having the NSA evaluate the TRNG, is to test the TRNG with a FIPS- or NIST-approved random number test.[23]

4.13.1 True Random Number Generation

A non-deterministic randomizer uses an unpredictable process, rather than a specified algorithm, to produce an output. The output of a non-deterministic randomizer cannot be determined even if an attacker possesses or determines the complete randomizer design, including internal state information at any particular time. From a conceptual view, a non-deterministic randomizer consists of one or more sources of entropy and a combining function to produce a uniformly distributed random output, as shown in Figure 4.25.

As shown, the N-bit random output is totally unpredictable; the probability of producing a specific random number is 1 / 2^n, where n is the number of bits in the random output. The probability should remain at 1 / 2^n even if an attacker has complete knowledge of the previous random number produced by the randomizer and has complete knowledge of the randomizer design and implementation.

The entropy source is the basis for the non-deterministic operation of the randomizer. Many physical components and processes can serve as acceptable entropy sources. Examples include ring oscillators, noise diodes, radioactive decay, and high-bandwidth signal noise in electronic devices. Entropy sources that can be implemented using digital logic are very useful since they can be synthesized with programmable arrays (FPGAs) or custom application-specific

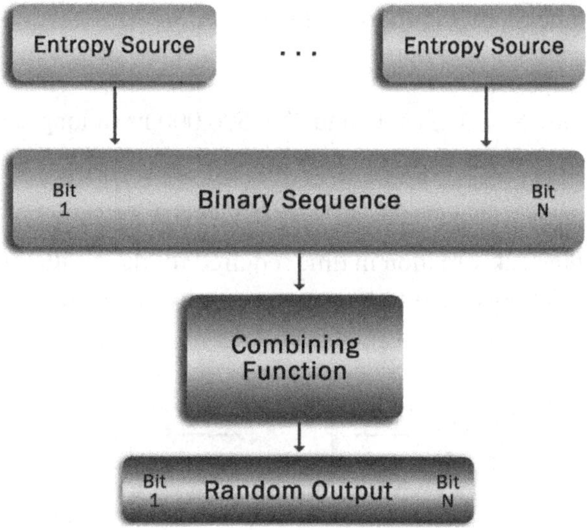

Figure 4.25:
Entropy input to a combining function used to create true random output.

integrated circuits ASICs—and available for the embedded systems developer to use within an applications processor or adjunct co-processor. While most embedded systems developers reading this book can simply take advantage of this hardware and need not understand the details of a true random number generator implementation, the following case study is provided for those who wish to augment their base cryptographic knowledge and depth.

4.13.1.1 Case Study: Ring Oscillator-based Randomizer

Ring oscillators have been thoroughly studied and modeled. The entropy source of ring oscillators is the variation in delay, or *jitter*, across the circuit, causing the state of the ring to be unpredictable. Readers unfamiliar with the concept of a ring oscillator should take a moment to read any of the commonly found descriptions, such as Wikipedia, available on the Internet.

Figure 4.26 presents a ring oscillator with three delay elements: D_0, D_1, and D_2. Each delay element consists of a simple logic inverter, except for the first, D_0, which uses a NAND gate to multiplex an external enable line that holds the ring oscillator in a steady state prior to oscillation.

When enabled, the ring oscillator has 2N states, where N is the number of delay elements in the ring. A timing diagram of this three-stage ring oscillator is shown in Figure 4.27.

In the diagram in Figure 4.27, D_n represent the delay elements, P is the period, and d is the delay of one element. Only a single delay element changes state at a time; thus, the ring oscillator produces a square wave output that is regular when viewed over a short interval. However, over a longer period, drift and jitter cause the ring oscillator to be less predictable, especially if the jitter is greater than the period of the oscillator. Although not proven here, it can be shown that the next state of a ring oscillator S_1 after time T, discounting any jitter, is related to the initial state S_0 by the following equation:

$$S_1 = (S_0 + T/d) \bmod (2N)$$

For example, if d = 5 ns, N = 5, $S_0 = 0$, and T = 500,000 ns (a long sample), then S_1 is (0 + 500,000 / 5) mod 10 = 0. With jitter, the next state is

$$S_1 = (S_0 + (T + J)/d) \bmod (2N)$$

Jitter, J, is defined as the peak variation in time required for the oscillator to complete N cycles and varies randomly with noise due to temperature and voltage changes within transistors

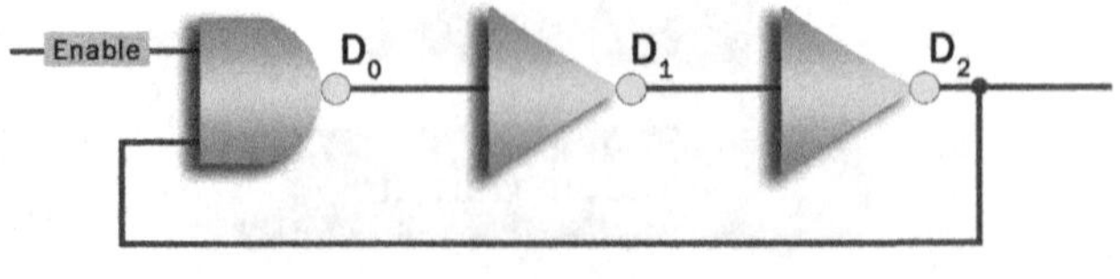

Figure 4.26:
Three-element ring oscillator.

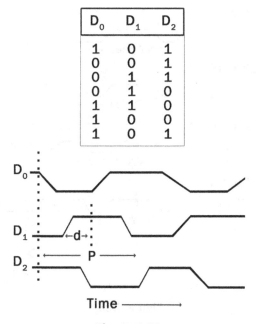

D_0	D_1	D_2
1	0	1
0	0	1
0	1	1
0	1	0
1	1	0
1	0	0
1	0	1

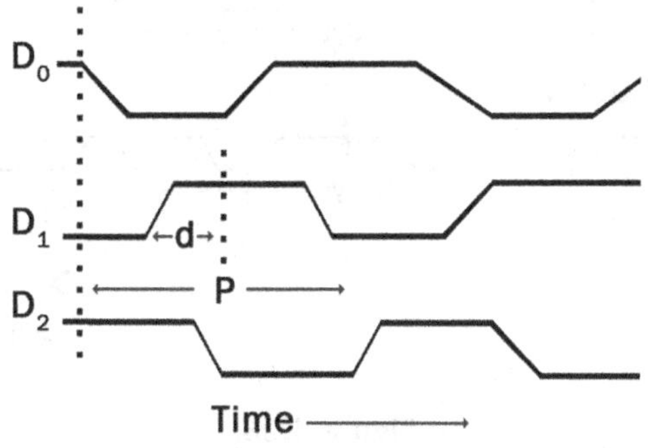

Figure 4.27:
Timing diagram for three-element ring oscillator.

(see Figure 4.28). The preceding equation shows that increased jitter causes the next state to be unpredictable.

In fact, the time T between samples has a direct effect on the amount of jitter that accumulates between samples. It can also be shown that depending on the peak jitter, J, after time T, the delay d, and the ring oscillator period P, if J is close to d and smaller than P, the next state would have small variations. But if J is greater than P, the output states exhibit large variation. We can deduce that a single ring oscillator could be used to produce random data as long as we sample it after a long period of time, making sure that J is greater than P. This approach would result in an unacceptably slow sample rate. In practice, a randomizer uses many ring oscillators summed together to form its output to avoid long sample rates. For example, Figure 4.29 shows two simple ring oscillators made up of two and three states.

A set of combined oscillators, each with an intentionally varied quantity of stages, is called a *leg*. Furthermore, multiple legs can be combined to create the final randomizer. When jitter is added to each leg of the randomizer and the magnitude of jitter is random and variable between legs, it can be easily observed that the combined output produces more oscillations, resulting in a random binary stream when sampled over time.

To use the preceding randomizer for classified or sensitive data, it must be secure and fail-safe. For example, if a randomizer composed of 20 ring oscillators with a varying number of delay

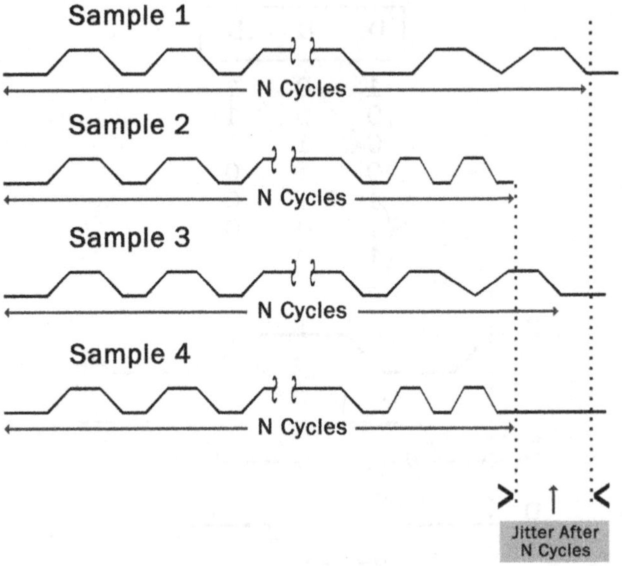

Figure 4.28:
Jitter.

elements suffers failure in a few legs, it could result in an output that is biased (no longer random). To ensure that randomness is retained in the midst of multiple failures, practical designs use a post-processing phase that takes as input the summed output of all the legs as well as the individual leg outputs, as shown in Figure 4.30.

Each leg has enough entropy to provide a random bit stream; therefore, even if a leg fails, the combined output would produce a random stream at the proper sampling rate. However, because a reasonably short sample rate is often required, each randomizer leg must be tested individually to rule out statistical biases.

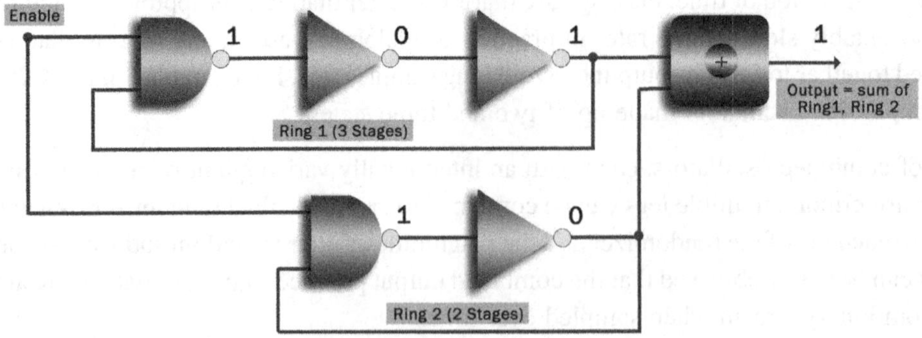

Figure 4.29:
Randomizer with two distinct ring oscillators.

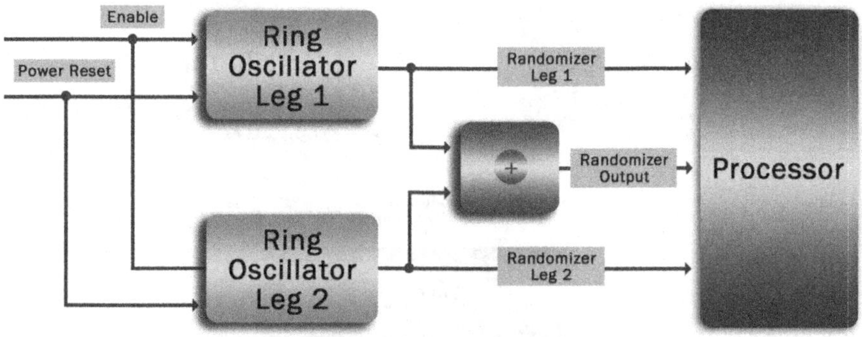

Figure 4.30:
Redundant oscillator outputs.

The processor sampling the outputs calculates the sum of the individual bit streams from each leg and compares to the randomizer output. If they do not match, a failure has been detected and an alarm is raised. In addition, the processor reads and aggregates N bits from each leg and from the combined output, and calculates the number of set bits in each N-bit stream. The processor uses probability theory to detect an inordinate quantity of set bits. For example, for a 256-bit stream, the allowable range of set bits may be 112 to 144. If all streams contain an acceptable number of set bits, then the test passes. Of course, there is a small probability of a false negative. When this occurs, a new set of data should be taken and tested. Based on probability theory, if numerous (e.g., six) sets of data fail the set bit range test, then a failure is deemed statistically assured and an alarm is raised; otherwise, the sampled data is used as the random bit stream.

The preceding test does not detect all types of bias failures in a randomizer. Randomness tests can be run at power up or periodically (e.g., once a year), sampling far more bits (e.g., a million) and looking for additional anomalies, such as improbably long strings of the same bit value.

Practical randomizers use both a non-deterministic randomizer (ring oscillators) combined with a linear function such as the SHA-1 hash to produce the final output, as shown in Figure 4.31.

This approach generates a smaller output given a set of random bits from the dual-leg randomizer. The hash output could be stored in memory for subsequent use in the creation of cryptographic keys. Using a hashed stream smoothes biases that might slip through the ring oscillators.

In summary, randomizers using ring oscillators must be carefully designed and implemented to make sure that they are truly random and incorporate built-in redundancy and error detection. The design of a randomizer must make sure that each leg has enough entropy (jitter) to produce

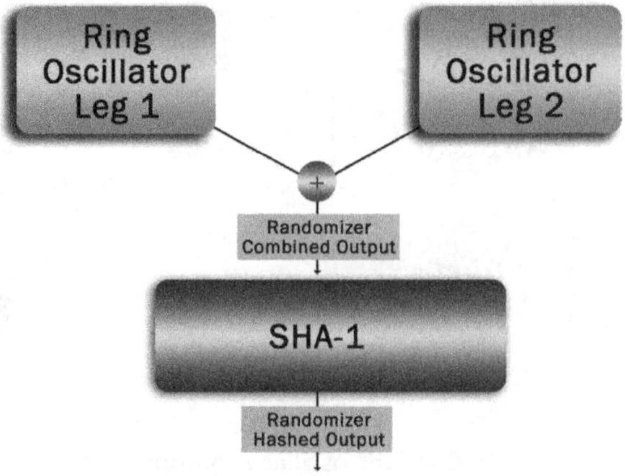

Figure 4.31:
Use of a hash function to narrow randomizer output.

a random bit stream. Ring lock-ups must be avoided, and each leg must produce an independent jitter with no affect on other legs. Adequate short-term and long-term tests must be performed to make sure the randomizer is fail-safe.

4.13.2 Pseudo-Random Number Generation

A deterministic randomizer produces an output based on a specific algorithm and not a probabilistic process such as a ring oscillator. As such, these types of deterministic randomizers or pseudo-random randomizers, while appearing to produce random bit streams, yield output that is totally predictable given a known initial state.

Key Point

Pseudo-random number generators, seeded by true random input, are required for applications that must generate cryptographic random material at a rate higher than can be created by a system's local true random number generator.

Key Point

A PRNG that is not properly seeded with true random input should never be used for the creation of private keys.

Periodically, during the operation of a TRNG, a new seed must be generated and input to the PRNG. If the algorithm is carefully designed, a deterministic randomizer will pass traditional statistical tests. Certification authorities (e.g., NSA, FIPS) will execute such randomness tests to validate PRNGs. PRNG algorithms can have a range in complexity, from linear congruential generators (LCGs) using modular arithmetic to combinations of encryption and hash algorithms.

Pseudo-random number generators must be carefully designed to avoid generating linear output that can be easily broken. PRNGS must use non-linear algorithms such as an AES-based keystream generator.

Key Point

A safe choice of PRNG is an FIPS-approved algorithm, such as those published in NIST Special Publication 800-90.

The list of FIPS-approved DRBGs can be found in Annex C of the FIPS 140-2 standard.[24]

4.13.2.1 NIST Special Publication 800-90

The most modern entries on the FIPS-approved list are the algorithms found in NIST Special Publication 800-90, published in 2007 and updated in May 2011 with a draft revision, SP 800-90A.[25] A couple of the more popular algorithms from this standard are discussed in the following sections. Note that the following algorithms' specifications match pseudo-random bit generators defined in ANSI X9.82, Part 3 Deterministic Random Bit Generators.[26]

4.13.2.1.1 HMAC_DRBG

HMAC_DRBG, as the name implies, uses the HMAC message authentication code algorithm to generate pseudo-random numbers. Variations of HMAC_DRBG use different HMAC algorithms (e.g., HMAC-SHA1 versus HMAC-SHA-256) and should be assumed to have commensurate relative strength. The typical interfaces to the DRBG are as follows:

- ***Instantiate:*** This method initializes the DRBG given true random input (entropy) of appropriate minimum bits.
- ***Generate:*** The core of the algorithm invokes the HMAC algorithm several times serially, using the entropy as input at various stages. Each invocation of the generate function outputs a string of pseudo-random bits of length equal to the length of the underlying hash algorithm. For HMAC-SHA-256, a 256-bit string of pseudo-random bits is generated.
- ***Re-seed:*** Most implementations will provide a re-seed operation that takes a new entropy input to refresh the internal state of the DRBG. The implementation may limit the number of bits that can be generated without a re-seed invocation. Periodic re-seeding is viewed as good policy to address the threat that the original entropy or internal DRBG state is compromised.

The HMAC_DRBG specification lends itself to efficient implementation, especially if the embedded microprocessor already has hardware acceleration for hashing. An implementation of HMAC_DRBG may be as small as 100 lines of code, making it easy to understand and verify against the specification.

4.13.2.1.2 CTR_DRBG

As its name suggests, CTR_DRBG uses a block cipher in counter mode to perform its bit jumbling. AES in Counter Mode (AES-CTR) is an obvious choice, especially if the system already incorporates AES-CTR for encryption. Similar to HMAC_DRBG, CTR_DRBG requires true random input and is typically invoked using instantiate, generate, and re-seed methods. One advantage of CTR_DRBG over HMAC_DRBG is that the algorithmic security of an implementation of CTR_DRBG using AES reduces to the security of AES-CTR itself. In contrast, pseudo-randomness is not an originally conceived property of hash functions. Therefore, the concern exists that a weakness could be found in HMAC_DRBG even if there is no weakness in the underlying hash. A weakness in CTR_DRBG would require a corresponding weakness in AES-CTR.

Cryptography without true random numbers should be considered little more than obfuscation that can be cracked by sophisticated attackers. Hackers allegedly compromised the Sony PlayStation 3 game console, claiming that the root of the subversion was caused by Sony's use of the following random number generator function:

```
int getRandomNumber()
{
    return 4;
}
```

Embedded systems developers: don't let this happen to you! The attackers, from an organization called failOverflow, presented their attack at the Chaos Communication Congress 2010. Because of this insecure random number generator, the attackers allegedly were able to compute the private key used to sign boot images and then boot Linux on a PS3.

4.14 Key Management for Embedded Systems

As previously discussed, the security of a cryptographic system depends on the secrecy of the entirety of the system's private keys. An adversary that obtains the key or a segment of the key will, in general, be able to obtain the plaintext from actively generated or previously captured ciphertext. At first glance, keeping keys secret seems simple: the source of the key (key generator) and the destination (user) are trusted agents instructed to protect the key as it travels from its source to its final destination within the embedded system. Many systems today rely on this process for securing the keys. The keys are generated in plaintext form—sometimes called *red keys*—and sent to the user by courier or by other secure means and then loaded into the embedded system.

Once the key is loaded, many embedded systems do not bother to periodically change the keys. It may be surprising to know that some critical banking systems in use today—for example, automated teller machines (ATMs) that use cryptography to protect card transactions—do not change the keys within the ATM unless a failure has occurred. And even then, the original keys are often reloaded!

4.14.1 Case Study: The Walker Spy Case

Another famous event in the annals of security and cryptography can basically be boiled down to poor key management. A U.S. government-classified system employed an overly simplistic key management system that proved disastrous. The basic attack was as follows: a trusted key administrator (John Anthony Walker, Jr.) replicated a set of red keys and gave them to an adversary, as shown in Figure 4.32.

Of course, the attack could not have been perpetrated without the insider threat. Without doubt, protecting against insider threats is one of the biggest challenges to all forms of security. However, we demonstrate how proper key management would have made this insider attack more difficult, if not impossible, to perpetrate.

This example describes how the U.S. government broadcasts messages to U.S. submarines at sea. Broadcast RF messages are sent from a ground-based broadcast station on U.S. soil to all the submarines. The broadcast approach enables submarines positioned all over the world to obtain critical information concurrently and rapidly. Of course, some messages are highly classified, describing submarine positions, nuclear command and control messages, and other secrets. Thus, the broadcasts must be strongly encrypted.

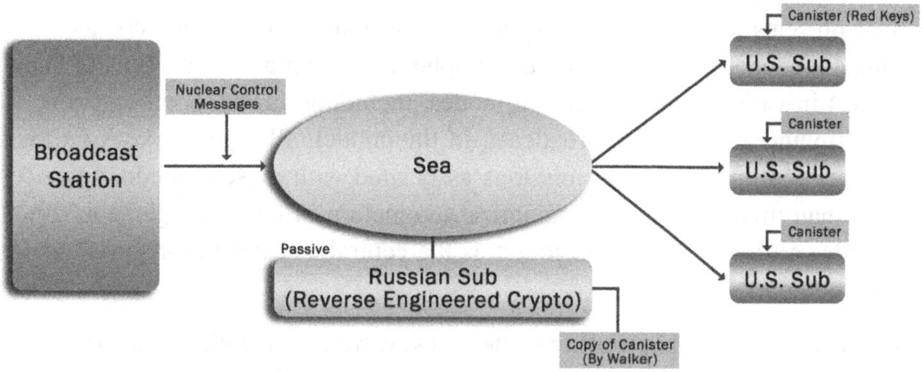

Figure 4.32:
Red key compromise.

Because it must remain underwater for long periods of time, each submarine is provided, prior to launch, a canister containing numerous red keys for use during a mission. The key list is printed on paper tape, and each key is used only for a predetermined, limited amount of time—for example, one day. The key is manually replaced with the next canister key at the end of the day. Cryptographic custodians or administrators are responsible for making sure that each key is properly loaded into the cryptographic communications units.

The system was compromised when one of the cryptographic custodians in a submarine, John Anthony Walker, Jr., made an extra copy of the paper tape key list and sold the copy to the KGB. With possession of the copied keys, the Soviets were able to monitor and decrypt all the messages sent by the broadcast station. Essentially, Soviet submarines were able to impersonate U.S. submarines since they had the authentication mechanism: the broadcast key used to decrypt all the messages. The daily key replacement policy was good: if a key were compromised, only messages encrypted and decrypted on the applicable day would be at risk. However, in this case, the threat was not an eavesdropper trying to obtain the key, but rather an insider threat that leaked lots of keys without detection for a long time.

This simple key management scheme of manually loading and reloading red keys by trusted personnel resulted in a major cryptosystem failure, what many believe to be the worst case of espionage in U.S. history and a loss of security that literally changed the balance of power between the United States and the Soviet Union. This example highlights the critical importance of a robust key management system for any product using cryptography to protect data, whether the data is classified, sensitive, or private. At a minimum, a key management plan needs to be articulated up front and presented, discussed, and approved prior to design or implementation.

4.14.2 Key Management—Generalized Model

Figure 4.33 presents a generalized infrastructure of embedded systems connected to a communications medium (RF, Internet, telephone switching system, private banking network, etc.) in various communication modes, including duplex point-to-point, duplex point-to-multipoint, and simplex broadcast. In this model, all types of keys are required for the various configurations and missions associated with these embedded devices. To minimize human involvement and maximize automation in key management, the model provides a trusted key server to act as the central key generator for all the required keys.

This model addresses the complete life cycle of a key, from its creation to its destruction. The process begins with embedded system registration and ends with key revocation (the embedded system is no longer trusted, is destroyed, or is no longer in inventory).

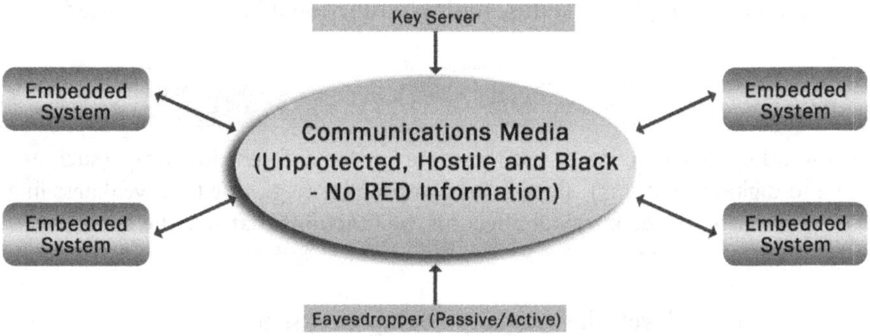

Figure 4.33:
Embedded computing infrastructure with centralized key server.

(1) Registration Endpoint systems need to register to the key server, and this normally requires an initial key load of some kind (symmetric or public key). If a symmetric key is required, it is usually black (it should not be red). For symmetric keys, initial key loads should be done either at the factory or in a trusted facility prior to shipping on site. It is best that the initial key loaded is used only inside the embedded device and used strictly to decrypt an operational key either sent by the key server or loaded initially in black form.

(2) Key Generation Some microprocessors provide a hardware-based one-time-programmable (OTP) key generation function. In other cases, the embedded systems firmware installation process can incorporate an internal key generation step. Key generation algorithms are included in most software or hardware cryptographic implementations.

Key Point

Ideally, private keys are born within and never leave the embedded system in which they are used.

Sometimes, however, private keys must be provisioned externally. For example, High-Definition Content Protection (HDCP) keys are licensed and distributed by Intel Corporation. How do these keys get securely transported from their originating trusted key generation server to the target embedded system? Either the embedded system must be manually keyed by a locally trusted administrator (not a scalable solution), or the device must have the ability to establish a secure connection (e.g., using IPsec or TLS) to the key provisioning server. This, of course, begs the chicken-and-egg question: how was the private key for that connection provisioned? An initial private key may be installed at manufacturing time, providing an ability to talk only to a provisioning server under the control of the embedded system customer. Regardless of how keys are provisioned, the embedded systems designer must have a plan and approach that provides adequate security and scalability to meet product requirements.

For most modern cryptographic systems, multiple functional keys are required.

> **Key Point**
>
> It is never a good idea or practice to use the same key for different functions (such as data encryption and digital signatures), since a key compromise (e.g., due to a weakness in a specific function or application of the key) will affect all the functions that use the same key.

Most practical, secure, and well-designed cryptographic systems use distinct keys for different functions. Some examples of key purpose are as follows:

Traffic Encryption Key (TEK): This key is used only for bulk encryption or decryption of messages or other traffic.

Key Encryption Key (KEK): This key is used to encrypt or decrypt TEKs; usually, the key server generates KEKs, and they are loaded into the embedded system once in its lifetime. The KEKs are loaded securely in a trusted facility, generally at the factory or an initial key load facility (KLF). Once KEKs are loaded, they are used by the embedded systems to load subsequent TEKs either manually or over the air from the key server. TEKs are loaded black (encrypted with a KEK) and decrypted by the embedded system when needed. KEKs are unique to each embedded device, even if a common TEK is loaded to many embedded systems. Therefore, a compromise of one embedded device and its KEK does not compromise any other systems. A key server, once notified that a KEK has been compromised, can provide a new TEK (if shared) to the uncompromised systems.

Transmission Security Key (TSK): A key used to provide transmission security as opposed to traffic security. For example, the link of a satellite ground station to a satellite may be protected at layer two with TSK-based authenticated encryption, even though the traffic or payload satellite data has been previously encrypted by a TEK before arriving at the ground station. This additional protection can be used as an anti-jam feature in some satellite systems. For ground systems, TSKs are sometimes used to protect internal source and target addresses of messages, offering traffic flow security and availability of the communication link. For example, in typical secure voice applications, the payload voice data is protected by TEK encryption, but the calling and called party telephone numbers are sent in the clear prior to establishing a secure connection. This weakness exists in most modern secure voice applications. An eavesdropper can easily determine who is calling each other as well as the traffic density. A TSK-encrypted link, when used, is often applied between major nodes of a communication system, such as a T1 line. An eavesdropper monitoring a transmission-secured T1 line sees continuous TSK cipher data even if payload voice traffic is low or non-existent.

Storage Key (SK): An SK is used to protect data that is at rest (DAR), stored either locally within the embedded system or remotely, such as a network attached storage (NAS) or enterprise storage area network (SAN). SKs can be generated internally by the embedded system or by the key server and sent in the black to the embedded device.

(3) Key Distribution Typical systems require both symmetric and asymmetric (public) keys for distribution. Once an embedded system is registered, additional keys are provided to it. All the keys sent from the key server to an embedded device are encrypted in a point-to-point mode, usually using the KEK of each embedded device.

(4) Key Change

Key Point

Robust and well-designed systems should change keys as often as practical to minimize the window of vulnerability of a compromised key.

The shorter the period of key use, the shorter the volume of compromised data. This period of use is called the cryptoperiod. For embedded systems that have been compromised and keys presumably obtained by an adversary, a key change is essential to the rest of the embedded devices that were potentially using the same TEK as the compromised system. The key change should be started as soon as the compromise is detected, even if it disrupts traffic for the duration of the key change. This type of key change is called a *supersession key change.*

Unfortunately, many autonomous embedded systems deployed in the field today do not effectively manage key changes, and these systems are often at risk of an undetectable compromise. In fact, many embedded systems are deployed with no ability for periodic or supersession key change. This can allow resources to be compromised without detection for months or years. Clearly, it is best to periodically change keys, even if it requires a manual new key load. We all suffer from this problem of changing keys; for example, we hate to change the passwords we use for online banking, social networking, and our PCs. But from a security point of view, not to mention peace of mind, frequent cycling of private security parameters is highly recommended.

(5) Key Destruction All keys whose cryptoperiod have expired or have been compromised should be destroyed within the embedded system and key server, and a record maintained, to avoid reuse of the compromised key.

(6) Key Protection

Key Point

The protection of keys, both from a physical as well as logical/software perspective, is one of the most important security features within an embedded system that employs cryptography, yet it is all too often ignored or poorly addressed.

If a system does not have adequate protection, an adversary can obtain the key and totally compromise any assets protected by that key. Unfortunately, there are many ways, some of which are very sophisticated, that an adversary can obtain the keys stored in an embedded system. While full-proof (or as near as possible) protection may be too costly for an embedded system, this does not mean that the issue should be ignored. At a minimum, the system designer must formulate a protection plan based on the threat environment and value of resources under protection.

Threat environment is often influenced by the physical location of the embedded system and whether there is some other form of physical protection beyond the embedded device itself. Unattended devices that are not located in a physically protected environment (such as a reasonably guarded building) are the most vulnerable for a passive and potentially active attack, including theft of the embedded system itself. Direct physical access of the system may result in detection or extraction of its keys. When a U.S. EP-3 military reconnaissance plane crash-landed in China, all the electronics, including cryptographic communications and storage devices, were at the disposal of the Chinese military; imagine the gold mine of information that was extracted. This event is known as the *Hainan Island Incident.*

There are many affordable techniques and design practices that can be embedded into embedded systems for key protection. Some are described next.

Non-volatile key storage: Avoid storing plaintext keys (and passwords) in non-volatile memory. This includes private encryption keys as well as the public keys used for public key cryptographic authentication. While the public portion of an asymmetric key pair need not be kept secret to maintain cryptographic strength, these keys are often used to prove authenticity of software, firmware, or other critical data. If an attacker is able to replace a plaintext public key with the attacker's key, then the authentication functions will incorrectly validate malicious firmware. Keys stored in permanent memory should always be encrypted. When the cryptographic subsystem is not operational (e.g., placed in a sleep state to reduce power), any internally managed keys must be encrypted before being stored into non-volatile memory.

In most embedded systems performing a security function, however, some plaintext keys must be physically present.

Key Point

Private keys stored in an embedded system's primary non-volatile memory (e.g., flash memory or hard disk) should be encrypted and decrypted with a plaintext KEK that is ideally held in non-volatile storage external to the main applications processor and protected against physical attacks.

These cryptographic storage systems are sometimes called *secure elements*. Protected storage is often coupled with cryptographic accelerators. These come in many forms, including

- Cryptographic engine within the same SoC as the applications processor
- Separate discrete microcontroller
- Trusted Platform Module (TPM)
- Hardware Security Module (HSM)

Without a secure element built into the embedded system, keys must be kept in flash or on disk, making them more susceptible to external attacks and software attacks resulting from weaknesses in operating systems controlling the platform.

In addition to common software-based side channel attacks using cache behavior analysis, attacks on keys include physical tampering and simple and differential power analysis (SPA, DPA). These attacks have proven quite practical in the wild. For example, in September 2011, cryptography researchers disclosed they were able to recover the private key from a commercial, fielded smart card microcontroller, the NXP DESFire MFICD40, in hours using a power analysis attack that required a few thousand dollars' worth of equipment to perpetrate.[27]

The goal of secure key storage is to prevent such attacks. The ability of hardware-based key storage approaches to prevent such attacks varies. For commercial secure elements, the FIPS 140-2 certification level is an important measurement of this robustness. FIPS 140-2 certification is discussed later in this chapter.

Attended State: For embedded subsystems that are attended, a good practice is to provide a user authentication mechanism, either through a password or a physical token, to enable use of the cryptographic subsystem. When the system is not in an active, authenticated, attended state, the cryptographic subsystem enters a protected mode in which all keys within are encrypted. Inserting the token or reentering the password unlocks the cryptographic subsystem, permitting non-volatile memory decryption of keys for operational use.

Unattended State: For autonomous embedded systems, robust key protection is desired since we must assume that an adversary will have unlimited time to mount either a passive or active attack to recover the keys. Protection should include key zeroization upon detection of physical probing (hardware or software) and zeroization upon detection of power loss to a certain level. There are many other attacks that an adversary may attempt, and upfront decisions must be made to either include or exclude countermeasures in the design. For example, an adversary may include in her repertoire an active attack that forces failure of the cryptographic subsystem microprocessor. To counter this threat, consideration must be given for the cryptographic subsystem design to be fail-safe. In other words, to prevent the leakage of keys to external interfaces when a hardware failure occurs, the device must contain alarm circuitry that inhibits all operation and zeroizes all volatile keys upon detection of a cryptographic subsystem failure.

An advanced approach to fail-safety is to use redundant circuits that can detect functional mismatches, providing early warning against systematic failure.

(7) Key Revocation Revocation is often a common life-cycle event when a public key certificate's lifetime ends.

> **Key Point**
>
> For embedded systems that are lost, stolen, or are known to have their cryptographic subsystems compromised, provision must be made to revoke all the keys associated with the compromised or lost unit.

Other devices within the compromised system's worldview (e.g., communication peers on a network) must be told that the keys in question have been revoked. Each embedded system in turn should zeroize all the affected keys.

Keys corresponding to embedded systems with short field lifetimes may not support revocation at all. When the lifetime is complete, the entire embedded system is disposed, including its private keys.

In other cases, revocation will be impractical due to a lack of communication capability to a revocation server that provides up-to-date revocation lists. In this case, the only recourse for a revoked key may be to re-install potential peers with a static list of revoked keys.

4.14.3 Key Management Case Studies

4.14.3.1 Case Study: Secure Wireless Push-to-Talk Radio Network

This case describes a secure wireless handheld voice radio network and the various options of managing its keys. As shown in Figure 4.34, each radio is equipped with an embedded cryptographic subsystem to secure voice traffic. This network is characterized by low-power wireless communications across a relatively small number of endpoints (radios), i.e., tens or hundreds, not tens of thousands. The wireless network may be deployed as part of a secret military mission to hunt down terrorists, a crime sting, or a private operation of some kind that requires a secure wireless network.

For the military application, at least some of the cryptographic algorithms used within the cryptographic subsystem are likely classified. For the other applications, the cryptographic subsystem most likely uses AES configured as a keystream cipher in OFB mode. Depending on the radio frequency (RF) channel bandwidth (e.g., narrowband or wideband), the voice compression algorithm is either a government standard such as Mixed Excitation Linear Prediction (MELP) or a commercial standard like G.729.[28] For

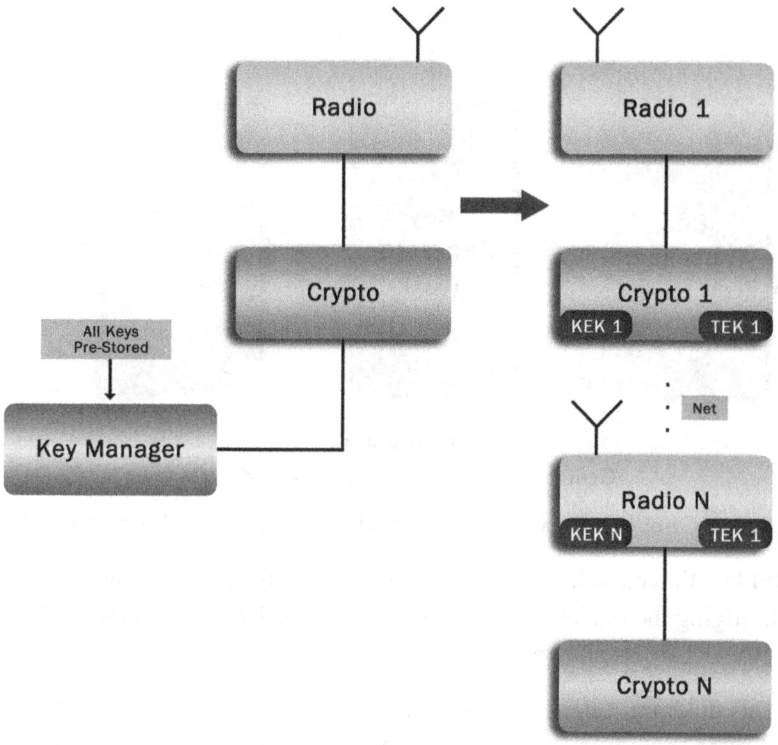

Figure 4.34:
Secure wireless push-to-talk radio network.

narrowband channels, we would not use pulse code modulation (PCM) for analog-to-digital conversion and voice compression. A generalized architectural diagram of the radio is shown in Figure 4.35.

Secure network operation is initiated by a user pressing the radio's push-to-talk (PTT) button. A PTT initiation interrupts the network and alerts all network subscribers that voice communication is about to take place. For example, upon a PTT initiation, the radio sends a plaintext pattern of alternating ones and zeros on the line, alerting all other receivers that a cryptographic synchronization is about to be initiated. For burst or noisy channels, the crypto synchronization message sent by the PTT initiator uses a proper error-encoding scheme to assure perfect reception of the IV by all the receivers. Release of the PTT button results in an encrypted end-of-message (EOM), alerting all receivers that all traffic should be dropped until a new cryptographic synchronization is transmitted. Once a channel is released, any other user can initiate voice communication using her PTT.

There are many options available for key management of the PTT network and its radios; they are described next.

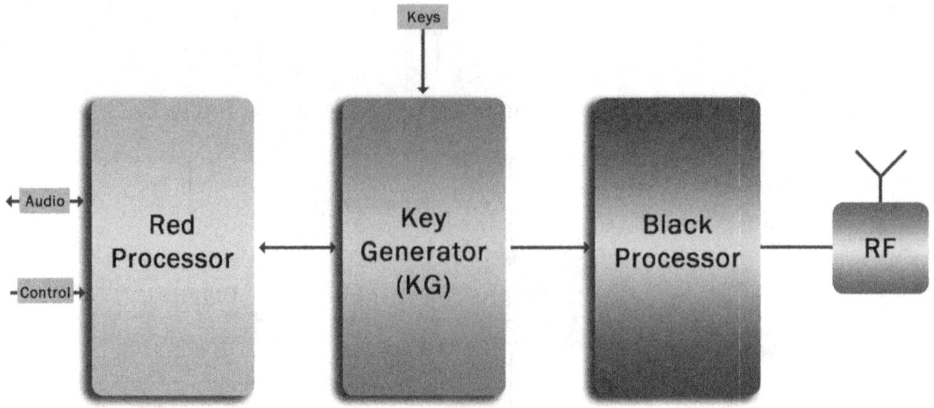

Figure 4.35:
Generalized secure military radio diagram.

4.14.3.1.1 Option One: Load a Common Red TEK into Each Radio at a Key Load Facility

Users must not be allowed to load their own red TEKs. This policy avoids the insider threat of an end user divulging the red TEK in her possession. It is best to load the red TEK in a trusted key load facility (KLF), as shown in Figure 4.36.

As shown in this diagram, a common method to provision a set of radios is to connect each of their key load ports to an installation network that also includes a key server. The key server may be a simple PC. The installation network may be a local Ethernet network that is used for the key provisioning process. It is good policy for the radio to have a distinct physical port that is used specifically for key provisioning. This reduces the possibility that an improper connection is established with the radio's internal key loading software. Once this red TEK is loaded over the installation wire, it is used as the common TEK of the deployed tactical network and not changed during the mission. At the end of the mission, when the radios are returned, a new red TEK is loaded at the KLF for the next mission.

This option is viable and simple to implement. However, if one of the radios is lost, compromised, or captured in a hostile environment, communications security is lost. The only

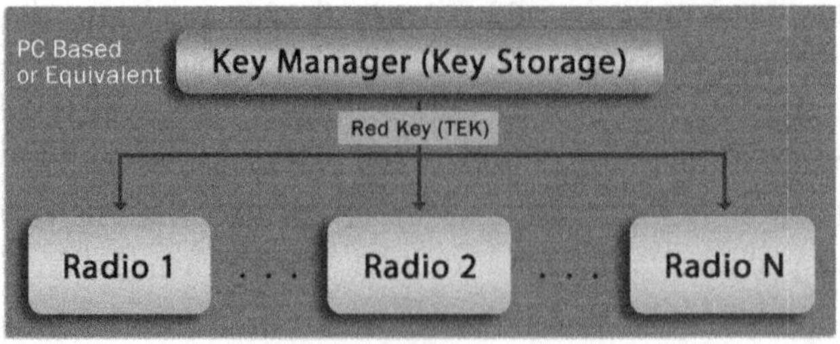

Figure 4.36:
Key load facility.

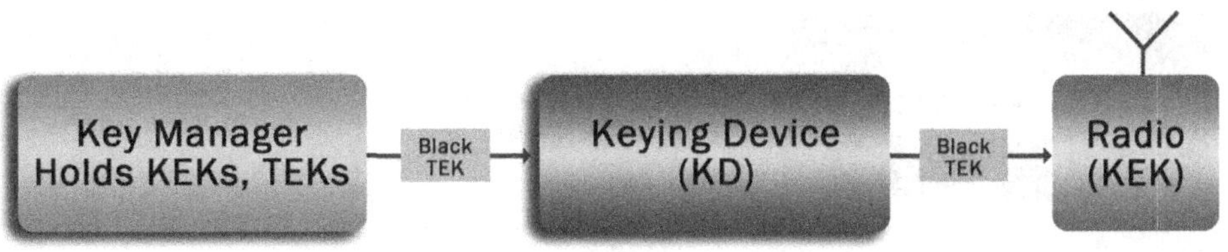

Figure 4.37:
Black TEK load into radio.

way to re-establish secure communications would be to abort the mission, returning the radios to the KLF for reloading.

4.14.3.1.2 Option Two: Load a Black Key in the Field

Instead of a TEK, a unique per-radio, red KEK is loaded into each radio at the KLF. In addition, the common network TEK is encrypted with each unique KEK and loaded into a portable keying device (see Figure 4.37), issued to each user to take into the field along with the radio. The user then loads the black TEK into the radio using the keying device. Finally, the radio decrypts the TEK using its unique KEK, and the radio is now available for network communications.

This option allows the user to manually reload a new TEK into her radio. Each keying device can be provisioned at the KLF with a large set of black TEKs, enabling dynamic rekeying in the field for extended periods of time.

The common network TEK is encrypted in the unique KEK using an electronic codebook scheme called a key wrap, described earlier in this chapter. All the KEKs are generated and loaded at the KLF and never exposed or used in the field.

Another method of provisioning the keying device is to load a copy of the per-radio KEKs instead of the black TEKs. Then the owner of the keying device (the field key provisioning officer) uses a trusted interface, such as a button, to dynamically generate and encrypt a TEK that can be installed into all the radios. The keying device should require strong multi-factor authentication, such as password and smart card insertion, from the provisioning officer to enable the load function. This approach enables a set of fielded radios to be rekeyed an infinite number of times, as long as the keying device remains protected. Because it contains red keys for all the radios, the keying device becomes a single point of failure in which compromise of the internal keys will expose all communications. Tamper protection of the keying device is indicated when it contains red keys.

4.14.3.1.3 Option Three: Key Load Using Public Key Cryptography

Option three is similar to option one, except that instead of loading a red TEK into the radio at the KLF, the TEKs are loaded over the KLF's installation network by first establishing a session key through a key exchange between a master key server and each of the radios, in

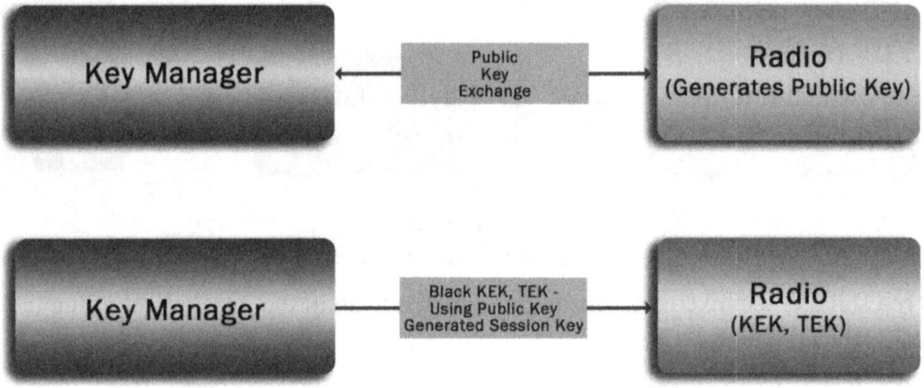

Figure 4.38:
Black key load to radio using public key cryptography.

turn. This session key is used to send the TEKs to each radio, as shown in Figure 4.38. For option three, no red keys are ever exposed, even at the KLF. This protects against some insider threats within the KLF's installation network. The use of public key exchange assumes that all radios are pre-provisioned, at the manufacturer, with private static public keys and a root certificate for the installation network prior to transfer to the KLF.

To simplify the installation process, the key server and radios connected on the KLF's provisioning network can derive the network TEKs from shared secret data established during each server-radio key exchange. In fact, multiple TEKs, to last multiple missions, can be created and installed using this process.

Of course, a public key exchange can be applied to option two, e.g., to provision the keying devices and radios with KEKs. Then the keying devices can be used to install TEKs periodically, as needed, to the radios in the field.

4.14.3.1.4 Option Four: Over-the-Air Rekey Capability

In this fourth scenario, option two or three is first used to load a KEK and an initial TEK into each of the radios at the KLF. Then, as needed in the field, the master key server remotely sends to each radio a new common TEK, as shown in Figure 4.39.

The key server interfaces to the radios to establish a session and remotely send the new TEK. Of course, this approach works only if the radios are within communications range of the key server. The key server first establishes the connection using the initial TEK and the PTT for synchronization. Once a session is established, the key server utilizes the following rekey protocol:

1. The key server sends an encrypted rekey alert message, notifying all radios that a rekey is about to take place.
2. Each radio leaves the traffic state and waits for a key load from the server.

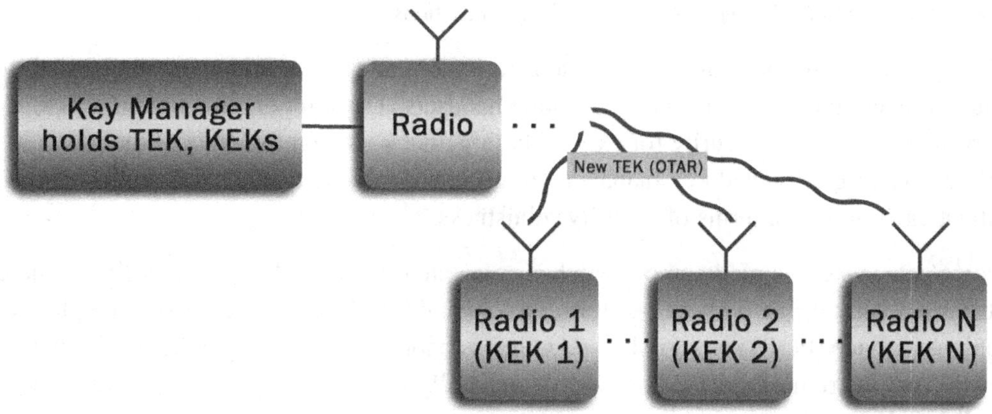

Figure 4.39:
Over-the-air rekey (OTAR) of new TEK.

3. The key server establishes cryptographic synchronization via a public key exchange with each radio and then sends the new common TEK.
4. This rekey sequence is followed for all radios until all of them receive the new TEK.

This rekey process is used for a periodic or a supersession rekey, the difference being that for supersession, the key server would only rekey the uncompromised radios. The key server destroys the KEK of the compromised unit. Secure communications continue once the compromised radio is logically eliminated from the network through this rekey process. Even if an attacker succeeded in recovering the TEK and KEK of the compromised radio, she can never join the network, since all radios were rekeyed with a new TEK.

If OTAR communication is possible in the deployed tactical network, this approach is streamlined yet robust. However, because it is imbued with a capability to send critical keying material over a wireless network, the key server must be completely trustworthy. In addition to running trustworthy software, the key server must be physically protected. When not in use, or left unattended, the key server must be unavailable for overt or covert key extraction.

One way to protect access to an unattended key server is to lock it using a token, such as a smart card. This token contains one-half of a storage key that is used to encrypt all KEKs in the key server's non-volatile memory. The other half of the split storage key is contained in the key server's cryptographic subsystem. In the locked mode, all keys are encrypted, and even if the key server computer is stolen, an adversary will not be able to recover any keys. Only when the key server is unlocked are the splits reconnected to recover the storage key and make the encrypted radio keys available for loading.

4.14.3.2 Case Study: Generalized Phone Communications

This case study describes how to secure sensitive voice and data communications for all types of human-operated equipment, such as mobile handsets. The solution discussed is scalable and can provide end-to-end security for extremely large networks, on the scale of the Internet. We explore several generalized key management options available to system designers and contrast each option in terms of security robustness.

A voice/data system is made up of switches connected to each other by trunk lines. The end instruments can take on a variety of form factors and interfaces including wired—plain old telephone service (POTS) or digital—and wireless phones. The cryptography resides within or near the end instruments, providing what is generally termed end-to-end encryption and avoiding link-to-link encryption (see Figure 4.40). End-to-end encryption avoids "red–bridge" compromises that could potentially occur at the junctures between the encrypted segments in a multi-link network.

For our generalized case, we are not precluding use of any physical network media. The switching network can be analog (e.g., POTS), digital (e.g., ISDN), the Internet (e.g., Voice-over-IP), wireless RF, or any combination thereof. The embedded cryptography residing at or near the end instruments is applied only at the applications layer of the OSI model. Voice/data is encrypted at the end instruments and traverses a non-trusted, potentially hostile communication network, whose availability cannot be guaranteed. We describe, however, mitigation approaches that improve availability of the network by the use of transmission security techniques. We also describe techniques that protect both voice/data traffic as well as signaling (indication of the calling and called telephone numbers), sometimes referred to as *secure dial.*

As shown in the system diagram in Figure 4.41, end instruments are connected to local switches. Calls initiated by an end user are routed either locally or through trunks to remote

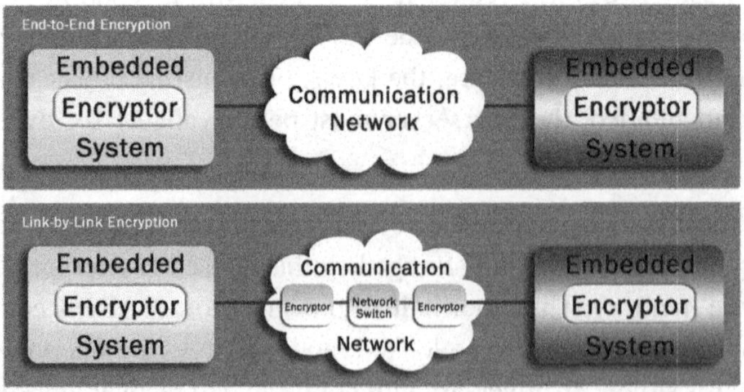

Figure 4.40:
End-to-end and link-to-link encryption comparison.

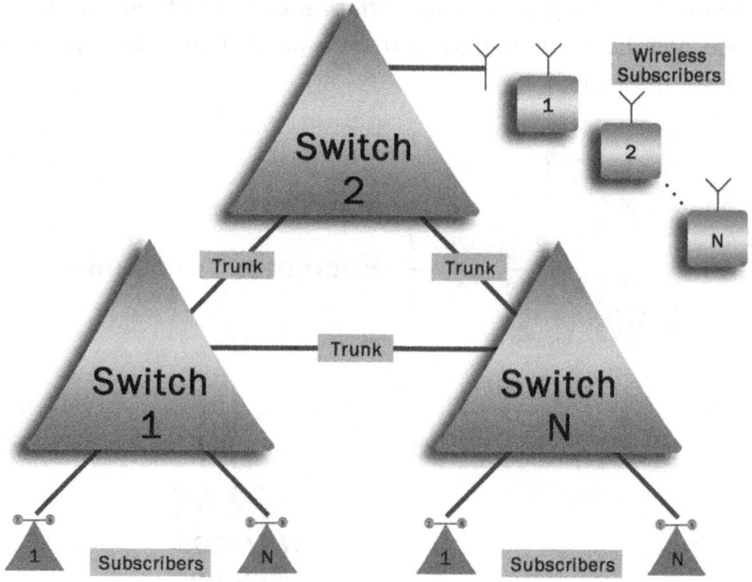

Figure 4.41:
Phone network system diagram.

switches and connected point-to-point to the called party. For this case study, we cover all types of communications devices, from secure phones developed from scratch to unmodified, commercial-off-the-shelf (COTS) wired or wireless phones with cryptography as an applique.

4.14.3.2.1 Cryptographic Subsystem Embedment Options

There are many options available for cryptographic subsystem embedment; they are described next.

4.14.3.2.1.1 Option One: Cryptographic Subsystem Embedded within the End Instrument Option one is shown in Figure 4.42.

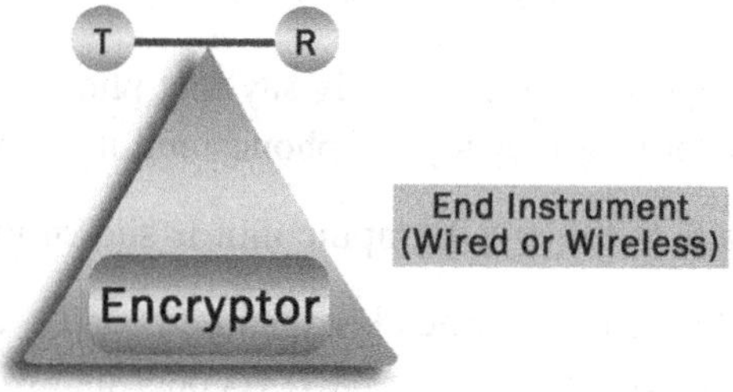

Figure 4.42:
Cryptography Embedded within Instrument.

4.14.3.2.1.2 Option Two: Bump-on-the-Line (Attached to End Instrument) Option two uses an unmodified phone, and the cryptographic element is an applique connected to the output of the phone, as shown in Figure 4.43.

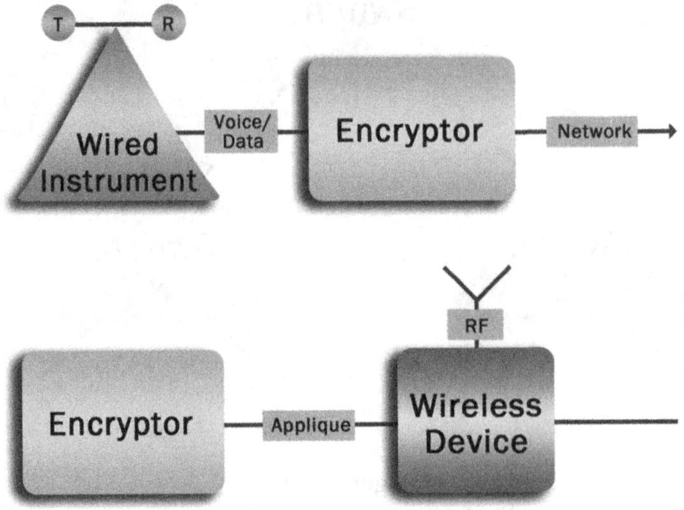

Figure 4.43:
Bump-on-the-line (attached to instrument).

4.14.3.2.1.3 Option Three: Bump-on-the-Handset Similar to option two, in option three, the cryptographic unit (encryptor) is connected between the handset and the base of the instrument (see Figure 4.44).

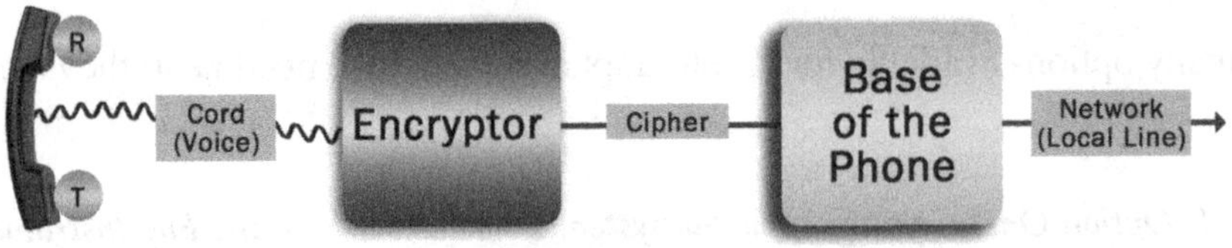

Figure 4.44:
Bump-on-the-handset.

With all these embedment options, we can safely say that phone voice/data security can be deployed without replacement of all legacy telephones and instruments.

A generalized block diagram of the cryptographic unit is shown in Figure 4.45.

Voice compression, depending on the traffic classification, can be a standard low-rate vocoder (codec) specified in G.729, a government-specified vocoder like MELP, or any other vocoder. The cryptographic unit can store several (or many) vocoders, and after cryptographic synchronization through an established protocol, determine a common vocoder contained in

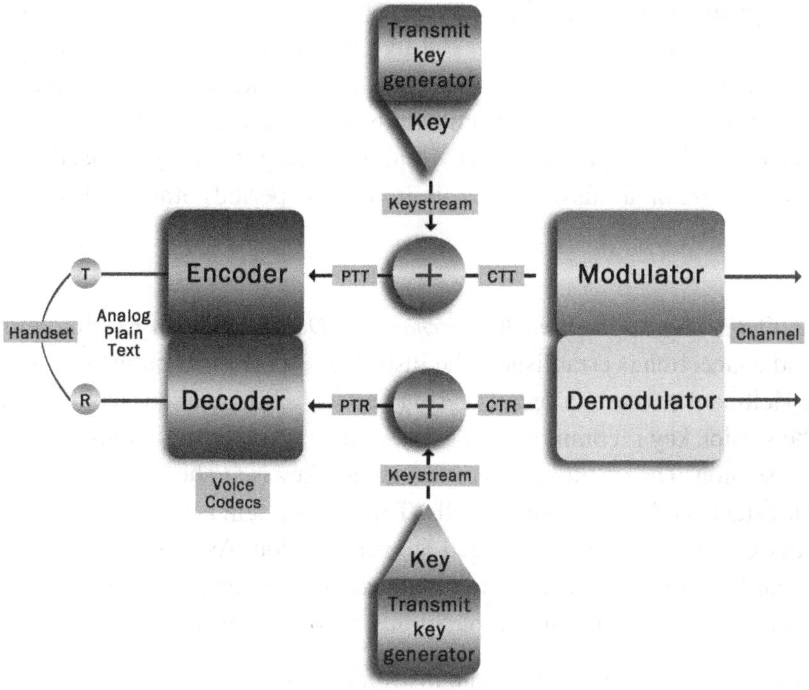

Figure 4.45:
Embedded cryptographic unit (full-duplex key generator) block diagram.

both end instruments. Note that encoding and decoding have historically been incorporated into the cryptographic unit even though these functions are not involved in cryptography. This is done primarily for performance; in addition, in some cases, there may be a desire to keep the codecs themselves secret.

For our generalized secure voice/data system, a typical call is established as follows:

1. The calling party goes off-hook, obtains a dial tone from the switch, and enters the called party's telephone number.
2. The switch receives the called party telephone number and either "rings" the called party if local or routes the number to a remote switch that then "rings" the called party.
3. When the called party answers, the switch infrastructure establishes an end-to-end connection between the calling and called parties.
4. End-to-end secure voice/data traffic is established by the end instruments.

4.14.3.2.2 Key Management Options

Once the end-to-end call is established, encrypted voice/data is exchanged, provided the end instruments possess a shared secret session key. Of course, there are options for establishing this session key.

4.14.3.2.2.1 Option One: Pre-Load a Common TEK to All Instruments The first option is a common, low-cost, and simple option. But by now, it should be obvious that this option is not robust. The risk of a key compromise for a widely shared key is relatively high. Once the key is compromised, the entire communications system is compromised. As explained in the previous case study, the pre-shared key option could potentially be used for a small number of subscribers on a mission for a limited cryptoperiod; otherwise, it is not recommended.

4.14.3.2.2.2 Option Two: Use Public Key Exchange to Derive a Session Key In option two, after the end-to-end connection is established, the instruments perform a public key exchange (e.g., using Diffie-Hellman) to create a secret session key for subsequent encryption and decryption of traffic. The session key is common only between the two end instruments and is destroyed at the end of the session. The cryptoperiod of the session key is as short as the phone call. A new session key is established on subsequent calls. Thus, the system is not vulnerable to exhaustive key space attacks, since the key is changed at every session. Assuming the endpoints are utilizing standards-recommended key lengths and algorithms, an active eavesdropper could not break a session key even with vast computational power at her disposal.

A public key is initially loaded into the end instruments by a trusted key server or certification authority (CA). The public key is encapsulated in a certificate, signed by the CA. Only certified public keys are honored by the calling and called parties. If an end instrument is compromised, the CA revokes the instrument's public key certificate and places it in a revocation list that is sent to the end instruments. During call setup, the end instruments do not honor any certificate on the revocation list. The CA may distribute revocation lists to end instruments periodically. Alternatively, the CA can push revocation lists to a set of nearby instruments, and those instruments can disseminate the lists to other terminals in a distribution chain.

While key management options one and two provide end-to-end security, the key exchange method is a better solution and is scalable to a large number of subscribers. Both solutions are point-to-point solutions. Point-to-multipoint solutions would require a common TEK for the multipoint network. Neither option one nor option two is able to do this, since session keys are assigned to only two parties. A common TEK requires a key server, similar to the earlier push-to-talk case study, for distribution of a common TEK using pre-established per-instrument KEKs.

These two key management options provide protection only of the voice/data information; they do not protect addressing information. Calling and called party numbers are dialed in the clear. An eavesdropper monitoring the communication links is unable to decrypt traffic but is able to determine the volume of traffic as well as who is calling each other. For systems that require addressing confidentiality, a modified approach is needed.

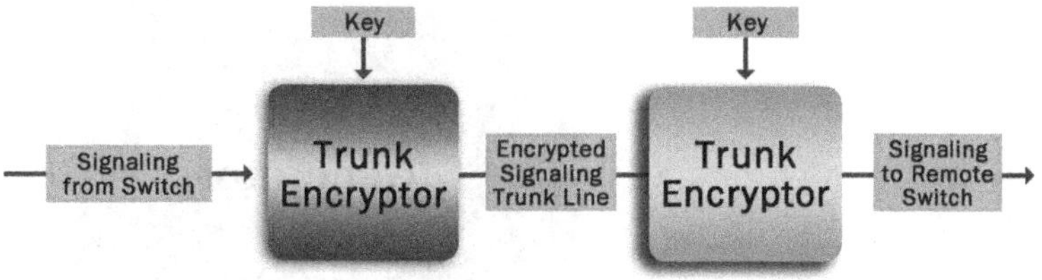

Figure 4.46:
Trunk protection using encryption.

4.14.3.2.3 Addressing Protection

4.14.3.2.3.1 Option One: Bulk Encryption between Trunk End Points Trunks carrying many subscriber telephone numbers between central switches can be protected by link-to-link encryption of the trunk lines, as shown in Figure 4.46.

This solution protects the addresses of instruments within the trunk segments but does not protect the addresses of local switch subscribers. Nevertheless, this approach is viable and provides significant system-wide transmission security. An active eavesdropper, monitoring the trunk lines, observes a continuous random stream (the cipher of the link), whether or not any signaling or activity is taking place between the trunks. Jamming the trunks is very hard, since it would require a very powerful wideband noise jammer.

4.14.3.2.3.2 Option Two: Local Protection of Addresses To eliminate the vulnerability of option one, per-subscriber addressing protection is required. This is a hard problem. One solution that is costly and complex is to provide a key distribution center (KDC) for each switching center. The KDC would provide a local key generator that connects to subscribers during dial or ring tone, securing the link during these phases.

The KDC, as shown in Figure 4.47, consists of a key controller, where all the symmetric keys are generated and stored, and a set of loop key generators (LKGs), to which the switch can connect during the dial/ring phase.

Each subscriber is pre-loaded with a unique KEK generated at the KDC. Secure call establishment would follow the protocol shown in Figure 4.48.

When the switch detects off-hook from the calling party, and before the terminal receives a dial tone, the switch connects an LKG to the calling party after the LKG receives the unique KEK of the subscriber. Note that the LKG is physically protected within the KDC, along with the primary key generator and key storage. The KDC also generates a per-session TEK and sends it to the LKGs of the caller and callee. The LKG and calling party establish a cryptographic synchronization using the unique KEK, and the LKG sends the TEK, encrypted by the KEK, to the calling party.

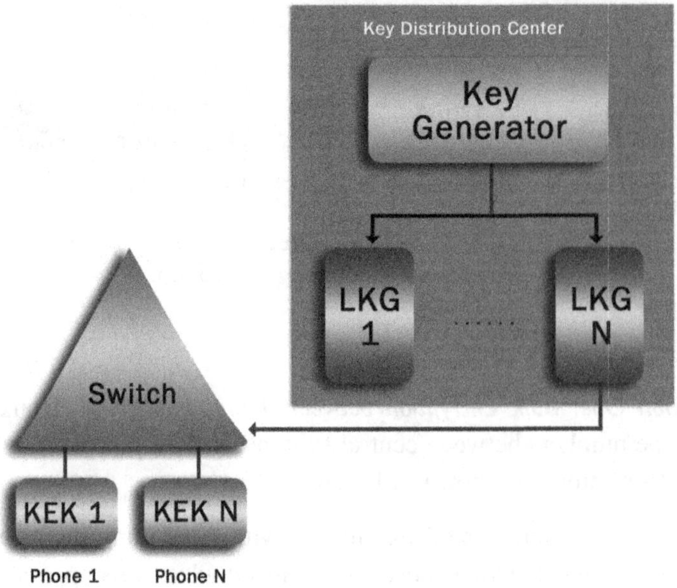

Figure 4.47:
KDC for secure dial.

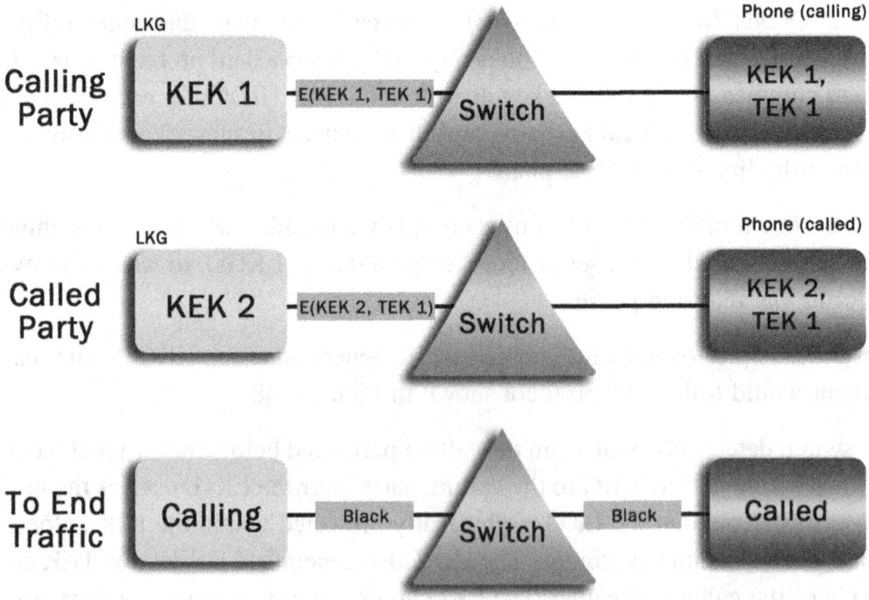

Figure 4.48:
Secure call establishment.

Once it receives the TEK, the calling party initiates a new cryptographic synchronization with the LKG using the TEK. The switch then sends the dial tone through the LKG. The caller in turns sends the called party's number securely. Once the switch obtains the dial digits, the LKG is dropped from the connection and made available for other calls.

The switch rings the called party, and when the called subscriber goes off-hook, the LKG assigned to the called party is first given the same TEK (session key) that was originally generated by the KDC for the calling party. Once the TEK is sent to the called party securely via the LKG, the LKG is dropped and an end-to-end connection is established between the calling and called parties. The calling party initiates traffic, secured with the TEK, by sending the cryptographic synchronization sequence.

A new random session key is generated for each call. After each call, the session key is destroyed at the KDC and in the end instruments. Exhaustive key attacks are infeasible.

This solution is secure but expensive and not well scalable for millions of subscribers, since each KDC would have to store the KEKs of all its subscribers. The interface between the KDC and the switch has a red component, namely plaintext to and from LKGs. Thus, each KDC has to be collocated with the switch in a secure facility.

Cryptographic systems today, in general, do not protect signaling information (addresses, telephone numbers, signaling protocols). The Internet does not protect addressing information. Consequently, active, intelligent eavesdroppers can create havoc with this vulnerability.

4.15 Cryptographic Certifications

Due to its importance in practically every facet of security technology, cryptographic implementation is one of the most heavily regulated components in the computing world. While this section does not attempt to cover all cryptographic-related certifications (for example, the financial and payments industries perform some types of device-specific approvals), we briefly discuss the ubiquitous FIPS 140 certification as well as NSA's U.S. government cryptographic certification and approval processes. We discuss when developers may be affected by these certifications, provide a general overview, and point readers to other sources of useful information and guidance.

4.15.1 FIPS 140-2 Certification

4.15.1.1 FIPS 140-2 Standard

FIPS publication 140-2 specifies security requirements (see summary in Table 4.6) for cryptographic hardware and software implementations.[29] FIPS 140-2 certification acts as a benchmark for base-level commercial quality and is mandated by certain organizations, such as the U.S. government for all its sensitive but unclassified communications (certification of

Table 4.6: FIPS 140 Requirements overview.

Requirement Category	Description
Cryptographic Module Specification	Documentation explains cryptographic boundary, algorithms used, key establishment methods, hardware/software/firmware architecture
Cryptographic Module Interfaces	Documentation explains all cryptographic interfaces and data paths
Roles, Authentication, and Services	Documentation of cryptographic system roles (software and users); role, identity, and multi-factor authentication requirements based on security level
Software/Firmware Security	Integrity protection requirements based on security level
Operational Environment	Protection and access control of cryptographic boundary within system
Physical Security	Invasive and non-invasive tamper evidence, detection, and resistance, depending on security level
Security Parameter Management	Random number and key generation, zeroization, handling, storage, I/O
Self-Tests	Pre-operational and runtime integrity protection and algorithmic validation tests
Life-Cycle Assurance	Configuration management, specification, design, and testing rigor depending on security level

classified communications systems is handled by the NSA). Because many embedded systems target government end customers, the use of FIPS 140-certified software and/or hardware is often a critical requirement for embedded systems designers. In addition, many commercial areas, such as the financial community, have adopted FIPS 140-2 as a requirement.

In some cases, using a, FIPS 140-certified software library or add-in network card within a design may not be sufficient. Government policy states that any product that implements cryptography must be FIPS 140 certified. Thus, equipment such as a network security appliance must itself be FIPS 140 certified.

It is important to note that FIPS 140 primarily specifies functional requirements and very little in the way of assurance requirements; correspondingly, validation is geared toward black box functional testing. More rigorous assurance requirements, such as source-code-level penetration testing, are required neither by the standard nor NIST's validation program. A salient example of the relative lack of assurance rigor is found in the following functional requirement (section 4.6.1 of the FIPS 140-2 standard):

> *For Security Level 1 only, the cryptographic module shall prevent access by other processes to plaintext private and secret keys, CSPs, and intermediate key generation values during the time the cryptographic module is executing/operational.*[30]

This requirement would seem to imply that a cryptographic software implementation would necessarily be instantiated as a separate process, independent and memory protected from any other processes in the system, using a message-based API to invoke cryptographic services. Yet practically all FIPS 140-2 validated software modules are implemented as dynamic link libraries (DLLs) that, by definition, share the same memory space as the applications that use them.

FIPS 140-2 specifies four qualitative levels, one through four, with relatively increasing functional requirements. Level 1 covers non-physical requirements, such as validation that algorithm implementations meet their specifications. Level 2 adds tamper detection requirements. Level 3 requires tamper resistance. Level 4 extends the tamper protection requirements but also adds protection against environmental anomalies such as voltage or temperature fluctuations. FIPS 140-2 does not, however, explicitly require protection against certain sophisticated attacks, such as those based on DPA, timing, and TEMPEST.

4.15.1.2 FIPS 140-3

FIPS 140-3 is expected to eventually supersede FIPS 140-2. Ratification of the new standard is anticipated no earlier than 2012. A short grace period will follow wherein FIPS 140-2 will still be accepted by the U.S. government. Embedded systems designers planning to field cryptographic products should be aware of FIPS 140-3 and plan to comply with the new standard. A good place to start is the current draft of the standard that can be found on NIST's public draft's website.[31]

FIPS 140-3 compliance will have limited impact on embedded systems designers targeting FIPS 140 Levels 1 and 2. One of the main drivers of FIPS 140-3 is to add explicit requirements for some of the aforementioned sophisticated attacks, which kick in at Levels 3 and 4. Specific types of attacks (including DPA) are called out in what is currently labeled Annex F of the draft standard.[32] However, the standard remains non-prescriptive with respect to how to defeat these attacks. The validation program will most likely rely on approved testing laboratories, such as SAIC, to apply well-known attack vectors and validate the operation of well-known countermeasures. For embedded systems developers who wish to learn more about FIPS 140 certification (either 140-2 or 140-3), good sources of information as well as hands-on training are provided by companies such as SAIC, which has an FIPS consultancy business independent (firewalled) from its government-testing laboratory.

Developers of security appliances that may benefit from the marketing value of an FIPS 140 certification are advised to take a training class, provided by SAIC and other FIPS certification experts, that will introduce designers to the general process of FIPS certification. In general, the documentary and testing requirements for a FIPS certification, assuming that the product already meets what are a reasonable and unsurprising set of functional requirements at any of the FIPS security levels, are not extreme. In fact, an FIPS consultancy company can provide templates for policy and user guides that will get you most of the way there.

Perhaps the most painful part of the FIPS certification process is the length of time—nine months to a year or sometimes longer—it takes to obtain the certification once the product has been submitted for evaluation. However, some government organizations may accept a product that is under evaluation. When an organization is working with an experienced FIPS consultancy, the probability of a product not passing an evaluation when the consultancy has been deemed the product ready for submission is close to zero.

4.15.2 NSA Certification

4.15.2.1 Cryptographic Product Classification

Information can be protected in a variety of ways, including physical protection by armed guards and structures such as buildings, vaults, or safety deposit boxes. Information, at rest or in transit, protected by cryptographic means is usually implemented by electronic hardware and/or software, in accordance with specifications developed by two U.S. agencies. The National Security Agency (NSA) specifies the design, documentation, assurance, and certification requirements for U.S.-classified or sensitive-but-unclassified information. The National Institute of Standards and Technology (NIST) specifies the requirements for private or commercial information.

U.S. government information is protected through the development of cryptographic elements or devices defined as

- ***Type 1:*** Cryptographic element or product that protects Top Secret (TS) or Secret (S) information.
- ***Type 2:*** Cryptographic element or product that protects sensitive-but-unclassified (SBU) information

Type 1 and Type 2 cryptographic devices use Type 1 or Type 2 cryptographic algorithms that have historically been specified by the NSA and are often classified. However, as we discuss further later in this chapter, this line has become quite blurred since the NSA launched its recent program to allow for some Type 1 classified devices to use layered NIST algorithms.

Unclassified information is governed by NIST under two additional categories:

- ***Type 3:*** Cryptographic element that uses NIST specified algorithms, including symmetric and hash algorithms. AES, DES, triple DES, Skipjack, and SHA are examples of NIST-specified algorithms.
- ***Type 4:*** Cryptographic algorithms that are registered with but not specified by NIST. Type 4 used to refer to exportable algorithms; however, this is no longer the case, and hence Type 4 has little applicability in modern embedded systems.

Cryptographic elements or products that want to obtain the NIST "seal of approval" must use Type 3 algorithms and undergo an FIPS 140 certification process. While commercial cryptographic elements do not necessarily require FIPS 140 certification, cryptographic elements for sale to U.S. government agencies require FIPS-140 certification.

Type 1 and Type 2 cryptographic devices are specified, tested, and certified by the NSA. The functional and security requirements for Type 1 and Type 2 devices are described in a set of classified and unclassified documents. An important governing document for the design, testing, and verification of Type 1 products is the Unified INFOSEC Criteria (UIC), a classified Secret document. Other documents describing cryptographic and associated protocols are classified For Official Use Only (FOUO).

The Type 1 specifications cover two categories:

- ***Top Secret (TS):*** Process that must be followed for securing TS information is covered in the sections for High Grade/High assurance of the UIC.
- ***Secret (S):*** Process that must be followed for securing S information is covered in the sections of High Grade/Medium Assurance of the UIC.

An embedded systems developer who designs Type 1 devices can choose to develop a single Type 1 device that covers both TS and S, or two separate devices, one for TS and one for S. A device designed to TS standards can secure TS, S, or unclassified information. A device designed to S standards can secure Secret or unclassified information (Secret-and-below or SABI). Type 3 and Type 4 devices (FIPS 140 certified or NIST registered) have historically not been allowed to secure U.S. DoD classified information (TS, S).

While Type 1 devices generally contain cryptographic algorithms that are classified, NSA approves the use of Type 3 algorithms (AES for example) for Type 1 applications. One example is allied interoperability where the NSA specified use of the AES suite of algorithms for use in Type 1 devices for use with U.S. allies. It is important to note, however, that while NSA allows the use of an unclassified algorithm suite (AES) for classified information, the device must still meet the NSA-specified Type 1 requirements stated in the UIC.

In fact, Type 1 devices can contain multiple algorithm suites—for example, one set of classified algorithms for U.S. only use and another set of unclassified algorithms for allied use.

4.15.2.2 Cryptographic Requirements for Type 1 Devices

The NSA specifies the Type 1 design and assurance requirements of a cryptographic device, using the generalized model shown in Figure 4.49. NIST follows a similar process for FIPS 140 certification.

In NSA terminology, the cryptographic element is often referred to as the End Crypto Unit (ECU), not to be confused with the automotive industry's Electronic Control Unit or Engine

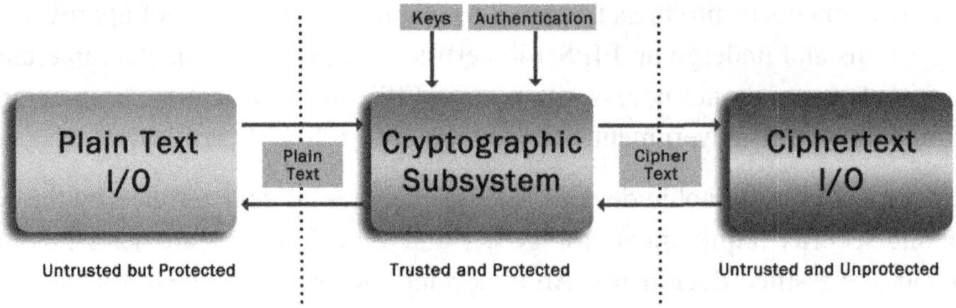

Figure 4.49:
Cryptographic security boundaries.

Control Unit. The ECU is the cryptographic element or subsystem that is designed, documented, and tested in accordance with the UIC. The ECU is the only "trusted" element in this model. All the security requirements in this model reside in the depicted security boundary.

NSA UIC requires ECU implementations that are standalone, physically separate units or boxes to meet advanced security requirements such as anti-tamper or anti-TEMPEST (electronic extraction of classified information). The model does not, however, preclude embedding the ECU into another subsystem or another embedded device, provided the security boundary is maintained as shown in Figure 4.49.

When an ECU is embedded into an embedded system, the security-critical interfaces may need to change. For example, if private key plaintext enters the embedded system from an I/O interface prior to loading within the ECU, then the portions of the embedded system that process the plaintext and pass it to the ECU must now be included in an expanded security boundary. Imagine that a Linux kernel manages the I/O interface and pathway to the ECU; suddenly, hundreds of thousands of open source code lines are brought into the security boundary! This could convert a simple certification process into an insurmountable one.

Ignoring this additional surface area could potentially result in a key compromise, since we do not know what else the untrusted software is doing with the key. Once a key is compromised, the system is compromised. What is even worse, we do not even know when the compromised occurred; it could have been on the first day of system use.

For any cryptographic application (classified or unclassified), the security boundary must be carefully described and drawn. Security boundary "fuzziness" must be avoided because it will ultimately lead to a design of a system that is **not** secure.

While the NSA and NIST specify the design of an ECU using the model shown previously, they do not guarantee that a system using the ECU is secure. For example, suppose we develop a secure packet network to handle classified information. Use of an NSA-certified ECU to protect the information flowing across the packet network is necessary but not sufficient to guarantee the security of the network itself. Rather, we require a security system certification. Functional and assurance security requirements must be specified and met at the system level in accordance with a system security policy. This policy includes key management and other aspects of system security that fall beyond the scope of the ECU itself.

Unfortunately, systems are usually specified for general functionality, and the secure elements are bolted on later. This usually results in vulnerabilities. One obvious example is the Internet. When the Internet was first conceived, specified, and implemented, security requirements were not formulated. Now that we use the Internet for many security-critical services, including financial transactions of all types, security is being bolted on. However, the Internet suffers from tremendous vulnerabilities in practically every aspect, from routing and key management to many of the core Internet communication protocols. It is safe to say that the Internet will never really be secure regardless of how many security features are retrofitted. There are just too many untrusted entities, and the security boundary of the Internet is "fuzzy" at best.

4.15.2.3 NSA's Cryptographic Interoperability Strategy

The sustained and rapid advance of information technology in the twenty-first century has provided motivation for the NSA to develop a new, adaptable cryptographic strategy for protecting national security information. NSA hopes that through its interoperability strategy, developers will make available strong cryptographic solutions that are easily adopted and shared between warfighters, first responders, and federal and international partners.

Chapter 5 discusses Suite B and its impact on common network security protocols such as SSL and IPsec. The promulgation of Suite B is part of a recent NSA initiative to use commercial-off-the-shelf cryptographic hardware and software for information classified at Secret-and-below (SAB), including sensitive but unclassified. In addition, the NSA has created new categories of cryptographic certifications to augment the traditional Type-1 certification discussed earlier. By varying requirements according to the value of information and its threat environment, developers and certifiers are able to meet security objectives while controlling cost and improving time to market. The certification categories for SAB are discussed next.

4.15.2.3.1 OTS for SECRET

Federal contractors build government-off-the-shelf (GOTS) products specifically for the U.S. government. GOTS-for-secret certification is intended to protect data up to Secret using the Suite B algorithm set. For these systems, the NSA has revised the set of security and test requirements and has minimized certification deliverables, resulting in decreased time to market.

Since Suite B algorithms are unclassified, a GOTS-for-secret device is not considered a Controlled COMSEC Item (CCI). CCI devices have rigorous handling requirements, including specialized accounting controls and markings. By avoiding CCI, total cost of ownership is significantly reduced.

4.15.2.3.2 COTS for SECRET

Commercial-off-the-shelf (COTS) refers to products created for the general market, such as a FIPS 140-2 certified VPN concentrator. COTS-for-secret allows, through the NSA-sponsored Commercial Solutions Partnership Program (CSPP), vendors to propose solutions consisting of the composition of multiple COTS products. The COTS-for-secret program is also known as Commercial Solutions for Secret (CSfC). COTS-for-secret stipulates certification to FIPS 140-2 and select U.S. government Common Criteria protection profiles. The Common Criteria assurance levels used are minimal since their purpose in this context is to ensure interoperability.

A major advantage of COTS-for-secret over GOTS-for-secret is that the equipment itself, prior to the loading of any secret keys, is unclassified even though it may be used for classified information protection. The use of COTS hardware and software eases the logistical burden for developers and opens up the classified communications market to more suppliers.

Nevertheless, the certification still requires a rigorous design and documentation process. Usually, the approval process sequence for COTS-for-secret certification is as follows:

1. For various applications, such as Wireless LAN, VPN, and disk encryptors, the NSA publishes a set of specifications describing the solution architecture that should be followed by the vendor.
2. The vendor develops a composed solution from the published architecture and presents the solution to the NSA for review and approval.
3. The NSA approves the proposed implementation.

It is important to note that NSA approval is quite different from the traditional NSA certification. While approval indicates that the developer has followed NSA guidelines, final responsibility for use of the device is left to the Designated Approving Authority (DAA) of the end-user organization, such as a military or intelligence service.

As we discuss in Chapter 5, the NSA has been working with both NIST and IETF to ensure that industry standards and protocols include the Suite B algorithms. In addition, the NSA has undertaken many infrastructure initiatives that will incorporate Suite B, including

1. DoD Public Key Infrastructure (PKI) project to move from legacy RSA to Suite B-preferred elliptic curve certificates by 2012
2. Key Management Infrastructure (KMI): an NSA program to provide key management services for Type 1 and SAB products
3. Suite B-compliant commercial PKI for sensitive but unclassified agencies

The NSA has also licensed many technologies and makes them freely available to vendors building Type 1 or Secret-and-below products. For example, the NSA licensed the rights to 26 elliptic curve technology patents and made them freely available to industry developing products for the U.S. government.

The recent major developments in NSA cryptographic strategy will enhance secure information sharing to many enterprises including DoD networks, federal and state agencies, and commercial-enterprise networks. In addition, the NSA also hopes that by developing Suite B with its associated scalability for high-speed data and efficient key sizes critical for wireless systems, legacy algorithms such as DES, 3DES, and SHA-1 that perform poorly or exhibit weak security will be discarded.

4.16 Key Points

1. It behooves embedded systems developers to understand NSA cryptographic guidance, even though NSA certification is not required for most embedded systems.
2. NSA Suite B includes a suite of unclassified cryptographic algorithms providing a variety of cryptographic services required in modern cryptography, including encryption, hashing, digital signatures, and key exchanges.
3. Because of its proven, unconditional, and simple security, the one-time pad is often considered the basis of all good encryption systems.
4. The lifetime of a key's use within a cryptosystem is called the key's cryptoperiod.
5. For the purposes of determining cryptographic strength, proprietary or secret algorithms are always assumed broken (fully understood and reproducible by an attacker) either through reverse engineering or other means.
6. Cryptographic modes enable the application of encryption to specific use cases that have varying requirements: some modes enable simultaneous origin authentication and data protection; some modes have been created simply for improved execution time efficiency; some modes propagate bit errors and thus thwart spoofing attacks; other modes sacrifice anti-spoofing in exchange for bit error tolerance.
7. A desirable trait of secure communications algorithms and protocols, traffic flow security (TFS) aims to prevent an eavesdropper from deducing private information from patterns in communication rather than the communicated encrypted data itself.
8. Block ciphers must be designed such that knowledge of large amounts of matched input and output blocks is insufficient to determine the key other than by exhaustive search of all possible keys for the applicable key length.
9. AES is the preferred block cipher in NSA Suite B.
10. For key encryption (otherwise known as key wrapping), variable-width ECB is the preferred encryption mode.

11. Authenticated encryption can be more computationally efficient than executing independent encryption and authentication algorithms.
12. Galois Counter Mode (GCM) is the Suite B-recommended authenticated encryption cipher.
13. Unfortunately, in most realistic uses of public key cryptography, the key distribution problem is no simpler than the symmetric key-sharing problem due to the need to distribute and perform a trusted installation of a common authority certificate.
14. The cryptography community traditionally estimates the cryptographic strength corresponding to key lengths of public key and other algorithms as a measure of bits of strength in a symmetric key algorithm, such as AES.
15. NIST special publication 800-57 provides estimates on the length of keys for particular algorithms based on the length of time the data needs to be protected.
16. Because of its computational inefficiency relative to symmetric key, public key cryptography is generally not used for encryption and decryption, particularly for the bulk processing of data, such as in a VPN session or full-disk encryption device. Public key cryptography, however, is preferred for digital signatures (providing origin authentication) as well as for the secure distribution of ephemeral symmetric keys.
17. Diffie-Hellman is an important algorithm because it enables perfect forward secrecy to be achieved by network security protocols: session keys cannot be compromised even if other session keys or one of the long-term private keys used for origin authentication is compromised.
18. Digital signature functionality must be added to a protocol that employs Diffie-Hellman to use its key agreement for secure peer-to-peer communications.
19. NSA Suite B specifies the use of X.509 v3 certificates using elliptic curve cryptography (ECC)-based digital signatures.
20. In some cases, the embedded systems designer may benefit from a cryptographic vendor who can provide custom, resource-optimized certificates.
21. While the performance difference between RSA and Elliptic Curve algorithms with their respective contemporary key sizes is not dramatic, ECC provides better security strength efficiency per bit and therefore is deemed by the NSA and most cryptographers as the superior long-term choice.
22. While some implementation options for ECC are covered by patents, all major cryptographic toolkit vendors (as well as the open source OpenSSL developers) have created ECC implementations that follow public specifications and are believed to be free from patent risk.
23. For embedded systems that operate on networks using legacy RSA-based PKI, ECC cannot be practically used until these infrastructures are upgraded to use ECC.
24. Of critical import to embedded systems designers is the requirement that they use an embedded processor with a true random number generator (TRNG).

25. Pseudo-random number generators (PRNGs), seeded by true random input, are required for applications that must generate cryptographic random material at a rate higher than can be created by a system's local TRNG.
26. A PRNG that is not properly seeded with true random input should never be used for the creation of private keys.
27. A safe choice of PRNG is an FIPS-approved algorithm, such as those published in NIST Special Publication 800-90.
28. Ideally, private keys are born within and never leave the embedded system in which they are used.
29. It is never a good idea or practice to use the same key for different functions (such as data encryption and digital signatures), since a key compromise (e.g., due to a weakness in a specific function or application of the key) will affect all the functions that use the same key.
30. Robust and well-designed systems should change keys as often as practical to minimize the window of vulnerability of a compromised key.
31. The protection of keys, both from a physical as well as logical/software perspective, is one of the most important security features within an embedded system that employs cryptography, yet it is all too often ignored or poorly addressed.
32. Private keys stored in an embedded system's primary non-volatile memory (e.g., flash memory or hard disk) should be encrypted and decrypted with a plaintext KEK that is ideally held in non-volatile storage external to the main applications processor and protected against physical attacks.
33. For embedded systems that are lost, stolen, or are known to have their cryptographic subsystems compromised, provision must be made to revoke all the keys associated with the compromised or lost unit.

4.17 Bibliography and Notes

1. Kahn D. *The Codebreakers: History of Secret Communication from Ancient Times to the Internet.* New York, NY: Scribner; 1967.
2. Kahn D. *Seizing the Enigma: The Race to Break the German U-Boats Codes, 1939–1943.* Barnes and Noble; 1991.
3. Deavours CA, et al. *Cryptology: Machines, History, and Methods.* Artech House, Inc.; 1989.
4. Fact Sheet No. 1 for the National Policy on the Use of the Advanced Encryption Standard (AES) to Protect National Security Systems and National Security Information (CNSSP-15) (June 2003).
5. NSA Suite B Cryptography. www.nsa.gov/ia/programs/suiteb_cryptography
6. Gilbert VS. Secret Signaling System. U.S. Patent 1310719, issued July 22, 1919.
7. Claude S. Communication Theory of Secrecy Systems. *Bell System Technical Journal* **28**(4): 656–715. 1949.
8. Song D, et al. Timing Analysis of Keystrokes and Timing Attacks on SSH. *Proceedings of the 10th Conference on USENIX Security Symposium* 2001;vol. 10.
9. AES Key Wrap Specification. http://csrc.nist.gov/groups/ST/toolkit/documents/kms/AES_key_wrap.pdf
10. Counter with CBC-MAC (CCM). Network Working Group, Request for Comments: 3610 (September 2003).

11. NIST Special Publication 800-38C. Recommendation for Block Cipher Modes of Operation: the CCM Mode for Authentication and Confidentiality (May 2004).
12. Vinodh G, et al. Optimized Galois-Counter-Mode Implementation on Intel Architecture Processors (April 2010).
13. Kleinjung, Thorsten, et al. Factorization of a 768-Bit RSA Modulus, version 1.4 (February 18, 2010), http://eprint.iacr.org/2010/006.pdf
14. NIST Special Publication 800-57. Recommendation for Key Management—Part 1: General (Revised) (March 2007).
15. Internet X.509 Public Key Infrastructure Certificate and Certificate Revocation List (CRL) Profile, Internet Engineering Task Force, Request for Comments: 5280 (May 2008).
16. Suite B Certificate and Certificate Revocation List (CRL) Profile, Internet Engineering Task Force, Request for Comments: 5759 January 2010.
17. SEC 4: Elliptic Curve Qu-Vanstone Implicit Certificate Scheme (ECQV), Version 0.97, Certicom Research (March 2011).
18. Xiaoyun W, Lisa Y, Hongbo Y. Finding Collisions in the Full SHA-1 (February 2005). Proceedings of Crypto Core.
19. Xiaoyun W, Andrew CY, Frances Y. Cryptanalysis on SHA-1, Presented by Adi Shamir at the Rump Session of CRYPTO 2005.
20. Secure Hash Standard. Federal Information Processing Publication 180-2 (August 1, 2002).
21. HMAC: Keyed-Hashing for Message Authentication, Internet Engineering Task Force, Request for Comments: 2104 (February 1997).
22. NIST FIPS PUB 198. The Keyed-Hash Message Authentication Code (HMAC) (2002), http://csrc.nist.gov/publications/fips/fips198/fips-198a.pdf
23. NIST Special Publication 800-22—Revision 1a. A Statistical Test Suite for Random and Pseudorandom Number Generators for Cryptographic Applications (April 2010).
24. Annex C: Approved Random Number Generators for FIPS PUB 140-2, Security Requirements for Cryptographic Modules (DRAFT) (June 14, 2011).
25. NIST Special Publication 800-90A—Revision 1. Recommendation for Random Number Generation Using Deterministic Bit Generators (Revised) (May 2011).
26. ANSI X9.82-3-2007. Random Number Generation—Part 3: Deterministic Random Bit Generators (September 2007).
27. Oswald D, Paar C. Breaking Mifare DESFire MF3ICD40: Power Analysis and Templates in the Real World. *CHES 2011, Nara*; September 30, 2011.
28. International Telecommunications Union. Telecommunication Standardization Sector, G.729: Coding of Speed at 8 kbit/s Using Conjugate-Structure Algebraic-Code-Excited Linear Prediction (CS-ACELP) (January 2007).
29. Federal Information Processing Standards Publication 140-2: Security Requirements for Cryptographic Modules, National Institute of Standards and Technology (May 2001).
30. Federal Information Processing Standards Publication 140-2: Security Requirements for Cryptographic Modules, (January 11, 1994).
31. Federal Information Processing Standards Publication 140-3: Security Requirements for Cryptographic Modules (Revised DRAFT) (September 11, 2009).
32. Annex F: Non-Invasive Attack Methods for FIPS PUB 140-3 (Draft), Security Requirements for Cryptographic Modules, NIST Computer Security Division (September 10, 2009).

CHAPTER 5

Data Protection Protocols for Embedded Systems

Chapter Outline

Embedded Systems Security. DOI: 10.1016/B978-0-12-386886-2.00005-9

5.1 Introduction

Many modern embedded systems have data-in-motion (network security) or data-at-rest (encrypted storage) protection requirements, driven by embedded intellectual property protection, secure remote management, digital rights management, financial transactions in Internet-connected devices, and more. This chapter focuses on the policies of data confidentiality, integrity, and origin authentication, since they are common to almost all data protection protocols. Notably absent from this discussion is a data access control policy, which tends to be application- or media-type specific and has less general applicability from a protocol perspective.

This chapter provides an overview of the latest and most important network security protocols such as TLS v1.2, DTLS, and IKE v2. In addition, the chapter discusses the far less standardized approaches to protecting data-at-rest, such as the use of full disk encryption (FDE). The chapter demonstrates how to make practical use of these tools in embedded systems and provides useful advice regarding performance, footprint, and certification standards such as NSA Suite B. This chapter is intended for embedded systems engineers (software, hardware, systems) who either have minimal experience in using data protection protocols or have a general understanding but lack practical experience integrating these features into embedded systems. The material is also useful for more experienced network security engineers who want to brush up on the latest standards revisions and expert guidance.

This chapter assumes basic understanding of cryptography, including symmetric encryption algorithms, public key cryptography, and key management concepts. These topics are covered in Chapter 4.

5.2 Data-in-Motion Protocols

Data-in-motion (also referred to as data-in-transit) security protocols are involved in the establishment and use of protected channels over which a pair (point-to-point) or set (point-to-multipoint or multipoint-to-multipoint) of entities can communicate securely. We concern ourselves primarily with Internet Protocol (IP)-based network security.

5.2.1 Generalized Model

5.2.1.1 Point-to-Point

Let's assume Alice wants to establish a protected channel with Bob. At a high level, Alice and Bob both employ a control program to initially establish the channel and a data program to subsequently send bulk information across the established channel (see Figure 5.1). When Alice and Bob are finished with their communication session, the channel is torn down (this cleanup is considered part of the control program).

To drill down a bit further, let's assume that public key cryptography is used for session establishment, and that prior to any sessions, Alice already holds Bob's validated long-term (static) public key, Bob holds Alice's validated long-term (static) public key, and they want to

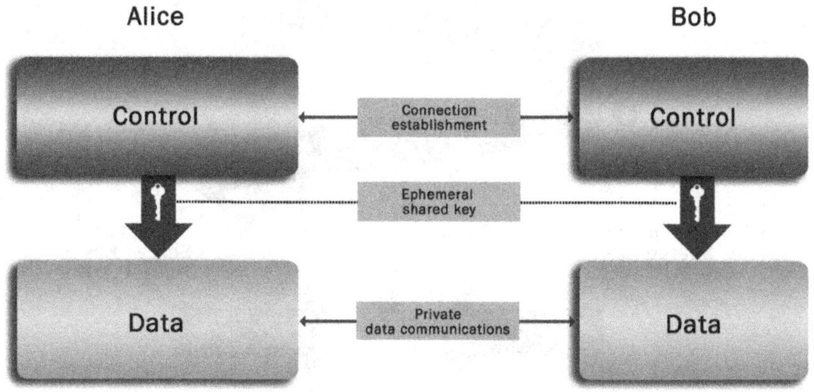

Figure 5.1:
High-level generalized model of point-to-point secure communication subsystems.

communicate only with each other (they will never obtain any other public keys). By "validated," we mean that Alice and Bob share a common certification authority (CA) whose pre-installed certificate is used to validate each other's public keys in advance. For example, Alice and Bob may receive each other's public key certificates from a trusted certificate server on the network. Note that in some embedded systems, the communication peers' public keys may be pre-installed at the factory in a trusted manner. In this case, Alice and Bob need not validate each other's public keys at all, and no shared common certificate authority public key need be pre-installed.

Figure 5.2 shows a generalized authentication and key agreement sequence used to establish a channel between Alice and Bob for secure bulk data transmission.

In step one, Alice creates a random, unpredictable ephemeral public key pair (e.g., RSA or elliptic curve) and sends the public part, P_A, to Bob. Alice keeps the other part, S_A, private. Bob receives P_A and also creates a random ephemeral public key pair (S_B, P_B). In step two, Bob sends P_B and the digital signature of a concatenation of the two exchanged public keys, P_A and P_B. Bob creates this signature, referred to as $Sign_B$ in Figure 5.2, using his static private key (remember that Alice already holds the matched public portion). The signature algorithm corresponds to the types of static public keys used (e.g., RSA or ECDSA). Alice receives this message and validates Bob's signature using Bob's static public key. In step three, Alice signs a concatenation of P_B and P_A and sends the result to Bob. Bob receives and validates the signature using Alice's static public key.

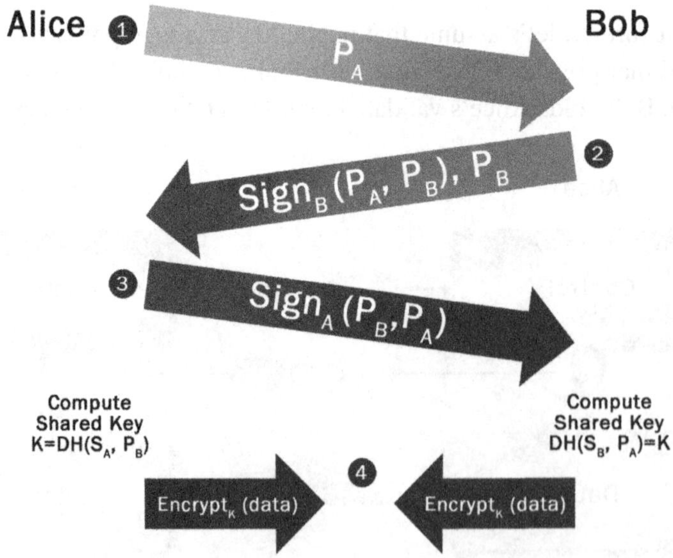

Figure 5.2:
Generalized point-to-point secure data-in-motion protocol.

We're almost done with the session establishment. Alice and Bob have exchanged ephemeral public keys and can therefore use the elegant and powerful Diffie-Hellman algorithm (refer to Chapter 4) to create a shared secret computed from their retained ephemeral private key and the received public key. The specific form of Diffie-Hellman depends on the type of ephemeral keys used, e.g., standard Diffie-Hellman or Elliptic Curve Diffie-Hellman (ECDH) when using elliptic curve ephemeral keys. Because it is computed from private keys held only by Alice and Bob, the shared secret can now be used as a symmetric key for bulk-authenticated encryption (e.g., AES-GCM). Alice and Bob can communicate confidently using this key, knowing that no one other than Alice and Bob can possibly read the data.

Let's look at the common goals of a data-in-motion protocol and how the preceding sequence satisfies each:

Mutual authentication: Bob's signature on P_A, which he sends in step two, provides assurance to Alice that she is indeed communicating with her intended recipient, Bob. Alice's signature on P_B in step three provides the same assurance for Bob. Thus, Bob and Alice have mutually authenticated.

Liveness: Because Alice's P_A is unpredictable (random), the signature of P_A returned by Bob demonstrates to Alice that Bob is actively participating in the conversation (not a replay of an earlier communication). Similarly, Alice's signature of P_B proves Alice's liveness to Bob.

Key Agreement: Bob and Alice have computed the same fresh, private key by which they can subsequently securely communicate. A corollary to key agreement is *key confirmation* in which Alice and Bob actually have evidence that the other party has computed the shared secret key. In our generalized scenario, confirmation is not achieved until bulk data is sent and then successfully received (authenticated and decrypted). Because the whole point of data-in-motion protocols is to use the secret key for communications, an explicit key confirmation step during the establishment process is not strictly necessary.

Key Point

Identity anonymity (part of transmission security) is often a desired goal of data-in-motion protocols.

In the preceding sequence, the key exchange messages and their corresponding signatures are sent in the clear. An eavesdropper who holds a copy of Alice's and Bob's public keys (we usually must assume this is possible since the keys are public) can validate the messages and determine that Alice and Bob are communicating. Identity anonymity can easily be added to the preceding sequence. Because he has Alice's ephemeral public key, Bob can compute the shared secret prior to sending the return message in step two. Therefore, Bob encrypts the signature in this message, thwarting the eavesdropper. Similarly, Alice can compute the shared secret prior to sending the message in step three. She too encrypts the signature.

In practice, depending on the underlying transport medium (e.g., Internet Protocol on top of Ethernet), identity information may be deduced from those transport packets if the communicating identities can be mapped to their network locations. As we discuss later in this chapter in the example of IPsec Tunnel mode, addressing anonymity may need to be addressed independently.

The simplified model discussed so far assumed that Alice and Bob can only ever talk to each other. In practice, Alice and Bob are but two constituents in a larger network with more entities. In a closed network with a manageable number of entities (for example, a corporate intranet), each endpoint can be pre-installed with the list of potential communicating partners, and identity information must be exchanged in the protocol to inform Alice and Bob which public keys to use for message validation. In an open network with a virtually unlimited number of entities (e.g., the Internet), public key certificates are usually sent as part of the handshake. In other words, before Alice can talk to Bob, Alice will send not her simple identity but rather her entire public key certificate, and Bob will need to validate that signature during the establishment process. Similarly, Bob will respond with his certificate.

The improved sequence, including encryption of the signatures and transmission of certificates, is shown in Figure 5.3.

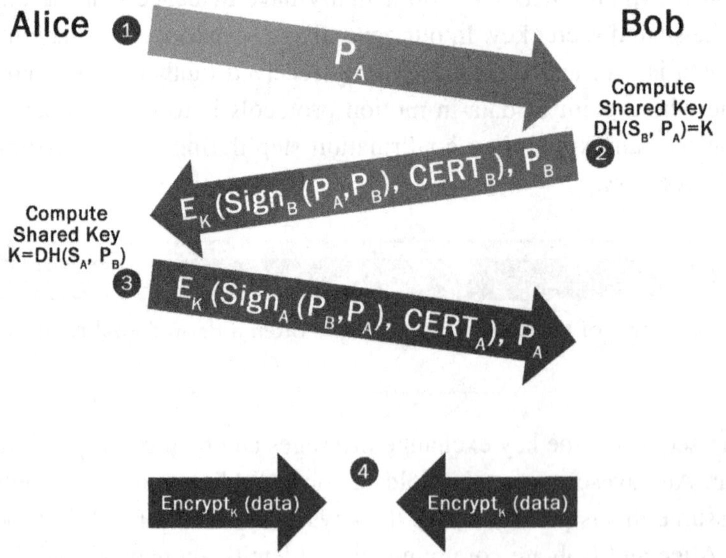

Figure 5.3:
Generalized point-to-point network security protocol with improved transmission security.

While other details are elided in this generalized model of data-in-motion point-to-point secure communication, it forms the basis of modern public key point-to-point network security protocols, such as IKE/IPsec, SSL, and DTLS, discussed later in this chapter.

For adventurous readers wishing to deep dive on the history, variety, and formal proofs of key establishment methods, we recommend the excellent book *Protocols for Authentication and Key Establishment*.[1]

5.2.1.2 Point-to-Multipoint

Suppose that Alice wants to communicate securely to a large number of receivers. Establishing a point-to-point session with each receiver may not be practical and would certainly be inefficient, requiring the same data to be sent over and over. An example of an embedded systems problem requiring point-to-multipoint is communicating a remote management change (configuration modification, firmware update) to a large number of deployed systems. When multicast communications are used (the data is sent once and the underlying medium handles getting the data to all receivers in the specified multicast group), the control portion becomes rather complicated. Instead of a simple key exchange protocol, a group key management scheme must be used. This scheme must handle periodic rekeying across the group.

> **Key Point**
>
> Multipoint applications force key management and exchange challenges to be decoupled from the bulk data protocols: the developer must find a way to distribute a shared key across the group and then multicast packets protected with that key.

Readers interested in learning more about multicast with IPsec should consult RFCs 5374 and 4046.[2,3] The IPsec discussion later in this chapter focuses only on the traditional point-to-point case.

Broadcast is a special case of point-to-multipoint in which communication is strictly one way (e.g., a satellite transmission to terrestrial receivers). Broadcast makes traditional IP-based network security protocols even more impractical due the inability to exchange any messages. Once again, a pre-shared key must be established between broadcaster and receivers, and then the data encrypted with the pre-shared key can be signed and encrypted prior to broadcast. This is how paid cable and satellite content is sent between service provider and a large number of homes. Broadcast approaches are discussed later in this chapter.

Conferences are the quintessential example of multipoint-to-multipoint: all participants need to communicate with all other participants. However, information is generated individually by each participant; therefore, this scenario is simply a special case of multiple concurrent point-to-multipoint connections.

5.2.2 Choosing the Network Layer for Security

As shown in Figure 5.4, the embedded systems designer can choose among the different layers in the communications stack for applying data protection protocols. Due to performance efficiency concerns, most systems will want to apply protection at a single level.

> **Key Point**
>
> Multi-layered data protection may be used as a defense-in-depth strategy and can also occur when upper layers are unaware or distrustful of lower-layer protection capabilities.

As expected, the practical protocol choices vary depending on the selected layer.

> **Key Point**
>
> Choice of network protection layer can have significant impact on functionality, performance, interoperability, and maintainability.

Understanding these trade-offs is an important goal of this chapter.

The lowest available choice for network security is the data link layer of the OSI model and/or the link layer of the TCP/IP model. Each data link type—Bluetooth, Ethernet, Modem, ATM,

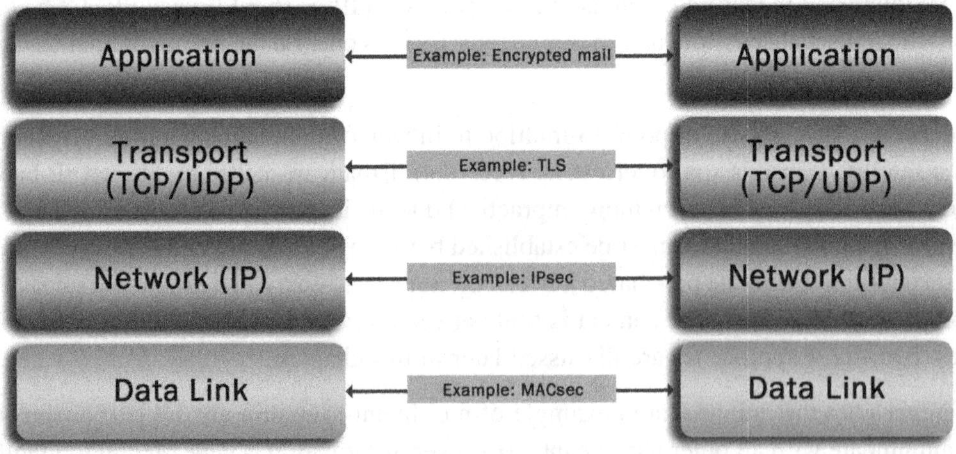

Figure 5.4:
Network security protection possibilities by layer.

IEEE 802.15.4, Frame Relay (just to name a few)—usually incorporates a distinct data protection approach tailored to its framing and control characteristics. Sometimes, data link encryption is built directly into the communications peripheral hardware.

> **Key Point**
>
> The efficient protocol handling and encryption in hardware and the lack of spare applications processing and storage resources force certain embedded systems to rely on data link layer protection rather than more flexible upper-layer choices.

A good example of this is an ultra-low-power wireless device such as a smart meter. Contemporary ZigBee-enabled devices rely almost exclusively on data link layer security to maintain optimal battery life and good network performance over low-rate radios.

Key management approaches may also differ widely based on layer. For example, data link layer protocols tend to use simplistic key management schemes. In some ZigBee networks, a single shared symmetric key is used for the entire network; periodic rekeying requires reflashing and may never be performed in practice; the key is factory installed into an administrative device on the network; and endpoint devices are sent the shared key over the air, in the clear. Those who have read Chapter 4 (or even those who have not), we hope, see lots of weaknesses with this key management scheme.

> **Key Point**
>
> The biggest drawback of link layer data protection is that, by definition, the scheme can work only on a network consisting of devices using the homogeneous data link media.

Therefore, a ZigBee-based smart meter is unable to obtain a direct, secured connection with the back-end power infrastructure that resides across the WAN. The only practical way to get an end-to-end secure connection between these entities is to go up to a higher level such as IPsec over 6LoWPAN (more on this later).

5.2.3 Ethernet Security Protocols

The remainder of this chapter focuses on the upper-level protocols, such as IPsec, due to their more general applicability. However, before we go there, one form of data link layer security bears mentioning: the key management and session establishment as well as the bulk encryption protocols, respectively, for Ethernet (both wired and wireless). Ethernet, especially wireless (Wi-Fi), is one of the most common data link types in embedded systems.

5.2.3.1 802.1X

Key Point

The IEEE 802.1X standard defines requirements for the secure joining of endpoints into Ethernet networks.

An 802.1X authenticator device acts as the gateway, rejecting all Ethernet traffic (except for the 802.1X connection requests) not originating from a properly authenticated endpoint. The 802.1X authenticator is often a relatively simple device, such as an Ethernet switch or Wi-Fi access point. The authentication decision logic is usually offloaded to an authentication server within the LAN, which may in turn take advantage of additional servers such as a user credential database. In 802.1X terminology, the requesting endpoint is called the *supplicant.*[4] The three-way supplicant, authenticator, authentication server architecture of 802.1X is shown in Figure 5.5.

A protocol framework for authentication, Extensible Authentication Protocol (EAP), is specified in 802.1X. However, specific authentication methods used on a particular network are not prescribed by 802.1X. EAP is defined in RFCs 3748 and 5247.[5,6] Authentication may be as simple as validating that the supplicant's Ethernet MAC address exists on a whitelist of allowed endpoints or as comprehensive as a dual scheme requiring valid username/password and mutual public key certificate-based authentication.

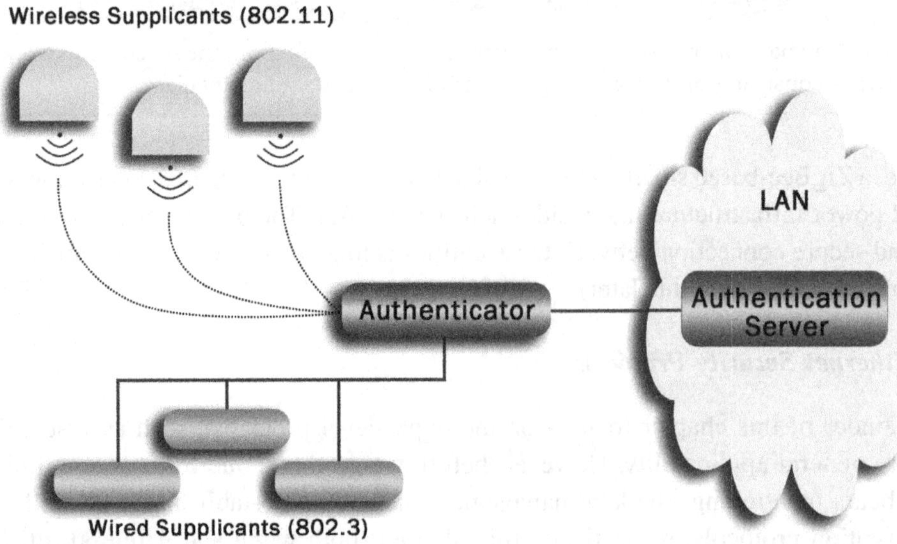

Figure 5.5:
IEEE 802.1X entities architecture.

More than 40 EAP methods have been defined over the years, and specific EAP methods are required to claim conformance to certain standards. For example, RFC 4017 defines the required EAP method requirements for IEEE 802.11 wireless networks.[7]

Plenty of detailed information about 802.1X is available in books and on the Internet. One suggested title is Jim Geier's *Implementing 802.1X Security Solutions for Wired and Wireless Networks.*[8]

5.2.3.2 802.11i

Key Point

The mapping of Extensible Authentication Protocol (EAP) requirements to wireless Ethernet LANs and other wireless Ethernet security requirements are defined in IEEE 802.11i-2004.

802.11i incorporates the 802.1X standard for authentication.[9] The generalized model for point-to-point connections presented earlier in this chapter, including strong mutual authentication, freshly agreed session keys, and liveness, are all covered by 802.11i. However, 802.11i data security is aimed at the link between the endpoint and the access point, not between the access point and the other end of the logical communication, behind the access point.

Key Point

The security functions specified in IEEE 802.11i protect against local eavesdroppers and hackers trying to gain access to a network via the access point but do not provide end-to-end protection.

Embedded systems designers must use a VPN or other higher-level protocol to connect the embedded system to a back-end network, even when 802.11i is in use. We must not trust the public networks (e.g., the Internet) that lie between the access point and the back-end systems. In fact, one could argue that a network that is strongly guarded with a VPN concentrator that enforces strong mutual authentication to all remote endpoints (that execute an appropriately configured VPN client) need not use data link security at all within those connections. This is a bit different from the use of mobile smartphones and laptops where the endpoint is used both for VPN connections as well as public Wi-Fi connections that both require data security.

At the time of this writing, 802.11i specifies the use of AES-CCM, an authenticated encryption mode discussed in Chapter 4, for data confidentiality and integrity protection. 802.11i's specification regarding how AES-CCM is applied to 802.11 Ethernet frames is called Counter Mode with Cipher Block Chaining Message Authentication Code Protocol (CCMP). It is interesting to note that AES-GCM, preferred by NSA Suite B in higher-level protocols, is not yet adopted in data link security protocols. At the time of this writing, the scope of NSA Suite B does not include data link security, implying a preference for IPsec and other higher-layer approaches. As wireless Ethernet speeds continue to increase, there is a good chance that AES-GCM will be added to IEEE 802.11 standards due to GCM's superior performance efficiency.

Data link security is generally more important in wireless than in wired LANs. Wireless networks are characterized by more dynamic topologies, where the network regularly is joined by endpoints that may not be well managed by the network's administrators. Wireless networks are more susceptible to eavesdropping and other man-in-the-middle attacks. Finally, wireless networks are simply replacing wired networks in most embedded applications due to their flexibility. The foundational standard for wireless Ethernet security, 802.11i is required reading for serious wireless Ethernet security implementers.

The Wi-Fi Alliance has been at the forefront of certification for security compliance and interoperability for wireless Ethernet (Wi-Fi) networks. The Wi-Fi Alliance bases its current certification testing on IEEE 802.11i.

5.2.3.3 WPA2

> **Key Point**
>
> Contrary to popular belief, Wi-Fi Protected Access 2 (WPA2) is not a definitive security standard, but rather a certification quasi-standard intended to provide industry with confidence that Wi-Fi-certified devices meet the requirements of IEEE 802.11i.

The Wi-Fi Alliance introduced WPA2 certification in 2004, concurrent with the release of IEEE 802.11i. WPA2 has been mandated for Wi-Fi-certified products since 2006 and replaces WEP and WPA, notoriously weaker technical approaches adopted by the Alliance certification program.

A good overview of the Wi-Fi Alliance's adoption of security standards culminating in today's WPA2 certification can be found on the Wi-Fi Alliance website.[10] The Wi-Fi Alliance's influence on wireless Ethernet security is so ubiquitous that most users of wireless Ethernet devices equate wireless Ethernet security with the WPA2 name rather than its constituent IEEE protocol standards.

> **Key Point**
>
> Embedded systems developers acquiring wireless Ethernet software for incorporation into devices should select stacks that not only claim conformance to WPA2, but have also have been successfully deployed in Wi-Fi-certified products.

Note that Wi-Fi Alliance certification is performed only on end products.

5.2.3.4 802.1AE

Key Point

Known as MACsec, the IEEE 802.1AE standard is the analog of IEEE 802.11i for wired Ethernet (IEEE 802.3) networks.[11]

Unlike 802.11i, which mandates use of 802.1X for joining networks, the scope of MACsec includes only the data plane, excluding network access authentication. However, draft standard IEEE P802.1af augmented 802.1X with the details necessary for application to 802.1AE. This draft has been incorporated into the next revision of 802.1X, 802.1X-2010, not yet finalized by IEEE at the time of this writing. The 802.1X standard will become the de facto wired Ethernet endpoint-to-switch authentication protocol since it enables back-end networks to implement a unified authentication server infrastructure (using RADIUS and EAP) for both wired and wireless endpoints.

The authors of 802.1AE were thoughtful enough about the rapidly increasing speeds of wired Ethernet (100 gigabit maximum as of the time of this writing, with some demand for 400 gigabit and even terabit[12]) to make AES-GCM, rather than AES-CCM, the mandatory default cipher for this standard.

In summary, data link protocols are appropriate for connecting embedded systems within homogeneous physical networks that require simple key management and efficient encryption processing. And data link security is better than no data-in-transit security if the embedded system is unable to sustain higher-level protocols such as IPsec. Ethernet security protocols are, of course, critical to developers of Ethernet switch and access point equipment.

Key Point

Developers looking for the most portable, flexible, and extensible data-in-motion security should look above the data link layer toward the network (IP) and higher layers.

5.2.4 IPsec versus SSL

The two most commonly used network security protocol families in embedded systems are IPsec and the Secure Sockets Layer (SSL). At a high level, these protocols fulfill the same security goals: confidentiality of transmitted data, integrity protection of transmitted data, and authentication of data origin.

The major difference between the two families is the level in the network stack at which they operate.

> **Key Point**
>
> IPsec operates at layer three (the Internet Protocol layer), making it invisible to both higher-level protocols (e.g., TCP) as well as applications.

> **Key Point**
>
> SSL and its modern variant, Transport Layer Security (TLS), operate on layer four, the transport layer; it is common for applications to use TLS by invoking a sockets-like API.

> **Key Point**
>
> A major advantage of TLS is that it can take advantage of TCP's guaranteed in-order packet delivery to simplify the protocol relative to IPsec that must handle dropped messages and out-of-order delivery.

> **Key Point**
>
> The use of TLS enables applications to select when they need the security (and are willing to pay the associated overhead).

For example, web browsers use TLS to secure web traffic. In contrast, once an IPsec connection is made, all applications and services running on that network connection will be forced to abide by the underlying policy.

> **Key Point**
>
> An important advantage of IPsec is that applications can derive security benefits without requiring any application-level decisions or code modifications (such as adjusting to use a modified-sockets API).

5.2.5 IPsec

5.2.5.1 Integrated versus Bump-in-the-Stack

IPsec is optional in IP version four (IPv4) and mandated in IP version six (IPv6). Most commercial network stacks, either third party or those built into operating systems, have IPsec already pre-integrated.

Some embedded software vendors supply a standalone IPsec that can run independently of an already-existing network communication subsystem; this architecture is called "bump-in-the-stack" (BITS) because the IPsec component is a separate layer ("bump") between the higher-level stack and the lower-level data link layer (e.g., Ethernet). Generally speaking, if an organization is already using a TCP/IP stack with integrated IPsec, then it should take advantage of that stack. Otherwise, using BITS may be a good choice due to simplicity, smaller footprint, and ease of integration, although it may not perform quite as well as a well-tuned stack implementation.

> **Key Point**
>
> An example of a good choice for bump-in-the-stack (BITS) IPsec is a virtualized environment where guest operating systems already have integrated network stacks, yet there is a desire to implement network security outside the virtual machine where flaws in guest operating systems could derail intended security.

The hypervisor can implement its own BITS IPsec layer (see Figure 5.6) below all the guest operating systems, even without the knowledge of the guest and its applications.

IPsec is often generalized to refer to the coordinated use of two separate protocols: IPsec and Internet Key Exchange (IKE). IPsec is concerned with the bulk flow of encrypted packets between network endpoints, whereas IKE deals with the thorny issues around key management and establishing the initial authenticated connection between nodes. IKE works with IPsec by establishing an ephemeral bulk encryption key (called a session key) for each new connection and then notifying the IPsec layer of the session keys to use.

5.2.5.2 IPsec RFCs

IPsec has been standardized for a long time with minor revisions. The original RFCs were published in 1998 and updated in 2005. RFC 4301 is the top-level IPsec standard, covering both IPv4 and IPv6.[13] While it can be somewhat painful, reading the RFCs mentioned throughout this chapter is highly recommended for developers who want to gain a reasonable understanding of the protocols instead of just using them.

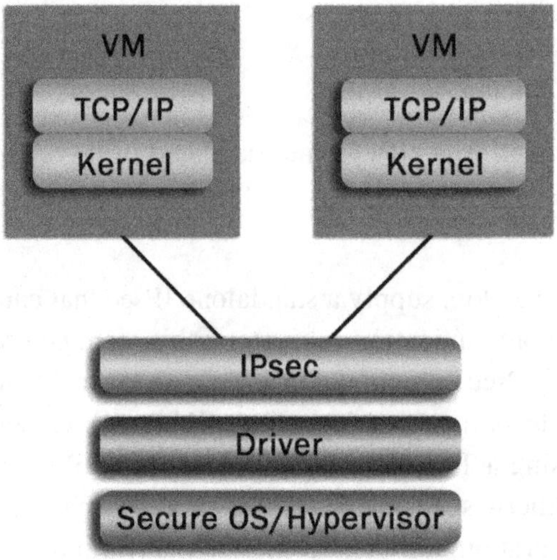

Figure 5.6:
Use of bump-in-the-stack ipsec in a virtual machine environment.

IPsec specifies two major sub-variants: Authentication Header (AH) and Encapsulating Security Payload (ESP). These protocols are covered by RFCs 4302 and 4303, respectively.[14,15] AH is used only for integrity protection and origin authentication. Because IPsec is usually employed for confidential communications, AH is rarely used; rather, ESP is the most interesting variant.

ESP has two modes of operation: Transport and Tunnel. Transport mode is intended for host-to-host connections (as opposed to host-to-gateway and gateway-to-gateway) and does not hide the original IP packet's header information.

In contrast, Tunnel mode fully encapsulates the original IP packet inside a new IP packet with completely new headers. Tunneling enables encrypted information to travel across open networks, such as the Internet; the outer headers and associated addressing are used to route the data between the VPN endpoints (client endpoint and server/concentrator) across the open network (see Figure 5.7).

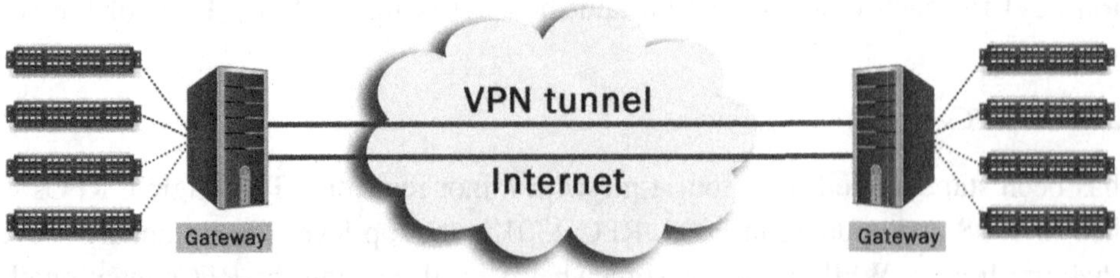

Figure 5.7:
VPN tunneling.

In addition to encrypting packet data, the original IP address information is hidden (encrypted) on the network.

Key Point

ESP Tunnel mode is the IPsec combination of choice when maximum security and confidentiality are required.

The encapsulation of an input IP packet into an IPsec ESP Tunnel mode packet is depicted in Figure 5.8. The trailing authentication data is used to verify data integrity of the encapsulated frame and the ESP header that helps to manage the security association between sender and receiver.

Tunnel mode adds more network overhead than Transport mode due to the new IP header. Transport and Tunnel modes are covered in detail within the aforementioned IPsec RFCs.

5.2.5.3 IKE

Internet Key Exchange follows the generalized model for secure connection establishment discussed at the start of this chapter. The peer who wants to establish a secure connection with a remote host sends the host its identification information (e.g., public key certificate) as well as some unique data used to validate that the messages are live (not replayed from an earlier session) and as a seed to a session key establishment algorithm. The initiator also signs data to assert its origin. The recipient peer will verify the signature to authenticate the initiator and then send back its signed data, random, and identification information for the converse operation. Each peer computes session keys based on the exchanged data and an agreed algorithm, typically a variant of Diffie-Hellman. These keys are passed down to the IPsec modules for the session communications.

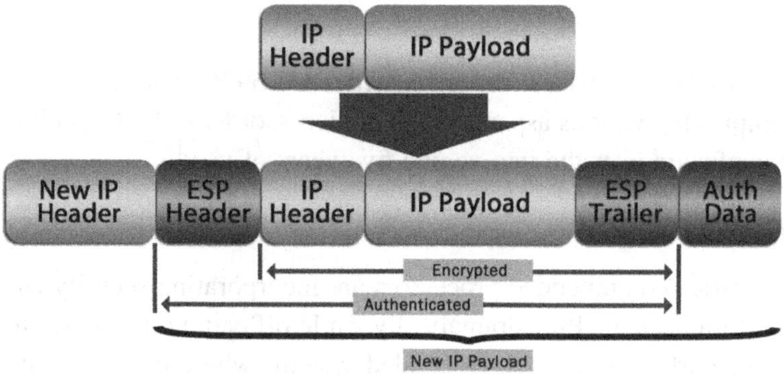

Figure 5.8:
IPsec ESP tunnel mode encapsulation.

IKE is also able to perform rekeying.

> **Key Point**
>
> In general, an IKE/IPsec or TLS session key's lifetime (called the cryptoperiod) should be limited and a new session key re-established to limit the damage should a session key be compromised.

The National Institute of Standards and Technology (NIST) provides cryptoperiod guidance in its special publication 800-57,[16] but ultimately, proper choice of cryptoperiod depends on the amount of data used for a single key as well as the time-prorated value of that data. For tactical data whose value degrades almost immediately after use, a cryptoperiod can be relatively long because a compromised key will give an attacker access only to stale information. For strategic data with a long lifetime (think nuclear weapons control codes), a cryptoperiod may be far shorter than in NIST guidance. Some network security implementations may not give the designer an ability to customize the cryptoperiod. When selecting network security products, the designer should consult the vendor and/or product documentation to understand whether this important characteristic is configurable.

IKE version 1 (IKEv1) is covered by RFC 2409,[17] and it is slowly being replaced by IKEv2 (RFC 4306), which fixes some important problems.[18] IKEv2 is not interoperable with IKEv1, making it difficult to use on its own unless the developer is confident that peers will always support the newer version. Yet the alternative is to use an older protocol that is known to have security flaws and inefficiencies.

> **Key Point**
>
> Generally speaking, if sufficient interoperability can be assured, IKEv2 is strongly recommended over IKEv1.

In late 2010, a new RFC for IKEv2 was published, RFC 5996, whose main purpose is clarifying and amplifying various aspects of the predecessor RFC 4306 specification that have met with some confusion over the intervening five years of use.[19]

5.2.5.4 IPsec Hardware Offload

Increasingly, sophisticated embedded processors are incorporating security engines that can accelerate the performance of IPsec dramatically while offloading this work from the primary CPU. This is particularly important in embedded systems where processing power is often tightly constrained. The security engine also decreases and simplifies the network security software content, reducing overall memory requirements.

For example, most modern multi-core network processors, including most of the Freescale QorIQ and the Cavium OCTEON family processors, have IPsec offload engines. Acceleration may include either of or both the public and bulk symmetric key cryptographic portions of IKE/IPsec.

5.2.5.4.1 Flow-through versus Lookaside

The QorIQ and OCTEON processors also provide good examples of the two different IPsec offload hardware approaches and programming models. The QorIQ has a single, feature-rich security engine that provides security services on behalf of all the general processor cores (eight PowerPCs in the case of the P4080 product). The P4080's security engine provides *flow-through* capability, which means that the offload engine can handle the bulk IPsec processing with little or no intervention of the CPU cores. Once IKE key material has been established and programmed into the engine, it will perform all the cryptographic algorithms (encryption, decryption, and integrity protection) as well as the IPsec header processing, such as adding the IPsec ESP packet headers and trailers. The P4080 provides offload support for other network security protocols, including TLS.

The Cavium OCTEON has a security offload processor built into each general-purpose core (up to 16 for the OCTEON 58xx family). However, the OCTEON provides *lookaside* processing instead of flow-through. This means that the engine offloads the heavyweight encryption processing but must rely on the main application cores for other IPsec protocol tasks, such as header and trailer processing and packet parsing and classification (identifying which tunnel's configuration, or security association, should be used for an incoming packet). The efficacy of either approach depends on the application and system requirements. Because its IPsec processing power scales with the number of cores, the OCTEON is able to achieve superior aggregate throughput. However, the P4080 is able to achieve its excellent performance without impacting the main cores.

Many Intel Architecture processors incorporate *AES-NI*, specialized instructions for accelerating AES. A lookaside approach, AES-NI claims the added benefit of avoiding local memory-based side channel attacks that have plagued software-only AES implementations.

Figure 5.9 illustrates how the hardware/software boundary differs between pure software IPsec, lookaside, and flow-through implementations. Note that some accelerators will have the line drawn in slightly different places. For example, a lookaside accelerator could handle portions of IPsec header processor or none at all. Furthermore, note that flow-through engines may require software key exchange via IKE, and others may actually handle the IKE negotiation as well. IKE is elided from Figure 5.9.

5.2.5.5 IPsec for Very Low Power Devices

The global smart energy initiative has raised concerns about the ability of power companies and their suppliers to provide adequate network security for a new generation of smart

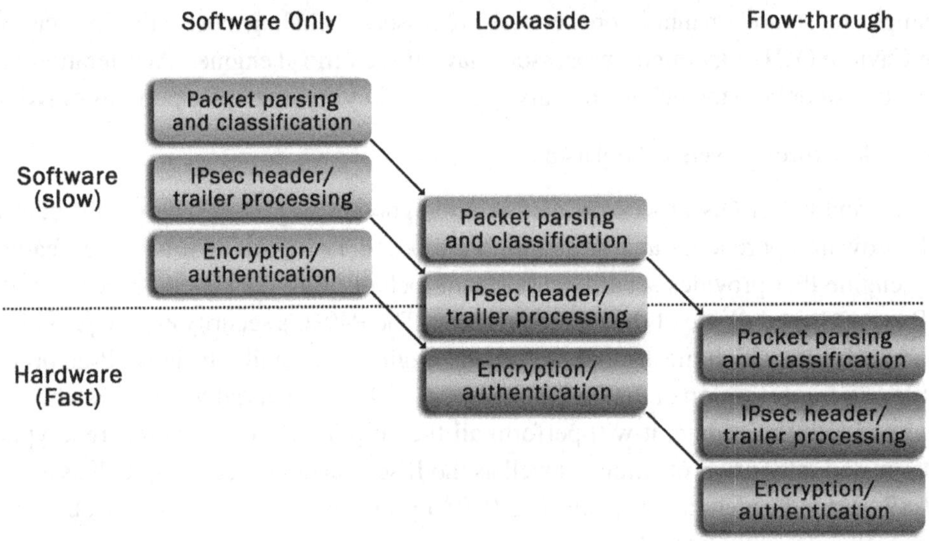

Figure 5.9:
IPsec offload comparison.

appliances, smart meters, gateways, and grid data centers as well as the home/personal area networks (HANs/PANs) and neighborhood area networks (NANs) that are being deployed to connect them. IPv6 is of enormous importance due to the desire to connect, ultimately, billions of smart energy network endpoints. However, a full-featured, RFC-conformant IKE/IPsec implementation is very sophisticated and not a good fit for a low-cost, battery-powered smart sensor armed with a microcontroller and its tens of kilobytes of flash and RAM talking to a low-rate, small-packet-size 802.15.4-based wireless network.

In fact, today's Zigbee-based devices use rudimentary data link layer security that has proven trivial to break (e.g., private symmetric keys provisioned over the network in the clear). In essence, designers are asserting that the security perimeter is best enforced at the gateway and that the threat to the HAN or perhaps even the NAN is not worth worrying about.

Yet some are looking past this stance to a future in which protecting the home from hackers will be critical. One promising standard for very low power devices is *6LoWPAN* (RFC 4944), which specifies a compression scheme for IPv6 (and by implication, IPsec) to enable its use over 802.15.4 networks.[20] Unfortunately, 6LoWPAN (and IP networking in general) is incompatible with the current generation of ZigBee; however, 6LoWPAN is specified in ISA 100.11a,[21] an open standard for wireless communication in industrial automation and process control systems, and the ZigBee Smart Energy Profile 2.0 Technical Requirements Document,[22] which is intended to result in a new ZigBee Smart Energy Profile specification.

Many 802.15.4 applications use UDP, rather than TCP, for transport. For example, an energy monitoring system might receive periodic, drop-tolerant datagrams sent from endpoint energy

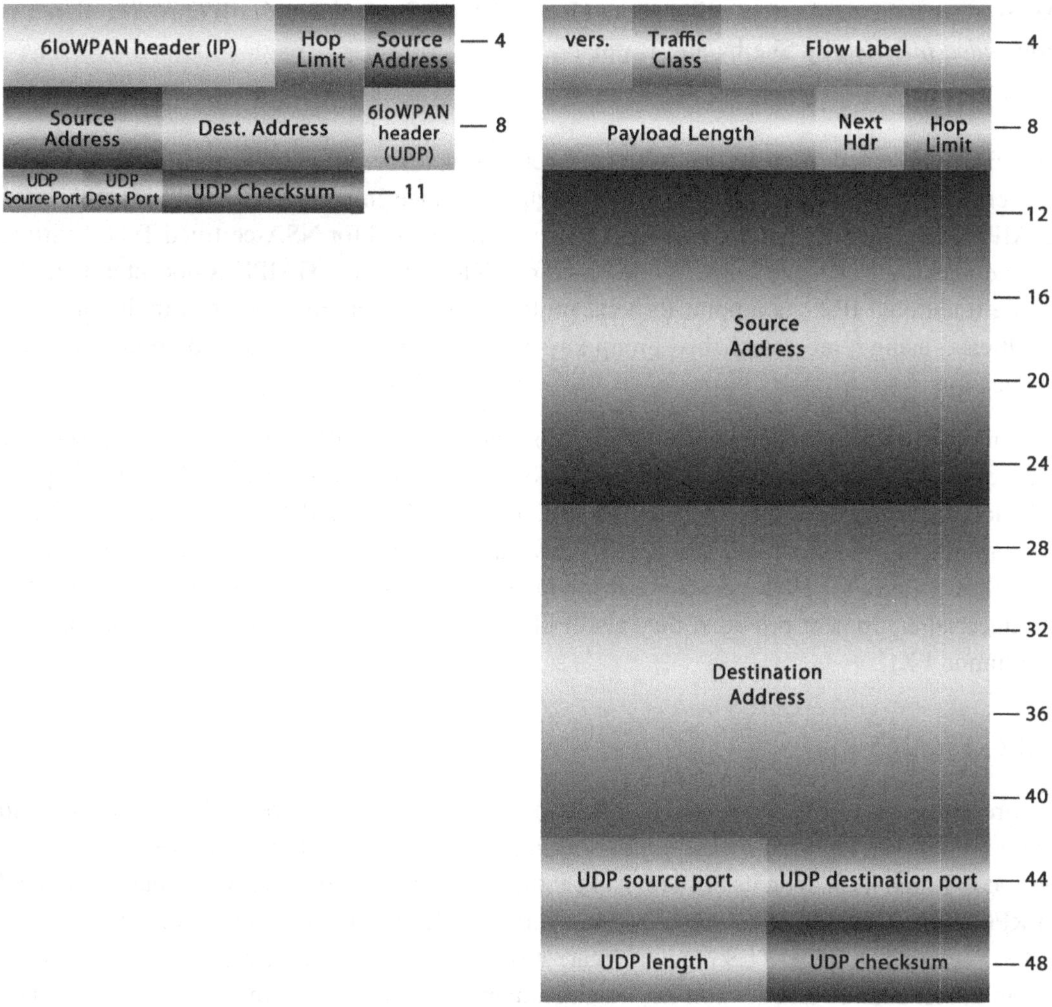

Figure 5.10:
Comparison of 6LoWPAN and standard IPv6 header sizes for routable packets.

sensors. As shown in Figure 5.10, 6LoWPAN-compliant UDP/IPv6 packet format enables compression of numerous fields, including the source and destination IPv6 addresses (at 128 bits each, the worst footprint offenders), IP version (six, implied), payload length (inferred from the data link layer), traffic class, flow label, and source and destination UDP ports.

Key Point

In the best case of a non-routable (local) packet, 6LoWPAN can reduce the UDP/IPv6 header size to just 4 bytes (relative to the usual 48); the best case for routable packets yields a header size of 11 bytes.

An excellent source of information about 6LoWPAN can be found in *6LoWPAN, The Wireless Embedded Internet*, by Shelby and Bormann.[23]

5.2.5.6 HAIPE

In 2004, NSA opted to create a customized version of IPsec protocol because of security concerns with commercial standards. The High-Assurance Internet Protocol Encryptor (HAIPE) standard (well, at least it's an NSA standard) is used for NSA-certified Type 1 virtual private network (VPN) systems. While based on IKE and IPsec, HAIPE is not interoperable. While traditional IPsec is a point-to-point protocol, HAIPE supports point-to-multipoint (multicast) using a pre-established group key in addition to its more common point-to-point application.

The initial HAIPE specification only called for classified algorithms (Suite A); however, subsequent revisions added a Suite B option to allow future interoperability between Type 1 products and Suite B (unclassified) products. Due to the IKEv2 and guidelines for use of Suite B cryptography in IKEv2 and IPsec, NSA has stated that HAIPE will eventually be replaced with IKEv2/IPsec for all Type 1 Suite B products. These products will interoperate with other commercial equipment provided they also follow the harmonized IPsec protocol and use a common PKI.

5.2.6 SSL/TLS

Secure Sockets Layer was invented by Netscape and had three major versions (SSL v1, v2, and v3) before it was replaced by the similar but improved Transport Layer Security (TLS) protocol, which itself is on its third revision at the time of this writing. TLS versions are defined in RFCs 2246 (Version 1.0, 1999),[24] 4346 (Version 1.1, 2006),[25] and 5246 (Version 1.2, 2008).[26] Similar to IKE, TLS aims to create a secure connection between two peers using origin authentication and session key establishment. Similar to IKE, the numerous SSL and TLS revisions have sought to improve security and usability based on feedback from widespread use. Nevertheless, fully RFC-compliant IKE or TLS products are sophisticated software packages that are prone to implementation error and misconfiguration. Once again, the best bet is to use the latest and greatest revision of the protocol—in this case, TLS v1.2. However, similar to IKE, the various flavors of TLS are not fully backward compatible to earlier versions. In particular, a TLS implementation that supports only version 1.2 will be unable to connect with an implementation that does not support version 1.2. Like IPsec, TLS cryptographic processing can be offloaded to security engines if present.

5.2.6.1 OpenSSL

OpenSSL is a popular open source SSL/TLS and cryptographic algorithm toolkit. Ports of OpenSSL exist for most modern embedded operating systems, including Linux and INTEGRITY.

At the time of this writing, TLS 1.2 is not yet supported in OpenSSL. Thus, if an organization's embedded device must connect with OpenSSL peers, then it may be forced to use TLS 1.1. OpenSSL should support TLS 1.2 soon, perhaps even by the time this book is published.

OpenSSL is a great example of why embedded systems developers should not assume that products are secure simply because their suppliers state they are or even because the products are dedicated specifically to security functionality. A search of U.S. CERT's National Vulnerability Database will show more than 100 software vulnerabilities in OpenSSL over the years, including five published in 2011. About a quarter of the software flaws published to date are classified as severe. For example, CVE-2007-4995: "off-by-one error in the DTLS implementation in OpenSSL 0.9.8 before 0.9.8f allows remote attackers to execute arbitrary code."

Embedded systems developers incorporating OpenSSL into products should verify that the version in use has the latest security patches applied. If a high robustness version of TLS is required, contact your embedded software supplier for commercially supported, certified options.

The OpenSSL toolkit consists of two major components: *libssl*, the implementation of SSL, TLS, and DTLS protocols; and *libcrypto*, a full-featured cryptographic library that can be, and often is, used independently of the SSL-related protocols. The OpenSSL toolkit builds these components into a combined library (usually a shared or dynamic link library) as well as a command-line tool for executing useful commands such as creating cryptographic keys, executing benchmarks, and encrypting the contents of a file.

For embedded systems developers of memory-constrained products, it is important to understand the memory footprint ramifications of incorporating network security protocol packages. In the case of OpenSSL, libssl's static footprint (code and data only, not including dynamic memory allocation) is approximately 200 KB. The cryptographic algorithm library, libcrypto, is approximately 1 MB in size. The OpenSSL executable, which incorporates other standard library functions, has a footprint of approximately 1.5 MB. These numbers change from one version to another as protocol changes are made, algorithms added, and so on. And, of course, numbers vary based on target CPU architecture and compiler used. In addition, because the source code is available, developers can customize the libraries to remove unneeded portions and reduce footprint.

Unfortunately, open standard APIs do not exist for SSL or TLS or practically any of the other data protection protocols discussed in this chapter. The OpenSSL API is documented on the www.openssl.org website.

5.2.6.2 GnuTLS

GnuTLS is another open source SSL/TLS library. Unlike OpenSSL's BSD-style license, GnuTLS employs the Lesser General Public License (LGPL). GnuTLS supports the latest standards, including DTLS and TLS 1.2, and includes a cryptographic library.

We use GnuTLS to illustrate one way that TLS is integrated into typical networking code. The general flow in GnuTLS for a client to establish a secure socket connection to a compatible remote TLS server is as follows:

1. Use the standard sockets interface to establish an unprotected connection.
2. Create a protected channel on top of the original unprotected socket using a TLS handshake.
3. Use the TLS socket API to send and receive packets on the protected channel.

The following IPv4 code fragments show how GnuTLS is used together with the standard socket APIs to create TLS-protected sessions. GnuTLS API calls and types are shown in bold:

```
int my_tls_connect(struct addrinfo *dest, gnutls_session_t session)
{
  int fd = socket(AF_INET, SOCK_STREAM, IPPROTO_TCP);
  ... /* some more initialization here */
  if (fd < 0)
    return -1;
  if (connect(fd, dest->ai_addr, dest->ai_addrlen) < 0)
    return -1;
  /* fd: standard, unprotected socket connection */
  /* associate the original socket with a TLS session */
  gnutls_transport_set_ptr(session, (gnutls_transport_ptr_t)fd);
  /* perform the TLS handshake */
  if (gnutls_handshake(session) < 0)
    return -1;
  /* Now session refers to an established TLS session */
  return 0;
}
/* The TLS analog to a socket send() */
void my_tls_send(gnutls_session_t s, void *data, uint32_t len)
{
  gnutls_record_send(s, data, len);
}
/* The TLS analog to a socket recv() */
int my_tls_recv(gnutls_session_t s, void *data, uint32_t maxlen)
{
  return gnutls_record_recv(s, data, maxlen);
}
```

While it is unfortunate that the lack of TLS API standardization ensures that the preceding code will not work with other TLS implementations, the code does show that the typical sequences are not complicated. In fact, developers can structure their code to abstract the implementation-dependent calls and types into a single portation module that is easy to swap in and out based on the selected library.

In practice, this TLS code will also need to include some global and per-session initialization calls as well as calls to specify authentication methods (X.509 certificate based, pre-shared key, etc.) and other configuration parameters. The detailed documentation for GnuTLS can be found on the public Internet at www.gnutls.org.

5.2.7 Embedded VPN Clients

Virtual private networks (VPNs) provide data confidentiality, integrity, and origin authentication across untrusted networks such as the Internet. We use VPNs to connect our laptops using a café's open Wi-Fi hotspot to our corporate intranets. VPN technology can be used to enable embedded systems to be secure managed remotely across heterogeneous networks (as long as these systems follow or can tunnel Internet Protocol). Given this description, it should come as no surprise that VPNs are frequently implemented with IKE/IPsec. A typical IPsec VPN client will incorporate the IKE protocol code while taking advantage of the IPsec functionality built into the underlying operating system. If an organization's underlying operating system does not support IPsec, then it must adopt a VPN technology that incorporates both IKE and IPsec.

Key Point

The lack of any standardized API or framework for IKE and IPsec has led to the creation of VPN client applications to encapsulate some of the tedious configuration, such as identifying a server IP address and authentication information, associated with these protocols.

A VPN application also gathers together numerous other auxiliary functions associated with IKE/IPsec, including support for digital certificate (e.g., X.509) parsing and verification, the behind-the-scenes use of system calls to communicate IKE-created keying material down to the IPsec stack, and interfaces to additional authentication methods and cryptographic hardware such as a smartcard.

Readers should not be fooled into thinking that incorporating a VPN application into an embedded system will be trivial. Developers should be prepared to read a lot of documentation regarding how to build and configure VPN software. Unfortunately, the suppliers of VPN clients have themselves made little effort to standardize on the APIs that embedded systems developers would use to incorporate them. For attended systems, a VPN client application may include a graphical user interface for configuration and session establishment. For unattended embedded systems, some form of development kit and documentation must be provided. Embedded systems developers should consider creating an independent API for invoking VPN services, insulating application software from the choice of VPN client package.

5.2.7.1 Openswan and Racoon2

Openswan is an example of an IPsec VPN package that can be used in either client or server modes. Openswan is a popular open source product for Linux operating systems. Originally,

Openswan included both the IPsec patches to the Linux kernel as well as the user-mode IKE daemon, called pluto. As of the Linux 2.6, however, IPsec is incorporated into the standard kernel distribution and well supported, reducing Openswan to the user-space IKE and associated connection establishment and configuration services. Openswan has been tested for interoperability with numerous open and proprietary VPN technologies, such as Cisco and Windows. At the time of this writing, Openswan information and software can be found on the Internet at www.openswan.org.

A modern Openswan implementation will require approximately 300 KB of code and data storage (static footprint), not including the cryptographic algorithms required by IKE.

Numerous VPN books have been penned, such as Jon Snader's *VPNs Illustrated: Tunnels, VPNs, and IPsec*.[27] A book on Openswan authored by its developers is also available.[28]

Most of Openswan is governed by GPLv2. Developers looking for a more friendly licensed alternative should consider Racoon2, which employs a BSD-style license. Racoon2 provides implementations of IKE (v1 and v2) and has been adapted for use with the IPsec stacks of numerous operating systems, including BSD, Linux, and MacOS. At the time of this writing, Racoon2 information and software can be found on the Internet at www.racoon2.wide.ad.jp.

5.2.7.2 OpenVPN

OpenVPN is another open source VPN technology. However, unlike Openswan, OpenVPN uses a variant on SSL/TLS to establish a tunnel and therefore requires a compatible OpenVPN peer. OpenVPN uses OpenSSL cryptography extensively. Because it consists entirely of user-space code, OpenVPN is considered easier to port to multiple operating systems, which has been done in practice, unlike Openswan, which has been relegated only to Linux systems.

OpenVPN is an example of what is termed an *SSL VPN* technology, of which numerous proprietary, commercial products exist. Because they use the SSL/TLS protocol (and many can support DTLS for UDP traffic, described in the following section), VPNs can be created at the application level, resulting in reduced privilege and overhead relative to an IPsec VPN that opens up an entire network pipe for all applications running on the system. In addition, SSL VPNs can be easier to install and configure.

Unfortunately, most SSL VPN implementations, including OpenVPN, are not interoperable. A decision to use a Juniper SSL VPN client likely requires adoption of Juniper SSL VPN concentrator equipment. OpenVPN clients talk only to OpenVPN peers in the back end.

5.2.8 DTLS

> **Key Point**
>
> Datagram TLS (DTLS) is designed to provide the same capabilities as TLS for UDP-based transport protocols and applications that use them.

DTLS was published in 2006 and is defined in RFC 4347.[29] It is gaining in popularity due to the increased use of voice and video streaming protocols that can tolerate dropped packets and prefer the lower overhead of UDP over TCP. Unfortunately, DTLS, to a large extent, must reinvent the wheel built by IPsec for handling dropped and out-of-order packets that can occur during the TLS handshake protocol. Therefore, DTLS adds complications to an already complex TLS protocol. Nevertheless, as with TLS, DTLS is a good choice when the system does not already incorporate an IPsec stack or when there is a desire to enable fine-grained application-level control of network security.

5.2.9 SSH

Secure Shell (SSH), as the name implies, was originally designed to provide secure remote access to computers by users or other computers. The original applications of SSH replace legacy protocols such as *telnet*, *rlogin*, and *rsh* that rely on plaintext transmission without cryptographic authentication. Essentially, SSH adds public key authentication and data-in-transit protection to a TCP socket session. The result is a more secure way for users and computers to create secure interactive sessions with another remote computer. A typical application of SSH in embedded systems is to run an SSH server within the fielded embedded systems and enable remote administrators to manage the embedded systems using the SSH protocol and system-specific shell commands.

> **Key Point**
>
> The history of SSH demonstrates the difficulty of achieving high levels of security in sophisticated data protection protocols.

SSH was originally developed in 1995 and within 5 years boasted millions of worldwide users. Yet 13 years after SSH's founding, a vulnerability was discovered in all versions of SSH. The vulnerability makes it possible for an attacker to recover up to 32 bits of plaintext from an arbitrary block of ciphertext when CBC mode (see Chapter 4) is used for encryption.[30] This vulnerability exists even in the latest version of SSH, SSH-2. There is no more recent version of the standard devoid of this vulnerability. The published workaround is to use CTR mode in place of CBC mode.

Several severe vulnerabilities in previous versions of SSH were found over the years prior to this CBC flaw. In fact, a bug fix made to address a vulnerability in the SSH-1 version protocol actually introduced a new, serious vulnerability that enables a remote attacker to execute arbitrary code within the SSH server, which executes with root privilege on many general-purpose operating systems.[31]

Unlike many other network protocols that are concisely defined in one or two RFCs, SSH-2 protocol is defined by five core standards: RFC 4250,[32] RFC 4251,[33] RFC 4252,[34] RFC 4253,[35] and RFC 4254.[36]

This standards complexity is perhaps indicative of the protocol complexity and its historical security difficulties.

5.2.9.1 Additional Protocols Based on SSH

SSH has been used as the basis to create secure versions of other popular communications applications that involve remote computer access by users and other computers. These include SSH File Transfer Protocol (SFTP), conceptually similar to File Transfer Protocol (FTP), and Secure Copy Protocol (SCP), a security-enhanced approach to remote copy protocol (RCP). SFTP is described in an expired draft RFC,[37] but no active RFC defines the protocol. Similarly, no RFC exists for SCP, leaving interoperability up to vendors (not usually a good idea).

5.2.9.2 OpenSSH

Perhaps one explanation for the lack of standardization around these uses of SSH is the ubiquity of OpenSSH, an open source implementation of SSH that includes both client- and server-side implementations as well as SFTP and SCP support. Development of OpenSSH is led by the OpenBSD community and enjoys a liberal BSD license. OpenSSH has its own fully contained cryptographic library and has been ported to a wide range of embedded operating systems. Developers should expect an SSH client to require on the order of 300 KB of ROM/flash storage for executable code and data.

5.2.10 Custom Network Security Protocols

Key Point

Due to the complexity and large footprint of both IPsec and TLS, a number of commercial software suppliers and university researchers have proposed alternative, lighter-weight network security protocols; for embedded systems designers more concerned about simplicity and footprint and less concerned about interoperability with arbitrary peers, custom protocols may be a viable alternative.

It is critical to have assurance that cryptographic experts have independently analyzed and validated the custom protocols.

Let's look at a sample set of simplifications to SSL based on areas of the protocol that have been found to be vulnerable. SSL allows for peers to negotiate a cipher suite to ensure interoperability. This makes sense: if a server supports 3DES and AES for bulk encryption and a client supports only 3DES, then the ciphersuite negotiation enables the peers to deduce the common 3DES algorithm so that communication can occur. However, security administrators deploying SSL on a server may desire 256-bit AES protection and not realize that a client may negotiate the server down to a weaker algorithm such as 3DES. Furthermore, man-in-the-middle attacks have successfully forced ciphersuite downgrades between peers. One example affects most shipping versions of OpenSSL and is referred to as CVE-2010-4180 in the NIST National Vulnerability Database.[38]

Other vulnerabilities of SSL affect its renegotiation, a protocol feature that allows a peer to request a re-establishment (authenticate and create a new shared key) within a previously established session. This vulnerability, referred to as CVE-2009-3555,[39] enables a man-in-the-middle attacker to inject malicious text during an otherwise valid established session between two unsuspecting peers. Because the attacker is unable to read the session's encrypted text, this vulnerability was thought at first to be relatively benign. However, researchers soon found a way to use injected text to cajole a web server into disclosing sensitive data. In particular, a successful attack on Twitter was performed: the injected text actually instructed Twitter to maliciously tweet the username and password information of a tweeting user. The attacker then grabbed the illicit tweet and could use the pilfered information to assume the identity of the unsuspecting user.

A potential simplification to SSL would be to remove the complexity surrounding the negotiation of cryptographic suites between peers as well as the ability to resume or renegotiate sessions. Imagine that an embedded system's designers want to be protected with a specific set of cipher suites, such as 384-bit elliptic curve digital signatures for authentication and key agreement as well as 256-bit AES-GCM for authenticated encryption of all data in transit. The client's SSL implementation can initiate a session specifying these suites and not permit anything weaker. The client will use a successfully established session encryption key for a client-defined cryptoperiod and then terminate the session permanently at the end of this period. No session resumption or renegotiation is permitted. If a peer needs to change the security association, it simply can close the socket and restart the protocol. While this approach may not be suitable for SSL VPN concentrators, it may be fine for embedded systems and have the benefit of simplifying the embedded system's code, reducing footprint, and avoiding weaknesses—those already discovered and perhaps others yet to be found.

With six versions and actively discovered weaknesses, SSL is a good example of a failure to rigorously employ the PHASE principle of complexity minimization. As with many sophisticated pieces of software, even those created to perform security functions, SSL's collection of authors simply did not have the discipline to sacrifice unnecessary features to create a simpler, cleaner, and ultimately more secure design. It is interesting to note that, in response to the renegotiation vulnerability, the workaround applied to web servers is to disable the renegotiation feature in the underlying SSL implementation. This workaround has been widely deployed, proving that the feature is unnecessary. Nevertheless, readers may find it humorous that the IETF's ultimate solution to this problem, rather than killing off renegotiation and simplifying the protocol, was to add yet another feature to SSL (RFC 5746) that links the renegotiation handshakes to their original sessions, preventing the man-in-the-middle attack.[40]

Let's consider a neighborhood of homes outfitted with smart meters that communicate with a neighborhood area network gateway that is designed and supplied by the same company as the smart meters. This is exactly the situation in which a custom, streamlined protocol, compliant but not necessarily conformant to the RFCs, may be appropriate. Alternatively, a committee tasked with creating the security standards for the smart grid could adopt this adjustment to the protocol. TLS is just one example; similar simplifications and improvements to the IKE and IPsec protocols could be made.

If an embedded design is tight on footprint or calls for a higher-assurance implementation and can tolerate a lack of full RFC conformance and interoperability, the developer should consult the operating system and/or systems software provider to discuss alternatives.

5.2.10.1 Traffic Flow Confidentiality

As mentioned earlier, IPsec ESP Tunnel mode provides improved transmission security by encapsulating the original IP packet's addressing information within a new header that typically corresponds to a gateway fronting a substantial network. Eavesdroppers can determine the origin and destination networks but not the identity of or the specific targeting of packets to individual nodes behind the gateways.

Another example of transmission security is the elision of traffic flow information. Traffic flow includes the rate and size of individual packets. Eavesdroppers may be able to gain unauthorized access to information about the nature of communications by observing traffic flow.

For example, a compromised, malicious node in a network can transmit sensitive information to a colluding eavesdropper by employing covert network timing channels. The compromised mode transmits a numeric value, N, by sending a rapid sequence of N packets preceded and followed by a long delay. A specific numeric value can be selected to signal the start and end of a new covert message, built from these packet sequences.

Another well-known approach involves the malicious application sending payloads of certain sizes to convey information. For example, the colluders may interpret payloads of less than 50 bytes as binary zero and payloads larger than 50 bytes as binary one. Special pre-arranged payload sizes can be used to signal the start and end of a new covert message, built from these

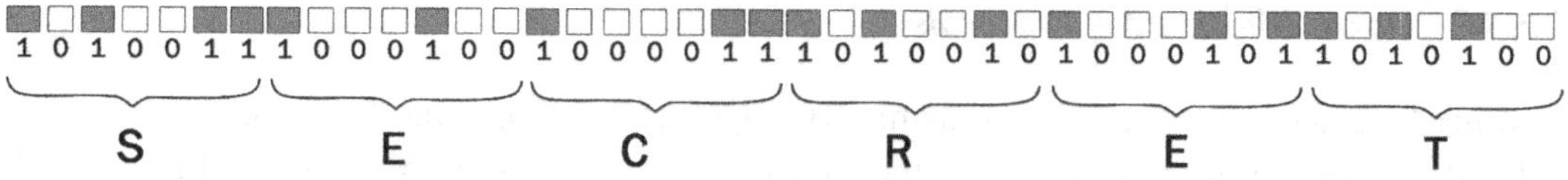

Figure 5.11:
A form of covert communication enabled by a lack of traffic flow confidentiality.

packet sequences. The packet-size attack is demonstrated in Figure 5.11, where the size of each packet communicates either a 0 or 1, generating the ASCII message "SECRET."

In both of these examples, the network security protocol's encryption of payloads does nothing to prevent the malicious insider from exposing sensitive information on the network.

> **Key Point**
>
> While the current open standards for network security protocols, such as TLS, DTLS, SSH, and IPsec, do not provide full traffic flow confidentiality, some embedded software suppliers provide versions of these protocols that are enhanced to provide this feature.

For example, the IPsec implementation can pad all payloads to a constant size, preventing the aforementioned packet-length attack. The additional padding overhead may be acceptable or even necessary for some environments. Furthermore, developers of embedded products that incorporate network security features may gain competitive advantage by offering their customers traffic flow confidentiality features.

5.2.11 Application of Cryptography within Network Security Protocols

Of course, network security protocols can be only as secure as their underlying ciphers. It is silly to use TLS to negotiate a weak symmetric key that can be easily broken with a brute-force attack. The breadth of public key, symmetric key, and hash/MAC algorithms and modes used in embedded systems today is staggering. Chapter 4 provides an overview of all these key areas of cryptography as well as government guidance for their general application.

5.2.11.1 NSA Suite B Guidance

A number of RFCs have been published that define the Suite B algorithm requirements for use with popular network security protocols. RFC 4869 covers Suite B requirements for IKE and IPsec. RFC 5430 covers TLS. A primary focus of these RFCs is to specify acceptable cryptosuites (e.g., AES in Galois Counter Mode with certain key lengths) for Suite B compliance when employing these protocols.

5.2.12 Secure Multimedia Protocols

Multimedia protocols govern the transmission of so-called real-time data, not in the embedded hard real-time sense, but rather in the sense that communication quality depends on minimized packet loss, low latency, and jitter. Examples include audio (e.g., voice), video, continuous environmental sampling, and certain kinds of remote control and management applications.

In some instances NSA has developed its own protocols because of security issues with existing commercial protocols or for lack of commercial protocols for certain applications.

For secure voice applications (wired and wireless), NSA developed a secure voice protocol, *Secure Communications Interoperability Protocol* (SCIP), that specifies secure association establishment procedures for voice calls using a wide set of voice codecs and for use in both point-to-point and point-to-multipoint applications. SCIP was initially deployed in 2001 and is now widely adopted for NSA Type 1-certified voice equipment and throughout NATO.

SCIP is not used in commercial products and hence is not useful for developers of embedded systems (e.g., radios) that are used beyond the sensitive government environment. Therefore, we do not discuss SCIP in any detail here. The use of this non-commercial protocol poses a problem for the NSA's recent cryptographic interoperability strategy (see Chapter 4) that strives to maximize use of commercial standards such as IPsec and published cryptographic algorithms. Therefore, we can expect the NSA's stance on secure voice to change rapidly, promoting a conversion from SCIP to open standards-based protocols.

In the commercial world, secure voice has become increasingly important as corporations have adopted the use of mobile cellular devices for sensitive communications and so-called softphones—phones that use the Internet for voice transmission. The general family of protocols for transmitting voice over Internet Protocol is called Voice over IP (VoIP).

VoIP encompasses numerous proprietary (e.g., Skype) and open standard protocols, a comprehensive discussion of which is not practical in this chapter. We cover just a couple of the more important ones, especially those that are general enough to include generalized multimedia (e.g., audio and video) transmission. Secure multimedia protocols can generally be broken down into two pieces, a signaling protocol (also called the control portion) that is responsible for locating the receiver (or receivers in a multicast application) and establishing a connection, and a data transport protocol responsible for moving encrypted multimedia data from sender to receiver(s). In some cases, multimedia compression codecs are ensconced within the data plane; these algorithms are beyond the scope of this discussion. Figure 5.12 shows the signaling and data plane paths in a local area network that connects to the outside world through a private branch exchange (PBX).

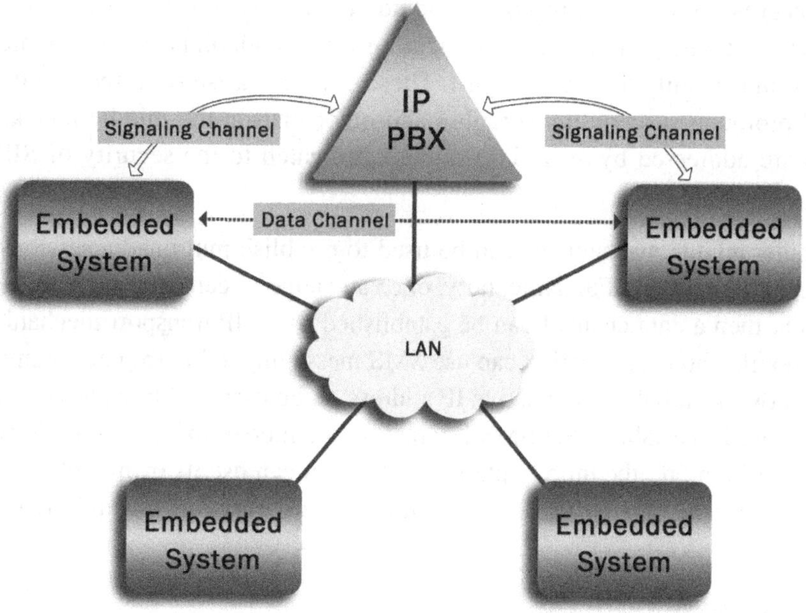

Figure 5.12:
Independent signaling and data channel paths are common in secure voice applications.

5.2.12.1 Signaling Protocols

Multimedia control plane protocols may be quite sophisticated, as they are responsible for locating the appropriate receiver(s), negotiating a common media format capability, crossing and tunneling across heterogeneous networks and communication media via routers and gateways, mutual endpoint authentication, and session teardown. The H.323 and Session Initiation Protocols (SIP) are the dominant open standard signaling protocols in use today. H.323 is defined by the telecommunications standardization sector of the United Nations agency for information and communication technologies (ITU-T). For detailed information on H.323, readers are encouraged to consult the ITU-T's specifications for H.323 and its subsidiary protocols.[41]

Unlike H.323, which was designed for traditional circuit-switched networks, SIP was originally designed for IP networks. Numerous comparisons of H.323 and SIP can be found in the literature.[42] Despite a large install base of H.323, SIP seems to be taking over as the dominant protocol. Like many other Internet Protocol-based standards, SIP is defined by the IETF—in this case, RFC 2543.[43] For readers interested in how complex these multimedia control protocols are, the SIP RFC is about 270 pages long as compared to the latest IPsec RFC, 4301, which is about 100 pages long. SIP is also the control plane protocol of choice for emerging mobile wireless networks, such as LTE, due to the desire to maximize compatibility with IETF Internet standards.

Security concerns for signaling protocols include the ability to steal an endpoint's identity, denial of service, transmission security (detecting which endpoints are communicating), and non-repudiation (inability to verify origin). These concerns arise from the fact that SIP is an unprotected protocol, susceptible to replay and other man-in-the-middle attacks. These weaknesses are addressed by an IETF standard dedicated to the security of SIP: RFC 3329.[44]

Numerous other ad hoc mechanisms can be used to establish multimedia sessions between endpoints on an IP network. Fundamentally, once an endpoint can determine the IP address of its destination, then a data channel can be established using IP transport mechanisms. For example, a mobile phone application can use SMS messaging to interrogate a named user over the cellular network and obtain the active IP address of the user's mobile phone. The details of SIP signaling used to establish the SMS interaction are hidden from the mobile phone application. Furthermore, the mobile phone application can use its own application-level authentication over the data channel, not relying on the security of the underlying cellular network.

5.2.12.2 Multimedia Transport Protocols

Once a session is established, multimedia data must be sent efficiently and confidentially over the link. As mentioned previously, data channel endpoint authentication is often desirable. Thus, one might think a standard TLS session would fit the bill. However, TLS runs over TCP, which introduces too much latency for multimedia applications. Because multimedia communication is generally tolerant of the occasional dropped packet, the low-latency UDP transport is often employed. Therefore, one might think that DTLS would fit the bill. In fact, DTLS may be just fine for high-bandwidth network links (e.g., video chat within a LAN). However, multimedia streams often run across the Internet and/or wide-area wireless networks (e.g., LTE) where bandwidth is limited and choppy. For this reason, an optimized secure transport protocol is indicated.[45]

5.2.12.2.1 SRTP

The Real-Time Transport Protocol (RTP) is a popular framework protocol for multimedia transport. The RTP framework is described in RFC 3550.[46] Application of the framework to specific information formats is covered in companion standards. For example, the application of audio and video conferences is standardized in RFC 3551.[47] This standard covers numerous audio and video encodings. One of the key differences between RTP and a simple UDP stream is the inclusion of quality-of-service information that is conveyed to endpoints for potential corrective action and sequencing information that is used to reconstruct the media stream from out-of-order packets. In addition, RTP handles multiple concurrent multimedia sources multicasting at varying rates (multipoint-to-multipoint), making the applications developer's job much easier.

> **Key Point**
>
> While proprietary, non-standard multimedia transport protocols exist, the majority of commercial and open source multimedia (especially voice and video) transport over IP protocols use RTP and its adjunct security protocol, Secure Real-Time Transport Protocol (SRTP).

SRTP is defined in RFC 3711.[48] At a high level, SRTP provides the same security services as the more generic secure transport protocols for IP (IPsec, TLS, DTLS): data origin authentication, integrity (including replay protection), and confidentiality. Unlike IPsec, which is agnostic to the choice of underlying encryption cipher, SRTP enforces the use of cipher modes friendly to multimedia. In particular, SRTP specifies variations of the counter and output feedback modes that, as discussed in Chapter 4, do not propagate errors when a packet is dropped. SRTP uses HMAC-SHA-1 for message authentication. Another critical difference of SRTP over tunneled IPsec is that SRTP retains the original RTP headers, enabling quality-of-service data to survive within the encrypted channel (obviously at the expense of some traffic confidentiality).

SRTP, like IPsec but unlike TLS, does not cover key establishment between communicating parties. Popular key exchange protocols used with SRTP include Multimedia Internet Keying (MIKEY) defined in RFC 3830,[49] Session Description Protocol Security Descriptions for Media Streams (SDES) defined in RFC 4568, and ZRTP defined in RFC 6189.[50] SDES piggybacks on SIP messages and is hence unsuitable to applications using other signaling mechanisms. Similarly, MIKEY is intended for signaling path key exchange. In contrast, ZRTP performs key exchange within the media channel and therefore is agnostic to signaling mechanism. There is some debate as to whether it is more sensible to perform key exchange within the signaling channel (which is already establishing connections) or to keep it decoupled. Given the variety of device types, formats, and networks that characterize the modern digital world, a decoupled approach, assuming sufficient efficiency, is desirable. Developers of secure multimedia applications can provide end-to-end authentication and encryption without worrying about how communicating parties are connected. Interested readers, however, are encouraged to review counterpoints for specific applications, such as VoIP.[51] Readers are also encouraged to examine RFC 5479,[52] an information document that discusses the various options of key management both within the signaling plane and within the media plane.

5.2.12.2.2 DTLS-SRTP

DTLS-SRTP, defined in RFC 5764,[53] presents a hybrid in which signaling path communicating parties are bound to certificates used in the media path key establishment. The reason this is important is that a Public Key Infrastructure (PKI) may not be practical between communicating parties in all instances. With DTLS-SRTP, the communicating parties use self-signed certificates bound to their signaling identities. If Alice dials Bob and gets connected by the signaling plane, then Alice trusts the signaling plane and now wants to ensure that her voice data is encrypted with a key that only Bob (as defined by his phone number) also shares.

The signaling plane forwards the certificates, and actual key agreement is performed by DTLS over the media channel. One of the desirable traits of DTLS-SRTP is that DTLS is a generic key exchange protocol used throughout the Internet for datagram-oriented data (like multimedia). SRTP is the de facto secure transport protocol for multimedia. Therefore, DTLS-SRTP represents arguably the best-understood, standardized combination of end-to-end secure Internet protocols for multimedia.

Perhaps surprisingly, current NSA Suite B guidance points to DTLS-SRTP. What makes this surprising is the protocol's inherent trust of the signaling domain. Imagine an intelligence analyst speaking classified information into a smartphone, communicating across bridged public cellular networks to another part of the world. Clearly, that would be a bad idea, since a compromised foreign phone company could direct the classified voice data to a malicious listener. The use of DTLS-SRTP in a classified domain, therefore, assumes that voice calls are connected over a private network, isolated from the public switched telephone network (PSTN) and the global Internet, or that the signaling plane enforces authentication deemed trustworthy by NSA. If signaling cannot be trusted, then a proprietary protocol or ZRTP would be indicated. Note that ZRTP has an option for using full PKI for authentication. Alternatively, the U.S. government is considering the use of mobile virtual network operators (MVNOs) that provide private signaling domains implemented as virtual cellular networks overlaid on established physical cellular networks.

At the time of this writing, Suite B guidance for the use of DTLS-SRTP is in draft RFC state.[54] The draft standard overrides SRTP's bulk data cipher with the Suite B preferred AES-GCM suite and requires elliptic curve cryptography, ECDSA and ECDH, for the authentication and key agreement algorithms. Embedded systems developers creating products with multimedia applications and targeting U.S. government use (autonomous vehicles is a salient example) would be well advised to consider the use of DTLS-SRTP.

5.2.13 Broadcast Security

So far, the network security protocols discussed in this chapter all rely on a duplex connection: transmitter(s) and receiver(s) both must send information to complete the protocol. Practically all IP-based protocols assume an underlying data link layer that inherently allows any endpoint to both send and receive data. Most middleware protocols, such as client/server CORBA, also assume a duplex communications medium. Broadcast, on the other hand, is an example of *simplex* communication: all information is sent in one direction, from sender to receivers. An example of a system that requires broadcast communications security is a digital satellite

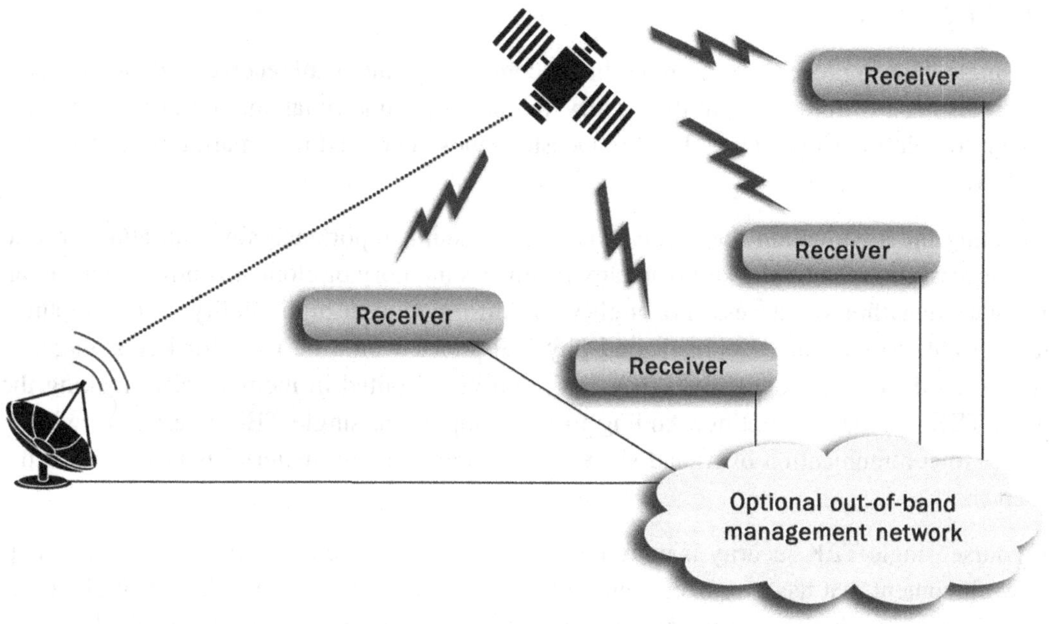

Figure 5.13:
Sample embedded systems-based broadcast network.

system that transmits protected multimedia content to widely distributed set-top box receivers (see Figure 5.13). The set-top box receivers are unable to transmit data directly back to the satellite.

To complicate matters further, broadcast systems are often characterized by a large and dynamic set of receivers and a dynamic set of information that needs to be transmitted, operate on physical networks that are tightly bandwidth constrained, and have limited physically protected storage (e.g., tamper-resistant smartcard) within each receiver. Like many other embedded systems, broadcast receivers are often unattended by authorized security administrators and cannot tolerate sophisticated key management schemes.

The conflicting requirements for dynamic content and receivers and simple key management present a significant challenge to embedded systems developers. Thus, we describe and compare several approaches for the use of information encryption keys in broadcast environments. We focus on two main types of encryption keys: key encryption keys (KEKs) and transmission encryption keys (TEKs). TEKs are the symmetric data encryption keys used for information content protection. For example, the information in transmission may be protected using an error-tolerant bulk encryption algorithm such as AES-CFB (described in Chapter 4). KEKs are used to encrypt (wrap) TEKs such that only a receiver that possesses the KEK can decrypt and use a TEK.

5.2.13.1 Single TEK

The single-TEK approach uses a single shared content key among all receivers, all of which are pre-provisioned with the TEK at the factory. Receivers need a trivial amount of local storage, enough to hold the single TEK. The broadcaster sends encrypted information protected by the TEK.

Protecting the keys within the receiver is of paramount importance since an attacker who can successfully recover the key from any receiver's memory or clone an entire receiver can then gain unauthorized access. For high-value information, a high-quality cryptographic element able to withstand sophisticated physical attack would be used for key storage. Since such devices are expensive, they are often very limited in memory size, making the single-TEK approach tempting. Adding to this temptation, single-TEK is efficient with respect to communication overhead since no additional keying material is broadcast to the receivers.

Of course, single-TEK security is extremely weak since a single compromised receiver will allow all content that has ever been sent with the TEK to be compromised. The single-TEK approach violates the important cryptographic key management principles described in Chapter 4 and should be avoided. The single-TEK approach is shown in Figure 5.14.

5.2.13.2 Group TEKs

Finite groups of TEKs can be used for collections of content to reduce the information compromise when a particular TEK is exposed while avoiding the need to update TEKs within the receivers. For example, a digital satellite system for pay TV could use separate TEKs for specific programs and/or specific times of day and/or groups of receivers. The broadcasted content includes metadata that enables the receiver to identify the TEK needed to decrypt a particular stream.

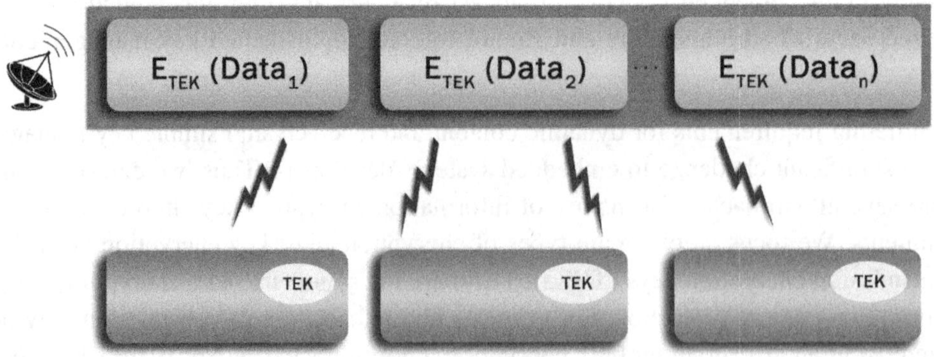

Figure 5.14:
Single, shared TEK.

This scenario implies that receivers must locally store lots of TEKs that still may be infeasible given hardware constraints. Even if we assume a significant number of usable TEKs, the group-TEK approach still suffers from cryptoperiod weakness: a compromised TEK will expose all information ever transmitted using the TEK.

The multiple TEK approach, shown in Figure 5.15, is communications bandwidth efficient.

5.2.13.3 Single KEK

The single-KEK approach uses a single KEK shared among all receivers to protect the TEKs. This also addresses the cryptoperiod problem since the broadcast can use unique TEKs for each transmission. A compromised TEK will expose only the limited content protected by that TEK.

Receivers need a trivial amount of local storage, enough to hold the single KEK. The broadcaster sends streams of information that include the wrapped TEKs and the information corresponding to those TEKs. The requirement of broadcasting the TEK impacts communications bandwidth. However, this trade-off is usually negligible—just a handful more bytes of data.

Given these favorable properties, the single-KEK approach seems like a reasonable compromise. In fact, DirecTV satellite television system has employed such a scheme. See Figure 5.16.

We hope that attentive readers are still horrified by this "compromise."

> **Key Point**
>
> The sharing of a single symmetric key for all protected broadcast content violates the most basic application of the least privilege principle and would be considered anathema to any cryptographer.

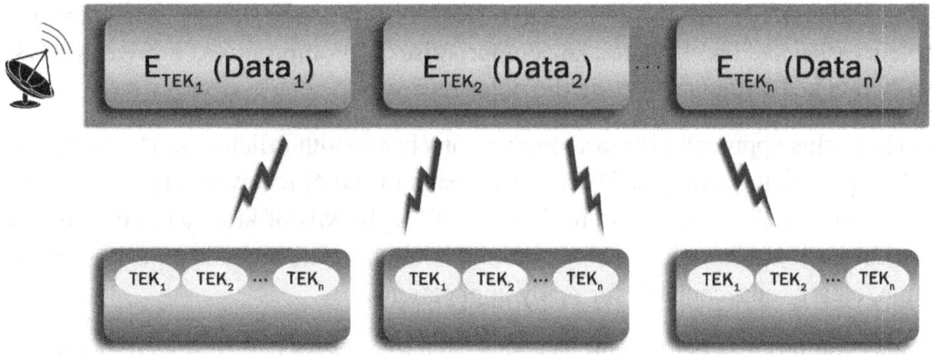

Figure 5.15:
Multiple TEKs.

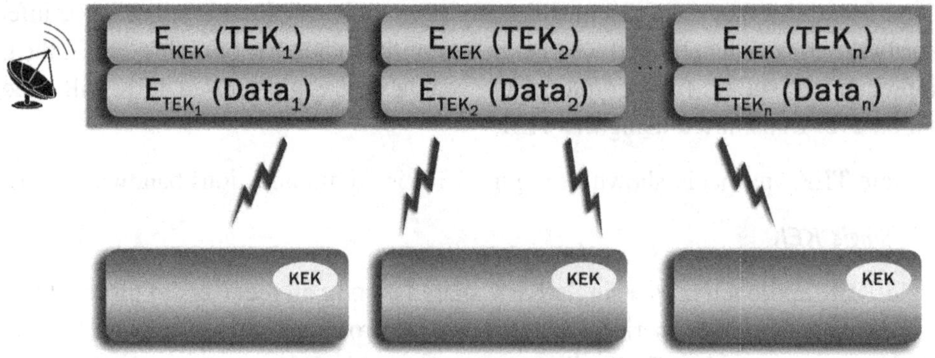

Figure 5.16:
Single KEK with broadcast TEKs.

Nevertheless, the existence of this approach in practice demonstrates how embedded systems developers will sometimes sacrifice security in exchange for development and maintenance simplicity, even for high-value information.

Hackers were able to break into DirecTV receivers to steal the KEK and then the content. It is believed that the compromise went undetected for years, causing immense value loss. When finally detected, the hacking led to a war between DirecTV and the thieves, with DirecTV attempting various patches, followed closely by increasingly sophisticated attempts to bypass. Eventually, DirecTV installed a patch that would accurately detect an illegal receiver and then permanently disable that receiver remotely.

However, one has to wonder how much pain would have been prevented if DirecTV had adopted a more secure approach from the beginning.

5.2.13.4 KEK Per Receiver

Clearly, we would like to have a unique KEK per receiver, used to wrap the per-transmission TEKs (see Figure 5.17). We would have excellent local receiver storage efficiency and avoid the problems associated with shared key compromises. A compromised receiver's KEK could be trivially revoked without affecting the other receivers.

Unfortunately, this approach kills communications bandwidth efficiency. The broadcast must include N copies of the wrapped TEK, one for each of the N receivers in the network. For a network of one million receivers and 128-bit TEKs, 16 MB of key material would be broadcast for each transmission. This corresponds to about 13 seconds on a 10-Mbit media connection, clearly unacceptable for many applications.

The obvious compromise is to group receivers such that KEKs can be shared, but in a limited manner (e.g., by geography or some other means) to reduce the bandwidth overhead without giving up entirely on least privilege.

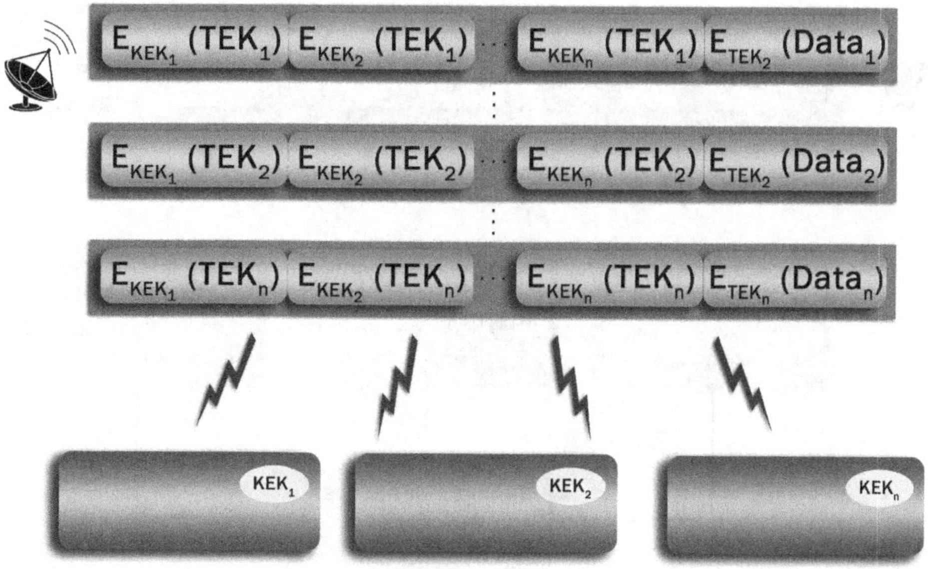

Figure 5.17:
KEK per receiver with per-KEK encrypted broadcasted TEKs.

Much academic research has been applied to the problem of creating efficient hierarchical groups of receivers and keys to strike the best possible balance between communications bandwidth and security. Enterprising readers who do not mind dense academic prose may be interested in the ACM paper "Key Management for Encrypted Broadcast."[55]

5.2.13.5 Single KEK with Group TEKs

In the preceding methods, we assumed a limited amount of physically protected, cryptographic element-based key storage. This constraint is not likely to change in the near future. However, what is also unlikely in modern embedded systems is a similarly tight constraint in other forms of non-volatile memory.

> **Key Point**
>
> A powerful technique for dramatically increasing the amount of protected key storage is to use the physically protected plaintext keys to wrap other keys stored in plentiful storage media, such as NAND flash or hard disk.

Therefore, we propose an approach (see Figure 5.18) that is bandwidth-efficient, is simple to manage, and takes advantage of significant local storage space. Each receiver has a unique KEK; however, the KEK is not used to broadcast TEKs. Instead, a large set of TEKs is wrapped by the KEK and provisioned to the receiver out of band by the service provider. The encrypted TEKs

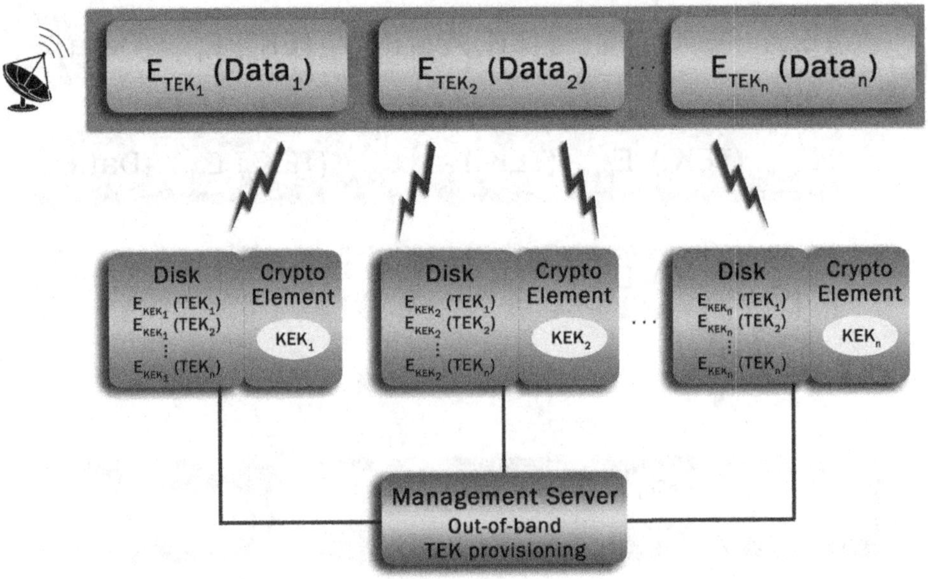

Figure 5.18:
Individualized KEKs with group TEKs provisioned out of band and wrapped on disk.

are stored in regular non-volatile memory. The same TEKs are provisioned to the broadcaster and used to encrypt selected groups of content streams, as described in the previous "Group TEKs" section. The lack of broadcasted TEKs provides maximum bandwidth efficiency.

The lifetime of the locally stored TEKs depends on the amount of storage, the number of individually keyed data streams that must be simulcasted, and the value of the protected information. The service provider must re-provision the receivers at a rate commensurate with the exhaustion of the TEKs' cryptoperiods.

While the rate of out-of-band provisioning may vary across different types of embedded systems, some form of access for this purpose is often possible. In Chapter 4, we discussed the scenario of a submarine that is deployed for a long time and receives mission orders via broadcast. The submarine must store enough TEKs locally until it eventually returns to port. A car must periodically be taken in for service. A smart meter can be occasionally serviced by a utility technician or may be connected to the home network for out-of-band communication. If no out-of-band provisioning were feasible, then grouped KEKs and broadcasted TEKs would be necessary.

5.3 Data-at-Rest Protocols

In 2010, the CBS network in the United States aired a program demonstrating how discarded office copiers are gold mines for private information, trivially harvested from disk drives within the machines. From copiers randomly selected from a used copier warehouse, investigators recovered lists of wanted sex offenders, drug raid targets, architectural design plans, personal identification information (name, address, Social Security number), and medical records—including blood test results and a cancer diagnosis.[56]

When asked whether this situation could be prevented, a copier company said that customers could purchase a $500 option that will erase copied images from the hard drive after use. Give the guy that wrote those couple lines of code a bonus! Another obvious solution is the topic of this section: data-at-rest encryption. Many embedded systems can benefit from data-at-rest protection.

Key Point

Essentially, any device that locally stores sensitive information is at risk of attack such that someone obtaining physical access to the embedded system can access the internal storage and read off the sensitive information.

Embedded systems may incorporate hard disk drives, flash memory, or attached USB thumb drives for this purpose. In the office equipment market, let's consider the following types of embedded systems that can and do store sensitive information during their normal operation:

- Fax machines
- Scanners
- Phone message recording systems
- Digital signage
- Video conferencing systems
- Security camera systems
- Smartphones and tablets
- Printers

Many embedded systems developers are simply unaware of the risks associated with stored data. And the customers of these products are in the same boat. Embedded systems developers can lead the way by first providing data-at-rest encryption features to improve the overall security of these products and then ensuring that end users take advantage of these facilities.

Key Point

Compliance regulations in certain industries require that sensitive stored data be protected with appropriate data protection protocols that include encryption.

In the medical sector, the Health Insurance Portability and Accounting Act (HIPAA) requires the protection of patient data stored within medical devices. In the financial sector, the Payment

Card Industry data security standard (PCI DSS) requires the protection of credit card information within financial processing systems. Data-at-rest protection within smartphones and tablets is a requirement for security-conscious enterprises and governments that allow those handhelds to be used for the processing of sensitive information.

5.3.1 Choosing the Storage Layer for Security

Earlier in this chapter we discussed the multiple layers in the communications stack from which developers may choose to apply data-in-transit protection protocols. As shown in Figure 5.19, an analogous choice exists for the protection of data at rest.

With *full disk encryption* (FDE), the entire media used for storage is encrypted. This has the advantage of ensuring that hidden files, such as operating system temporary files and swap space, are not exposed. When FDE is handled within the media peripheral itself, this is referred to as a *self-encrypting drive* (SED). Self-encrypting drives are common in the laptop market. The advantage of SEDs for the embedded systems developer is that little or no new software must be written to take advantage of the data protection facilities. Encryption is performed with specialized hardware within the storage device, offloading the main embedded applications processor.

> **Key Point**
>
> If self-encrypting storage media is feasible, it is an excellent choice due to ease of use, excellent performance, and the ability to hide the storage encryption key from the main applications processor and memory.

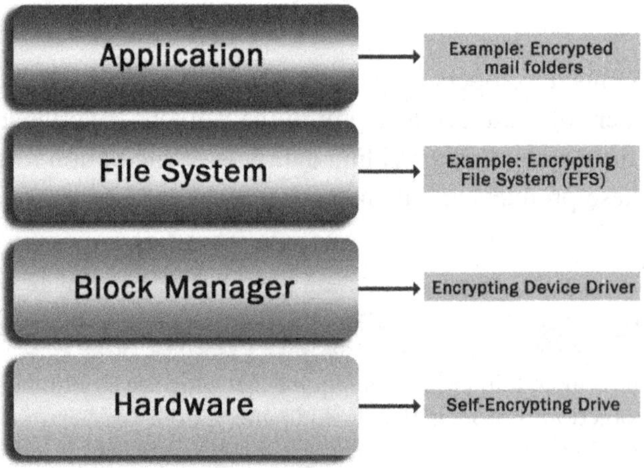

Figure 5.19:
Data-at-rest protection choices by layer.

Unfortunately, many embedded systems will simply be unable to use the available standalone SDE products due to form factor limitations.

Encryption can be performed at the next level up, the device management layer, typically a block-oriented driver. Protection at this level would cover the entire managed media device (FDE). The performance implications of this approach vary. If the embedded platform contains a symmetric encryption accelerator, then the overhead is likely to be reasonable. A pure software cryptographic implementation may cause a dramatic loss in performance. Whether this overhead is acceptable depends on the rate of storage media access requests and other system-level performance requirements. Embedded systems developers can architect the encryption facilities such that the device driver calls out to generic media block encryption routines, ensuring that software is easier to maintain across different generations of the embedded product that may use different types of storage.

Key Point

Some embedded operating systems provide board support packages (BSPs) that include encryption-aware storage device drivers.

The next level candidate for data-at-rest protection is the file system.

Key Point

The major advantage of implementing storage protection at the file system layer is to provide finer granularity over the choice of information that requires storage confidentiality.

Having this granularity is especially important if encryption is performed in software with minimal or no hardware acceleration. Depending on the file system implementation, developers may be provided options for encryption at the volume level or at the individual file level.

Finally, applications can add their own data protection, either using underlying file system encryption features or a custom implementation. For example, an audit logging application can encrypt its audit records prior to calling the standard file system output functions.

In the cases of volume-, file-, or application-level data protection, developers can employ separate keys for these groups of data rather than a single key for the entire system. This is a sensible application of least privilege principles.

Key Point

As with network security layer selection, there is no single right answer for selecting the layer for data-at-rest protection; in some cases, developers may want to use more than one layer for defense-in-depth.

Developers resorting to custom, application-level approaches will also need to take on key management responsibility, whereas users of encrypting file systems or SEDs can take advantage of the key management framework provided by the product supplier.

5.3.2 Symmetric Encryption Algorithm Selection

Data-at-rest presents some unique challenges for encryption algorithms relative to network security protocols. While developers may be able to rely on the operating system or file system vendor to handle the encryption details, it is critical to understand whether the supplier has selected a strong technical approach. Furthermore, readers who are implementing a do-it-yourself data protection system need to understand these details.

For readers who may have skipped Chapter 4, now is a good time to go back and become familiar with that part of the chapter covering symmetric encryption algorithms and modes. For the purposes of this discussion, we omit integrity protection, as most data-at-rest applications concern themselves only with confidentiality.

For data-at-rest protection, we generally want an encryption algorithm that can be performed without adding additional storage space: a plaintext media block is encrypted in place, generating a ciphertext block of the same size. The most basic encryption mode, ECB, would provide this memory conservation but is not suitable for data-at-rest encryption since any two same plaintext blocks will encrypt to the same ciphertext, making it easy for an attacker to find patterns in the data and potentially derive information. We must consider other modes, most of which require an initialization vector (IV). However, to avoid space expansion, the data protection system must include a means for implicitly deriving this IV.

Implicit IV derivation poses a surprisingly difficult challenge for common encryption modes. From our study of these modes in Chapter 4, we know that the common modes require uniqueness: the same IV must never be reused for a particular key. For example, with CTR mode, a predictable counter can be used, but the same number can never be repeated for a given key. For CBC mode, a unique and unpredictable number must be used. Network security protocols have the freedom to generate the IV and send it along as part of the transmitted data; for AES-CBC, each transmission can generate a new random number for the IV and transmit this IV to the receiver. But for data-at-rest, we have no room to store the IV for subsequent decryption.

The obvious source for an implicit IV would be the sector number and offset for a particular data block. Using this combination provides every disk block with a relatively unique input value. However, as data blocks are overwritten during the course of normal operation, the same sector and offset are reused for the same key. This implies a serious weakness in the applicability of common encryption modes for data-at-rest protection. Numerous other weaknesses of common modes, especially CBC, have been identified when applied to data-at-rest protection protocols. An excellent paper discussing these weaknesses is that by Clemens Fruhwirth, "New Methods in Hard Disk Encryption."[57]

5.3.2.1 Tweakable Ciphers

The good news is that cryptographers have worked diligently to address this encryption mode challenge. Liskov, Rivest, and Wagner introduced the concept of a *tweakable block cipher* in 2002.[58] The basic idea of a tweakable cipher is to apply the IV concept to the single-block cipher itself rather than to a chaining mode built on top of the block cipher. As shown in Figure 5.20, the block cipher converts a plaintext block to a ciphertext block, using both the traditional key as well as the tweak as inputs.

Key Point

The practical application of tweakable ciphers for the data-at-rest protection problem is the property that the cipher's security does not preclude reuse of the initialization vector; thus, media sector number and block offset within the sector provide a perfect fit for tweak selection.

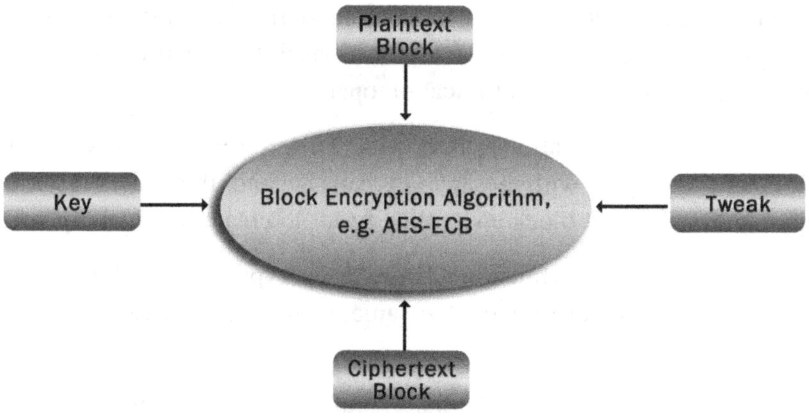

Figure 5.20:
Tweakable block cipher overview.

5.3.2.2 XTS-AES

In 2007, IEEE's Security in Storage Working Group (SISWG) published standard P1619.[59]

> **Key Point**
>
> The IEEE P1619 standard defines the XTS-AES cipher mode as a result of a thorough study of numerous potential tweak-based algorithms for use in data-at-rest protection applications.

This choice is further bolstered by NIST in its Special Publication 800-38E,[60] which approves the XTS-AES cipher mode and references its definition in IEEE P1619-2007. Finally, FIPS 140-2 has been amended to include XTS-AES as an approved cipher for validation.

As an example of commercial adoption, note that the latest Mac OSX release at the time of this writing, codename Lion, is the first Mac operating system to provide a native full disk encryption feature; this product uses XTS-AES.

The tweak algorithm found in XTS-AES is based on and almost identical to the one originally created by noted cryptographer Phillip Rogaway, called XEX.[61] In addition to strong security, XEX and therefore XTS-AES are also designed for efficiency when applied to storage of many sequential data blocks (as is common with file storage).

The XTS-AES block cipher is depicted in Figure 5.21.

An odd aspect of this cipher is the requirement for twice the keying material; for 128-bit security, 256-bits of key must be used. The first half of the key is used to process the plaintext; the second half is used to encrypt a 128-bit representation of the sector number, which acts as the primary tweak, as shown in Figure 5.21. The result of this encryption is then fed to a function that performs a Galois field multiplication (implemented as a sequence of shifts and XORs) of the encryption result with a Galois constant derived from the secondary tweak, the numeric index of the data block within the sector. Consult the IEEE P1619 document for a detailed definition of the Galois multiplication operation.

The result of this Galois multiplication is used twice. First, it is added (XORed) to the plaintext block, which is then encrypted with the first key half. Second, the Galois result is added (XORed) again to the plaintext block encryption result to create the final ciphertext block.

Decryption is similar; however, while the AES-ECB decryption algorithm is used to process the ciphertext, the tweak cipher remains the same, using the AES-ECB encryption algorithm.

In practice, data is stored to media in sectors. Therefore, the preceding block encryption algorithm must be executed in a loop across the entire sector. Note that while XTS-AES handles partial blocks, that part of the algorithm is often unnecessary. For example, the

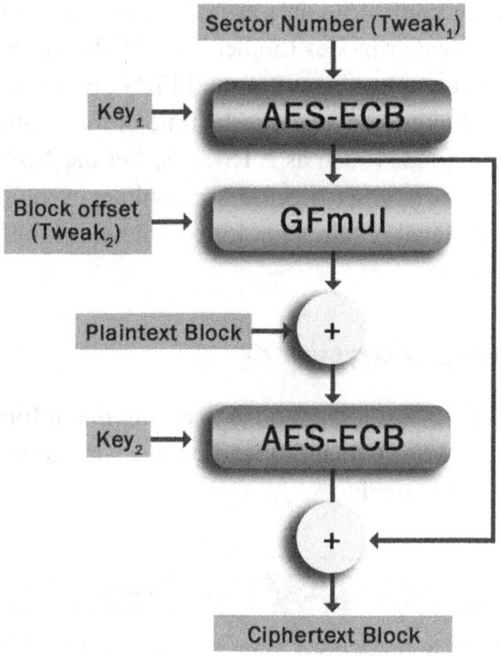

Figure 5.21:
The XTS-AES data-at-rest encryption cipher.

common sector size of 512 bytes will result in 32 block encryptions, and most media management layers will access a full sector at a time. For such a system, given a function, *xts_encrypt*, which takes the sector number and size in bytes, plaintext block, and encryption key as input, the following simple code sequence handles the sector encryption:

```
sector_encrypt(uint8_t *sector, uint32_t sector_num, uint32_t
    sector_size, uint8_t key[])
{
  uint32_t i;
  assert((sector_size % AES_BLOCK_SIZE) == 0); /* 512 % 16 */
  for (i = 0; i < sector_size/AES_BLOCK_SIZE; i++) /* 32x */
    xts_encrypt(sector+i*AES_BLOCK_SIZE, key, sector_num, i);
}
```

It is also easy to see from this code sequence that XTS-AES is parallelizable. If the embedded system contains an AES hardware accelerator (especially one that has direct support for XTS mode), the preceding implementation should be modified to take advantage of the accelerator's ability to process multiple AES blocks at once. Furthermore, if the media allows for sector size configurability, developers may want to vary the sector size to see if better throughput (potentially at the expense of slightly reduced space efficiency) can be achieved.

Embedded systems developers selecting data-at-rest protection products are advised to avoid legacy approaches that use weaker modes (numerous CBC-based implementations have been commercialized) and instead employ the NIST- and FIPS-approved standards. Perhaps surprisingly, NSA's current Suite B guidance at the time of this writing is reticent with respect to data-at-rest protocols and modes such as XTS-AES. Yet the NSA has expressed a strong desire for commercial products to employ Suite B-compliant cryptography and points to FIPS 140-2 validation as the entry-level requirement for commercial solutions. Thus, Suite B makes an indirect case for XTS-AES.

5.3.3 Managing the Storage Encryption Key

The primary purpose of data-at-rest protection is to ensure that information residing on lost or stolen media cannot be accessed by unauthorized parties who must be assumed to have complete physical access to the media.

> **Key Point**
>
> The symmetric storage encryption key must never be stored in the clear on the media.

However, it is often necessary to store an encrypted copy of the symmetric key on the media (or perhaps an attached Trusted Platform Module, if available). The key is unwrapped for active use while the system is executing in an authorized manner. For personal computers such as laptops and smartphones, unwrapping is triggered by successful authentication of the user (e.g., using a password, smartcard, biometric, etc., or multiple factors).

5.3.3.1 Generating the Storage Encryption Key

A typical method of storage encryption key establishment, for example, is to convert user credentials into a key using a key derivation function (KDF). A popular KDF used to convert passwords is the password-based key derivation function, version 2 (PBKDF2). PBKDF2 is defined in the RSA Laboratories specification PKCS #5;[62] it is duplicated in RFC 2898.[63] PBKDF2 applies a hash function to the password concatenated with a salt (random bitstring). To make password cracking more difficult, the standard recommends that the hash output be rehashed multiple times. The recommended minimum hash iteration count is 1,000, although the number is expected to increase over time. Apple's iOS 4.0 uses 10,000 iterations. In September 2010, RIM BlackBerry's encrypted backup service was determined to be vulnerable due to faulty application of PBKDF2: instead of following the standard, the BlackBerry software used an iteration count of one. The BlackBerry vulnerability is documented in the National Vulnerability Database as CVE-2010-3741.[64]

When the password is used to directly generate the storage encryption key, a change in password changes the encryption key and therefore forces re-encryption of the entire protected media. To avoid this problem, a permanent, unique encryption key is created when the media is initially provisioned, and then this key is wrapped (encrypted) with the password-derived key. With this two-level keying scheme, a periodic password change (good security practice) only requires re-wrapping of the encryption key.

The user authentication approach may be sufficient for limited types of attended embedded systems that can tolerate user intervention whenever the protected volumes must be unlocked. Imagine an enterprise printer that has FDE (sadly, most of them do not, but that is why we're here!). With such a high-availability system in a workplace environment, it may be acceptable to require a system administrator to log on to the printer to unlock the protected volumes in the case of a system reboot (planned or unplanned).

Nevertheless, this approach is not sufficient for large classes of unattended embedded systems. If the embedded system encounters a fault and automatically reboots, the encrypted volumes must be able to get back online without manual credential input.

5.3.3.2 Remote Key Provisioning

We can consider two classes of unattended embedded systems: those that have a remote management network interface and those that do not. For the latter, the embedded system simply lacks any mechanism for dynamic interaction that can unlock an encryption key. In this case, if information value demands data-at-rest protection, then the designer is advised to incorporate a cryptographic co-processor that provides physical tamper-resistant key storage and internal execution of the data encryption algorithm. The device driver sends in plaintext to this encryptor and receives ciphertext for storage on disk and similarly requests decryption of disk blocks as needed.

For network-enabled embedded systems, a remote management server holds a database of the provisioned data encryption keys. A server connection is initiated by the embedded system whenever a data encryption key must be unlocked (e.g., at boot time). The embedded system and server mutually authenticate (using some choice of the network security protocols described earlier in this chapter), and then the server provides a copy of the embedded system's provisioned data encryption key over the secured channel. This approach obviously depends on available network connectivity to enable the data protection services. The system designer must adopt a policy to handle the case in which the network is down when an unlock operation is required. The policy may disable certain sensitive data storage operations, allow for non-protected storage, or may force the device into a non-operational state.

5.3.3.3 Key Escrow

When we forget a website password, there's always a convenient "forgot your password?" link we can click to reset it.

Key Point

When implementing a data-at-rest protection system, developers must consider key escrow to guard against the possibility that the authentication information used to unlock the storage encryption key will be lost.

Furthermore, there are situations in which the system owner may need to extract the data from storage, such as after a system failure. Holding a copy of the data encryption key in an offsite secure location is advisable in most system designs to prevent loss of data when the data encryption key is no longer accessible. If the embedded system lacks a network management interface, then the internally stored key must be exportable onto media for offsite escrow storage (e.g., in a secure vault). If the system supports network management and the remote key provisioning described in the previous section, then developers simply need to ensure that remotely provisioned keys are retained on a secure server or copied to protected offline media.

This description begins to touch on the general subject of key life-cycle management. Many embedded systems developers adding in security functions, such as data-at-rest protection, for the first time may be ill equipped to handle the problem of managing keys across a large deployment. Furthermore, for an embedded system manufacturer, customers may expect that a key life-cycle management system be provided as part of its offering. For example, a developer of smart meters can sell not only smart meters, but also a back-end key management system for creating, distributing, auditing, and escrowing keys used for data-at-rest protection within the smart meters. While a comprehensive discussion of enterprise key life-cycle management is beyond the scope of this book, many embedded software suppliers that provide network security and data-at-rest protection solutions also offer a key life-cycle management system. Readers are encouraged to consult vendors to see what options are available.

5.3.4 Advanced Threats to Data Encryption Solutions

The authentication software that runs to unlock encrypted media must itself be trustworthy and tamper protected. For example, the embedded operating system may incorporate the authentication function directly. The embedded operating system image (and any preceding boot loaders) is not encrypted; only the rest of the media, which contains sensitive files, is protected. If the embedded operating system is not trusted (e.g., at risk of containing malware or vulnerabilities that would permit the loading of malware), then the authentication process could be subverted. For example, a key logger could record the user's password, enabling recovery of the storage encryption key and all the encrypted data.

If we assume the embedded operating system is trustworthy, then we still must ensure that anything executing prior to launch of the operating system is trusted. This is another good example of the need for secure boot (discussed in Chapter 2).

In some cases, the designer may want the embedded operating system to also be encrypted. When FDE is in use and a sophisticated operating system (e.g., Linux) resides on the encrypted disk, *pre-boot authentication* may be employed: a small portion of the encrypted disk contains a mini-operating system that is booted for the sole purpose of performing the authentication and unlocking of the media prior to booting the full operating system. If the embedded operating system is a secure microkernel (Chapter 2), then a separate pre-boot authentication module is not required.

Attacks against pre-boot authenticators have been successfully perpetrated. For example, the system is booted to a malicious operating system (e.g., alternative booting from an external USB drive) that tampers with the pre-boot code to steal the authentication credentials as they are input.[65] Secure boot can prevent this attack as well; the signature of the modified authenticator will fail to match the known-good version, aborting the boot process.

Another example of advanced threat is the so-called *cold-boot attack*. Unless the embedded system is using a self-encrypting hard drive where the keys are stored within the media and never exposed to the main processor, disk encryption requires that the storage encryption key be kept in memory (in the clear) while the system is operational, invoking the encryption and decryption algorithm to access data. When the system is turned off, RAM is unavailable, and the only copy of the encryption key is itself encrypted. Or is it? In some systems, RAM is not immediately cleared. An attacker boots the system using a malicious operating system that grabs the plaintext key in RAM. This attack has been performed successfully.[66]

Data-at-rest protection within an embedded system equipped with secure boot and a trusted operating system impervious to remote attack can still be defeated by removing the protected media and booting it on a different computer that lacks this secure environment. Binding the storage encryption key to its intended embedded system platform can prevent this attack. In this case, the permanent storage encryption key is derived (in whole or in combination with user credentials) from a platform-specific key, such as a fused one-time programmable key or TPM key (if applicable). Even if the user's credentials are stolen, the storage encryption key cannot be derived outside the targeted embedded platform. The downside of this extra level of defense is that a hardware failure that prevents access to the platform credential will render the data permanently inaccessible (unless the derived storage encryption key itself is securely escrowed).

Embedded systems developers looking to incorporate network security and/or data-at-rest protection into their next designs are faced with a plethora of design choices and constraints. This chapter has sought to provide designers with an overview of the key issues that need to be considered. At the software architecture level, key network security protocol choices include IPsec and TLS; while the latest revisions solve important security problems, designers must also consider requirements for interoperability with communication peers that speak older versions. From a hardware and systems perspective, designers need to consider the hardware security capabilities, including full protocol offload as well as cryptographic algorithm acceleration and secure key storage, of embedded processors. Footprint and latency constraints as well as certification requirements may further influence the road taken. Special considerations for data-at-rest protection include the use of government-approved symmetric encryption algorithms designed specifically for such applications and proper management of the long-term keys typically used for this purpose.

5.4 Key Points

1. Identity anonymity (part of transmission security) is often a desired goal of data-in-motion protocols.
2. Multipoint applications force key management and exchange challenges to be decoupled from the bulk data protocols: the developer must find a way to distribute a shared key across the group and then multicast packets protected with that key.
3. Multi-layered data protection may be used as a defense-in-depth strategy and can also occur when upper layers are unaware or distrustful of lower-layer protection capabilities.
4. Choice of network protection layer can have significant impact on functionality, performance, interoperability, and maintainability.
5. It is the efficient protocol handling and encryption in hardware and lack of spare applications processing and storage resources that forces certain embedded systems to rely on data link layer protection rather than more flexible upper-layer choices.
6. The biggest drawback of link layer data protection is that, by definition, the scheme can work only on a network consisting of devices using the homogeneous data link media.
7. The IEEE 802.1X standard defines requirements for the secure joining of endpoints into Ethernet networks.
8. The mapping of Extensible Authentication Protocol (EAP) requirements to wireless Ethernet LANs as well as other wireless Ethernet security requirements is defined in IEEE 802.11i-2004.
9. The security functions specified in IEEE 802.11i protect against local eavesdroppers and hackers trying to gain access to a network via the access point but do not provide end-to-end protection.
10. Contrary to popular belief, Wi-Fi Protected Access 2 (WPA2) is not a definitive security standard, but rather a certification quasi-standard intended to provide industry with confidence that Wi-Fi-certified devices meet the requirements of IEEE 802.11i.
11. Embedded systems developers acquiring wireless Ethernet software for incorporation into devices should select stacks that not only claim conformance to WPA2, but have also have been successfully deployed in Wi-Fi-certified products.

12. Known as MACsec, the IEEE 802.1AE standard is the analog of IEEE 802.11i for wired Ethernet (IEEE 802.3) networks.
13. Developers looking for the most portable, flexible, and extensible data-in-motion security should look above the data link layer toward the network (IP) and higher layers.
14. IPsec operates at layer three (the Internet Protocol layer), making it invisible to both higher-level protocols (e.g., TCP) as well as applications.
15. SSL and its modern variant, Transport Layer Security (TLS), operate on layer four, the transport layer; it is common for applications to use TLS by invoking a sockets-like API.
16. A major advantage of TLS is that it can take advantage of TCP's guaranteed in-order packet delivery to simplify the protocol relative to IPsec, which must handle dropped messages and out-of-order delivery.
17. The use of TLS enables applications to select when they need the security (and are willing to pay the associated overhead).
18. An important advantage of IPsec is that applications can derive security benefits without requiring any application-level decisions or code modifications (such as adjusting to use a modified-sockets API).
19. An example of a good choice for bump-in-the-stack (BITS) IPsec is a virtualized environment where guest operating systems already have integrated network stacks, yet there is a desire to implement network security outside the virtual machine where flaws in guest operating systems could derail intended security.
20. ESP Tunnel mode is the IPsec combination of choice when maximum security and confidentiality are required.
21. In general, an IKE/IPsec or TLS session key's lifetime (called the cryptoperiod) should be limited and a new session key re-established to limit the damage should a session key be compromised.
22. Generally speaking, if sufficient interoperability can be assured, IKEv2 is strongly recommended over IKEv1.
23. In the best case of a non-routable (local) packet, 6LoWPAN can reduce the UDP/IPv6 header size to just 4 bytes (relative to the usual 48); the best case for routable packets yields a header size of 11 bytes.
24. The lack of any standardized API or framework for IKE and IPsec has led to the creation of VPN client applications to encapsulate some of the tedious configuration, such as identifying a server IP address and authentication information, associated with these protocols.
25. Datagram TLS (DTLS) is designed to provide the same capabilities as TLS for UDP-based transport protocols and applications that use them.
26. The history of SSH demonstrates the difficulty of achieving high levels of security in sophisticated data protection protocols.
27. Due to the complexity and large footprint of both IPsec and TLS, a number of commercial software suppliers and university researchers have proposed alternative, lighter-weight

network security protocols; for embedded systems designers more concerned about simplicity and footprint and less concerned about interoperability with arbitrary peers, custom protocols may be a viable alternative.

28. While the current open standards for network security protocols, such as TLS, DTLS, SSH, and IPsec, do not provide full traffic flow confidentiality, some embedded software suppliers provide versions of these protocols that are enhanced to provide this feature.
29. While proprietary, non-standard multimedia transport protocols exist, the majority of commercial and open source multimedia (especially voice and video) transport over IP protocols use RTP and its adjunct security protocol, Secure Real-Time Transport Protocol (SRTP).
30. The sharing of a single symmetric key for all protected broadcast content violates the most basic application of the least privilege principle and would be considered anathema to any cryptographer.
31. A powerful technique for dramatically increasing the amount of protected key storage is to use the physically protected plaintext keys to wrap other keys stored in plentiful storage media, such as NAND flash or hard disk.
32. Essentially, any device that locally stores sensitive information is at risk of attack where someone obtaining physical access to the embedded system can access the internal storage and read off the sensitive information.
33. Compliance regulations in certain industries require that sensitive stored data be protected with appropriate data protection protocols that include encryption.
34. If self-encrypting storage media is feasible, it is an excellent choice due to ease of use, excellent performance, and the ability to hide the storage encryption key from the main applications processor and memory.
35. Some embedded operating systems provide board support packages (BSPs) that include encryption-aware storage device drivers.
36. The major advantage of implementing storage protection at the file system layer is to provide finer granularity over the choice of information that requires storage confidentiality.
37. As with network security layer selection, there is no single right answer for selecting the layer for data-at-rest protection; in some cases, developers may want to use more than one layer for defense-in-depth.
38. The practical application of tweakable ciphers for the data-at-rest protection problem is the property that the cipher's security does not preclude reuse of the initialization vector; thus, media sector number and block offset within the sector provide a perfect fit for tweak selection.
39. The IEEE P1619 standard defines the XTS-AES cipher mode as a result of a thorough study of numerous potential tweak-based algorithms for use in data-at-rest protection applications.

40. The symmetric storage encryption key must never be stored in the clear on the media.
41. When implementing a data-at-rest protection system, developers must consider key escrow to guard against the possibility that the authentication information used to unlock the storage encryption key will be lost.

5.5 Bibliography and Notes

1. Boyd C, Mathuria A. *Protocols for Authentication and Key Establishment.* Heidelberg: Springer-Verlag; 2003.
2. Multicast Extensions to the Security Architecture for the Internet Protocol. Internet Engineering Task Force, Request for Comments: 5374; November 2008.
3. Multicast Security (MSEC) Group Key Management Architecture. Internet Engineering Task Force, Request for Comments: 4046; April 2005.
4. IEEE Standard 802.1x-2001. *IEEE Standard for Local and Metropolitan Area Networks—Port-Based Network Access Control.* IEEE Computer Security; June 2001.
5. *Extensible Authentication Protocol (EAP).* Internet Engineering Task Force, Request for Comments: 3748; June 2004.
6. *Extensible Authentication Protocol (EAP) Key Management Framework.* Internet Engineering Task Force, Request for Comments: 4017; March 2005.
7. *Extensible Authentication Protocol (EAP) Method Requirements for Wireless LANs.* Internet Engineering Task Force, Request for Comments: 5247; August 2008.
8. Geier J. *Implementing 802.1x Security Solutions for Wired and Wireless Networks.* New York: Wiley; 2008.
9. IEEE Standard 802.11i. *Supplement to Standard for Telecommunications Information Exchange Between Systems—LAN/MAN Specific Requirements—Part 11: Wireless LAN Medium Access Control (MAC) and Physical Layer (PHY) Specification for Enhanced Security*; July 2004.
10. The State of Wi-Fi Security. *Wi-Fi Certified WPA2 Delivers Advanced Security to Homes, Enterprises and Mobile Devices.* Wi-Fi Alliance; September 2009.
11. IEEE Standard 802.1AE-2006. *IEEE Standard for Local and Metropolitan Area Networks—Media Access Control (MAC) Security.* IEEE Computer Security; August 2006.
12. Lawson S. Facebook Sees Need for Terabit Ethernet. *Computerworld*; February 3, 2010.
13. *Security Architecture for the Internet Protocol.* Internet Engineering Task Force, Request for Comments: 4301; December 2005.
14. *IP Authentication Header.* Internet Engineering Task Force, Request for Comments: 4302; December 2005.
15. *IP Encapsulating Security Payload (ESP).* Internet Engineering Task Force, Request for Comments: 4303; December 2005.
16. NIST Special Publication 800-57. *Recommendation for Key Management—Part 1: General (Revised)*; March 2007.
17. *The Internet Key Exchange.* Internet Engineering Task Force, Request for Comments: 2409; November 1998.
18. *Internet Key Exchange (IKEv2) Protocol.* Internet Engineering Task Force, Request for Comments: 4306; December 2005.
19. *Internet Key Exchange Protocol Version 2 (IKEv2).* Internet Engineering Task Force, Request for Comments: 5996; September 2010.
20. *Transmission of IPv6 Packets over IEEE 802.15.4 Networks.* Internet Engineering Task Force, Request for Comments: 4944; September 2007.
21. ISA-100.11a. *Wireless Systems for Industrial Automation: Process Control and Related Applications.* The International Society for Automation; 2009.
22. *Zigbee Smart Energy 2.0 Technical Requirements Document, draft version 0.7.* Zigbee Alliance; 2011.
23. Shelby Z, Bormann C. *6LoWPAN, The Wireless Embedded Internet.* New York: Wiley; 2009.

24. *The TLS Protocol Version 1.0*. Internet Engineering Task Force, Request for Comments: 2246; January 1999.
25. *The Transport Layer Security (TLS) Protocol Version 1.1*. Internet Engineering Task Force, Request for Comments: 4346; April 2006.
26. *The Transport Layer Security (TLS) Protocol Version 1.2*. Internet Engineering Task Force, Request for Comments: 5246; August 2008.
27. Snader JC. *VPNs Illustrated: Tunnels, VPNs, and IPsec*. Addison-Wesley Professional; November 2005.
28. Bantoft K, Wouters P. *Openswan: Building and Integrating Virtual Private Networks: Learn from the Developers of Openswan How to Build Industry Standard, Military Grade VPNs and Connect Them with Windows, MacOSX, and Other VPN Vendors*. Birmingham, UK: Packt Publishing; February 2006.
29. *Datagram Transport Layer Security*. Internet Engineering Task Force, Request for Comments: 4347; April 2006.
30. SSH CBC Vulnerability, US-CERT Vulnerability Note VU#958563.
31. SSH CRC32 Attack Detection Code Contains Remote Integer Overflow, US-CERT Vulnerability Note VU#945216.
32. *The Secure Shell (SSH) Protocol Assigned Numbers*. Internet Engineering Task Force, Request for Comments: 4250; January 2006.
33. *The Secure Shell (SSH) Protocol Architecture*. Internet Engineering Task Force, Request for Comments: 4251; January 2006.
34. *The Secure Shell (SSH) Authentication Protocol*. Internet Engineering Task Force, Request for Comments: 4252; January 2006.
35. *The Secure Shell (SSH) Transport Layer Protocol*. Internet Engineering Task Force, Request for Comments: 4253; January 2006.
36. *The Secure Shell (SSH) Connection Protocol*. Internet Engineering Task Force, Request for Comments: 4253; January 2006.
37. SSH File Transfer Protocol. Internet-Draft from the Secure Shell Working Group.
38. National Institute of Standards and Technology National Vulnerability Database. http://web.nvd.nist.gov/view/vuln/detail?vulnId=CVE-2010-4180.
39. National Institute of Standards and Technology National Vulnerability Database. http://web.nvd.nist.gov/view/vuln/detail?vulnId=CVE-2009-3555.
40. *Transport Layer Security (TLS) Renegotiation Indication Extension*. Internet Engineering Task Force, Request for Comments: 5746; February 2010.
41. Recommendation ITU-T H.323. *Packet-Based Multimedia Communications Systems*. Telecommunications Standardization Section of ITU; December 2009.
42. Papageorgiou P. *A Comparison of H.323 vs. SIP*; June 4, 2001.
43. *SIP: Session Initiation Protocol*. Internet Engineering Task Force, Request for Comments: 2543; June 2002.
44. *Security Mechanism Agreement for the Session Initiation Protocol (SIP)*. Internet Engineering Task Force, Request for Comments: 3329; January 2003.
45. Schulzrinne H. *Issues in Designing a Transport Protocol for Audio and Video Conferences and Other Multiparticipant Real-Time Applications*. expired Internet Draft; October 1993.
46. *RTP: A Transport Protocol for Real-Time Applications*. Internet Engineering Task Force, Request for Comments: 3550; July 2003.
47. *RTP Profile for Audio and Video Conferences with Minimal Control*. Internet Engineering Task Force, Request for Comments: 3551; July 2003.
48. *The Secure Real-Time Transport Protocol (SRTP)*. Internet Engineering Task Force, Request for Comments: 3711; March 2004.
49. *MIKEY: Multimedia Internet KEYing, Internet Engineering Task Force*. Request for Comments: 3830; August 2004.
50. *ZRTP: Media Path Key Agreement for Unicast Secure RTP*. Internet Engineering Task Force, Request for Comments: 6189; April 2011.

51. Orrblad J. *Alternatives to MIKEY/SRTP to Secure VoIP.* ftp://ftp.it.kth.se/Reports/DEGREE-PROJECTREPORTS/ 050330-Joachim-Orrblad.pdf; March 2005.
52. *Requirements and Analysis of Media Key Management Protocols.* Internet Engineering Task Force, Request for Comments: 5479; April 2009.
53. *Datagram Transport Layer Security (DTLS) Extension to Establish Keys for the Secure Real-time Transport Protocol (SRTP).* Internet Engineering Task Force, Request for Comments: 5764; May 2010.
54. *Suite B Profile for Datagram Transport Layer Security/Secure Real-Time Transport Protocol (DTLS-SRTP).* Internet Engineering Task Force, Draft RFC; April 8, 2011.
55. Wool A. Key Management for Encrypted Broadcast. *ACM Transactions on Information and System Security (TISSEC)* May 2000;**3**(2).
56. Keteyian A. Digital Photocopiers Loaded With Secrets. http://www.cbsnews.com/stories/2010/04/19/eveningnews/main6412439.shtml?tag=currentVideoInfo;videoMetaInfo; April 20, 2010.
57. Fruhwirth C. New Methods in Hard Disk Encryption; Institute for Computer Languages Theory and Logic Group. *Vienna University of Technology*; July 18, 2005.
58. Liskov M, Rivest R, Wagner D. *Tweakable Block Ciphers.* LNCS, Crypto: Santa Barbara, CA; 2002.
59. Security in Storage Working Group of the IEEE Computer Society Committee. IEEE P1619. *Standard for Cryptographic Protection of Data on Block-Oriented Storage Devices*; 2007.
60. NIST Special Publication 800-38E. *Recommendation for Block Cipher Modes of Operation: The XTS-AES Mode for Confidentiality on Storage Devices*; January 2010.
61. Rogaway P. *Efficient Instantiations of Tweakable Blockciphers and Refinements to Modes OCB and PMAC*, http://www.cs.ucdavis.edu/~rogaway/papers/offsets.pdf; September 24, 2004.
62. RSA Laboratories. *PKCS #5 v2.0: Password-Based Cryptography Standard*; March 25, 1999.
63. *PKCS #5: Password-Based Cryptography Specification Version 2.0.* Internet Engineering Task Force, Request for Comments: 2898; September 2000.
64. National Institute of Standards and Technology National Vulnerability Database. http://web.nvd.nist.gov/view/vuln/detail?vulnId=CVE-2010-3741.
65. Turpe S, et al. *Attacking the BitLocker Boot Process.* Proceedings of the 2nd International Conference on Trusted Computing. TRUST; 2009. Oxford, UK, April 6–8, 2009; LNCS 5471, Springer, 2009. DOI: 10.1007/978-3-642-00587-9_12.
66. Halderman AJ, Schoen SD, Heninger N, Clarkson W, Paul W, Candrino JA, Feldman AJ, Appelbaum J, Felten EW. *Lest We Remember: Cold Boot Attacks on Encryption Keys.* Proceedings of the 17th USENIX Security Symposium; August 2008. 45–60.

CHAPTER 6

Emerging Applications

Chapter Outline

Embedded Systems Security. DOI: 10.1016/B978-0-12-386886-2.00006-0

Welcome to the final chapter of the book and one that is perhaps the most fun for the authors to write and hopefully equally enjoyable for you to read. While this book includes numerous case studies, this chapter provides a small number of more detailed case studies of the issues, concerns, and guidance regarding the creation of secure embedded systems. In particular, we focus on some modern topics that may soon present unique challenges for your next embedded project: embedded network transactions, automotive security, and how to securely incorporate the popular Android and Linux operating systems into mobile and embedded systems.

6.1 Embedded Network Transactions

Key Point

Much of the world's commerce and critical infrastructure control depends on the security of network transactions.

The ability to provide end-to-end transaction security, encompassing both client and server, therefore is critical to enterprises, governments, and consumers. Increasingly, embedded systems act as the clients (and sometimes the server) in such transactions. Yet today's security posture for embedded network transactions is hopelessly inadequate. Critical infrastructure is saddled with vulnerabilities in both operating systems and applications, in both client endpoints and in servers. We employ filters, scanners, and Patch Tuesdays, but there are always new vulnerabilities that leave our critical resources exposed. And as we have become increasingly dependent on network transactions, the determination and sophistication of hackers has risen in tandem.

> **Key Point**
>
> For as little as $25,000, a sophisticated hacker will break into most any network, even the most strenuously protected enterprise data centers.

Hackers may bypass server security controls, or more often, subvert a comparatively weak client endpoint from which to access the server's crown jewels. Increasingly, those client endpoints are embedded systems.

While there are many security products and research projects that attempt to improve network transaction security, there has never been a solution that is both highly assured (i.e., certified to provide users with a very high confidence that the system will protect high-value information against sophisticated attackers) and deployable (cost effective and easy to use). This section explains the primary modern security threats facing network transactions and surveys some of the latest attempts (some weak, some promising) to address the problem. In particular, we identify security approaches that are scalable across all client-computing platforms performing commercial transactions over the Internet, including cell phones and deeply embedded devices.

6.1.1 Anatomy of a Network Transaction

In a network transaction, a client makes a request to a remote server. The client may be making a payment (on behalf of a cell phone user), modifying the state of a remote resource, or requesting a wide variety of other services. The server satisfies the request by transferring funds, modifying account state, and so on, and often sends back a response to the client, completing the transaction.

> **Key Point**
>
> Transaction security is defined by the ability of both client and server to have trust in the authenticity, integrity, and confidentiality of the messages that are sent and received by both parties throughout the transaction.

Attacks on transactions may target the transaction data—either in situ, within a browser, or during transmission—or the endpoint computers and resources to which they are connected.

6.1.2 State of Insecurity

The most common way for a hacker to gain access to critical information in servers is to hijack the client endpoint. Client identities can be stolen through a wide variety of techniques, some of which we discuss later. The hacker uses commandeered credentials and/or web sessions to perform illicit transactions and to gain unauthorized access to servers. While servers are usually protected by expensive security appliances such as firewalls, intrusion detection systems, intrusion protection systems, and unified threat management devices, embedded systems are generally much more weakly protected and are often situated beyond the server's security zone, providing a softer attack profile for hackers.

As reported by analyst firm IDC,

> *Transmission vehicles for malicious codes (malware) have progressed over the past decade from floppy disk, to email, to most recently, networks themselves. This is especially threatening because malware writers have discovered that clients attached to high-speed wireless and wired networks are powerful means to spread infections. Clients have become the source of distributed network attacks on servers and more importantly, applications.*[1]

The pernicious effects of these attacks on transactions have been well publicized. We discuss just a handful of recent events that demonstrate the increasing level of sophistication seen in these attacks. According to a *Businessweek* report, top NASA officials were socially engineered via fraudulent e-mail, resulting in the compromise of computers in the agency's Washington headquarters in 2006. One of the key findings of the subsequent investigation, as reported by the NASA Inspector General, was that "the scope, sophistication, timing, and hostile characteristics of some of the intrusions indicate they are coordinated or centrally managed."[2]

During the 2008 U.S. presidential campaign, CNN reported a sophisticated endpoint computer malware intrusion aimed at compromising presidential policy information. Agents from the FBI and Secret Service reported to the Obama campaign headquarters that "you have a problem way bigger than what you understand…you have been compromised, and a serious amount of files have been loaded off your system."[3]

InformationWeek reported that the well-publicized 2007 TJX breach that resulted in the theft of 45 million customer records originated, at least in part, from in-store kiosks that were used by hackers to gain access into the corporate LAN.[4]

Also in 2008, Wired.com reported how teen queen Miley Cyrus's e-mail (residing in the Google cloud), along with some private photos, was stolen when a hacker tricked a MySpace worker into turning over credentials enabling administrator access to MySpace's servers.[5]

This is, of course, but a smattering of the reported attacks on transaction-based computing. However, the daily deluge of such stories provides testament to the state of insecurity in which we transact business, control infrastructure, and socialize on the Internet. In the following section, we survey the major threats facing today's network endpoints.

6.1.3 Network-based Transaction Threats

6.1.3.1 Man-in-the-Middle and/or Eavesdropping

Man-in-the-middle (MITM) attacks are highly desirable by attackers because they do not necessarily require a successful client endpoint intrusion such as downloaded malware. Rather, the attacker intercepts client messages and injects messages back to the client (and in some cases, also back to the user's intended server) to redirect the client to a false server or otherwise impersonate a valid connection for nefarious purposes.

> **Key Point**
>
> Man-in-the-middle attacks are possible not only on unencrypted links, but also with encrypted links, such as an SSL-protected HTTP connection, by exploiting weaknesses in the cryptographic infrastructure or underlying operating systems.

One example of the MITM threat can be found in so-called "evil twin" Wi-Fi access points in which the attacker impersonates a public, unencrypted Internet link to obtain access to a mobile device, potentially without the user even being aware of the illicit connection.[6] MITM attacks can be used to steal private information directly from the user's computer (which can then be used to launch additional attacks) as well as for eavesdropping.

6.1.3.2 Phishing and Other Social Engineering Attacks

One of the most common criticisms against new technological solutions to security problems is that they are unable to counter the human factor. While it is true that human error can render impotent technological security controls, as we discuss later, common socially induced security problems can indeed be countered by technology.

Phishing attacks are attempts made by malicious websites or e-mails to fool users into providing sensitive information. In general, social engineering refers to the manipulation of people to cause them to perform actions or divulge information.

Another well-known social engineering attack baits users into installing infected media—for example, a USB flash drive that has been dropped on the ground in a corporate parking lot. The infected USB flash drive is believed to be one of the primary infiltration vectors of Stuxnet, discussed in Chapter 1.

> **Key Point**
>
> In general, social engineering attacks are increasingly prevalent and effective, due in part to the growing sophistication of hackers who are willing to mine the Internet for personal data that can be used to launch attacks and the increasing use of social networking sites like Facebook that draw more personal data into the public domain and make that data more readily accessible.

6.1.3.3 Malware Attacks

Data-stealing malware on an endpoint device can log keystrokes or scrape screen contents to obtain transaction information that can be used to corrupt future transactions or provide a means to attack servers. Malware can disable anti-malware software, launch other forms of attacks using the hijacked computer's networking capabilities, alter what the user sees on the screen, or perform a practically limitless range of subversions intended to steal information or cause users to perform activities that would result in the stealing of information.

6.1.3.4 Web Application Attacks

Vulnerabilities in server-side web applications may violate the security of transactions. For example, a cross-side scripting (XSS) vulnerability can enable a remote attacker to hijack a web transaction by stealing authentication credentials or billing information when a user browses a vulnerable website. Another common attack involves the exploitation of SQL injection vulnerabilities in a database to maliciously alter a website that interacts with the database.

6.1.3.5 Pharming Attacks

Pharming attacks redirect clients to a malicious server. For example, in Domain Name System (DNS) attacks, redirection is accomplished by modifying the server's IP address, within the client's applicable DNS server.

Key Point

Similar to man-in-the-middle attacks, DNS-based pharming attacks are particularly dangerous because neither the client nor the server must necessarily be subverted for the attack to succeed.

DNS cache poisoning is one example of a DNS attack in which attackers inject fraudulent addressing information into a DNS server, tricking it into directing clients to malicious websites. Improved versions of DNS can thwart poisoning attempts but are not always deployed.

Digital certificates are intended to protect the user from pharming attacks. Digital certificates are the currency of the Internet's Public Key Infrastructure (PKI), which enables client endpoints to authenticate and communicate securely with servers by relying on a trusted certification authority (CA). Fraudulent websites should be unable to present a valid certificate accepted by the user's browser. However, users and embedded applications do not always pay attention to browser warnings for invalid certificates, and there are weaknesses in Internet PKI that may enable attackers to obtain fraudulent certificates that appear to be signed by the certification authority and hence are accepted by browsers. One study demonstrated precisely this form of attack, taking advantage of a weakness in the MD5 algorithm that is sometimes used for certificate digital signatures.[7]

The ultimate DNS attack involves subverting the Internet's trusted certification authorities and domain name registrars—for example, Network Solutions. Such an attack can cause all

Internet communications to a server to be redirected. That is precisely what happened to CheckFree, one of the world's largest online bill payment companies. According to reports, the company's credentials, required to modify its Internet identity via the Network Solutions website, were stolen. Experts believe the credentials were stolen using a phishing attack on CheckFree's employees. Forty-two million customers access CheckFree's bill payment site. The company's Internet address was redirected for several hours. CheckFree warned more than five million customers of the breach.[8]

The CheckFree attack demonstrates the fragility of network transaction security: embedded systems developers and service providers who work assiduously to stay up to date with the latest security patches, anti-malware technologies, and social engineering behavior awareness are still totally exposed.

6.1.3.6 Combinations of Attacks

An obvious combination of attacks involves the use of phishing to install malware. For example, a phishing web link within e-mail could cause the user's browser to connect to a website that has been impregnated with a zero-day flash vulnerability that causes malware to be surreptitiously downloaded to the user's computer.

An attack on Citibank's Citibusiness service used a combination of phishing and MITM to defeat Citibank's two-factor authentication scheme. The phishing site asked for the client application's credentials but then acted as an MITM to forward these credentials to the valid Citibusiness site. Thus, someone attempting to test the site for authenticity would find that the site responded correctly to valid versus invalid logins.[9]

Another common example entails the use of a web application flaw, such as a cross-site scripting or SQL injection vulnerability, to launch a phishing attack within a compromised web page.

As alluded to earlier, the CheckFree debacle was most likely caused by the combination of a phishing attack to obtain credentials followed by the DNS subversion.

6.1.4 Modern Attempts to Improve Network Transaction Security

If there is one thing readers can deduce from the breadth and frequency of vulnerabilities and attack vectors that aim to compromise network transactions, it is the fact that developers and service providers have been fighting a losing battle trying to stay ahead of the black hats. There simply are too many avenues for attack. Current endpoint operating systems, middleware, and applications were simply never designed for a high level of assurance. This software is also tremendously complex and getting more complex all the time. As we discussed in Chapter 1, complexity leads to flaws that can be exploited. The combination of complexity and lack of a high-assurance development process can be thought of as the double whammy of insecure software.

Key Point

Without a dramatic improvement in secure design and assurance, the dual trend of increasing software defects and attacker sophistication converges to a total loss of confidence in shared networks such as the Internet as a medium for commerce or other critical network-based computing.

The Internet itself was not designed to provide private, secure transactions; rather, it was designed for the precise opposite: collaboration and sharing of information.

Let's look at the current state of the art with respect to mitigating network transaction security threats.

6.1.4.1 Anti-malware

Many transaction endpoint systems run some form of anti-malware software. Examples include anti-virus, anti-spam, anti-phishing, and firewalls. Anti-malware is built into operating systems, browsers, e-mail clients, and can be purchased as add-ons. This software is valuable because it can detect many forms of malware before they can do damage.

Key Point

The problem with anti-malware software is that it is a fundamentally a reactive technology; as new vulnerabilities and attack methods are discovered, the anti-malware protocols quickly become obsolete.

In addition, black-hat-created malware may be operating in the wild far long before being detected by the community trying to thwart it. A number of independent tests of leading anti-malware products have found all of them unable to cope with a significant number of modern malware specimens.

Secunia, a vulnerability management company, performed one test that found Symantec to be the best at detecting a suite of malicious files and web pages. The bad news is that Symantec detected only 21% (64 of 300) of the malware specimens.[10]

Information Security Magazine performed a test of 8,114 malware specimens against seven different anti-malware vendor products. The best performing product, from Trend Micro, was unable to detect 7.9%, or approximately 640, specimens.[11]

Finally, anti-malware products have themselves been found to be vulnerable. Anti-malware vendors are forced to change their products rapidly to catch up to the latest specimens. Anti-malware developers do not follow a rigorous high-assurance development process that requires independent evaluation or certification. The U.S. CERT National Vulnerability Database has posted approximately hundreds of security flaws in Symantec software.[12] A number of studies have found anti-malware vulnerabilities to be exploitable.[13]

Anti-malware is an important component of modern transaction security strategy. However, it is by no means something that can assure transaction security.

6.1.4.2 Secure Browsers

> **Key Point**
>
> An encouraging approach to transaction security involves sandboxing the embedded web client from the rest of the computing environment; the theory is that a sandboxed browser is less likely to be infected with malware and therefore more likely to be trusted for secure transactions.

There are a number of approaches to browser sandboxing.

One sandboxing approach uses a separately booted operating system (a version of Linux) invoked with a special button on the computer. A downside to this is that the user's regular environment is inaccessible while the sandboxed browsing environment is running. A reboot must occur to move between the two environments. Of course, since the alternate browsing environment runs on a general-purpose operating system using general-purpose browsers, it is certain that hackers would target the alternate environment if it were to achieve widespread adoption. Users are relying on security through obscurity.

Some vendors of secure portable storage products, such as USB thumb drives, offer a specially packaged Linux browser environment.[14] The browser is launched from and stores sensitive data to the attached storage device. The disadvantage, of course, is the extra cost and inconvenience of incorporating a separate physical device. In addition to the mobility across endpoints that a portable drive provides, this approach has one key advantage over PC-based solutions such as the alternate boot: physical security. Some of these portable drives are sealed against tampering and will self-destruct their internal cryptographic keys if an attempt to physically breach the device's casing is detected.

Another approach to secure browsing is to harden the browser itself. Google's Chrome browser is notable for its attempt to provide a more secure browser in which malicious web pages or plug-ins would be less likely to cause harm to unrelated web browser components. However, Google engineers did not use high-assurance development practices to implement Chrome. While Chrome may have some fundamentally good isolation concepts, the lack of commitment to high-assurance engineering ensures a product that cannot be trusted for high-value transactions. It should come as no surprise that in September 2008, just a few months after Chrome's release, security researchers discovered a buffer overflow vulnerability in the browser that could allow a hacker to take complete control of the endpoint.[15] Since this time, hundreds of vulnerabilities have been posted to the U.S. CERT National Vulnerability Database.

There have been a few academic attempts to create higher assurance web browsers. A promising recent effort from Samuel King at University of Illinois at Urbana-Champaign, called the OP web browser, also claims isolation between browser-level components. However, the OP researchers have gone further and employed some formal methods to verify browser behavior.[16] The increased level of assurance can provide OP users with confidence that the claimed sandboxing capability actually works. Other than the technical obstacles of completing the product and having it independently certified at a high-assurance level (both honorable goals), a major obstacle for something like the OP browser lies in achieving market acceptance. Developers and users are creatures of comfort; unless OP can be configured to look and feel exactly like the most popular browsers (e.g., Internet Explorer, Firefox, Safari) and can keep up with the ever-changing pace of browser functionality (scripting, plug-ins, look-and-feel), the global impact of such an offering is in question.

Another approach to browser security is to run the browser environment under a concurrently running virtual machine instead of a separate boot. Assuming adequate performance, this virtual browser appliance approach is favorable compared to a secondary boot. However, the virtual machine software must be trusted to provide the browser sandbox. Typical commercial hypervisors, such as VMware and Xen, were not developed using high-assurance methodology, have had numerous security vulnerabilities reported,[17] and therefore cannot be trusted to withstand attempts made by sophisticated attackers to break the sandbox.

6.1.4.3 Two-Factor Authentication

Two-factor authentication has been the predominant security response from banks and other security-critical network transaction enterprises. One approach is to require the user to enter an identification number in addition to the usual username and password. The identification number may be displayed on a specialized device in the key fob form factor. The number is generated cryptographically and changes after a short period of time. The bank will supply this identification device and run a complementary server that generates a matching number. RSA's SecurID is one popular two-factor authentication solution.

Two-factor authentication is beneficial because it increases the chances that the client and server can mutually authenticate; hackers in possession of only a valid password are unable to access a two-factor protected account. Two-factor authentication protects against some phishing attacks that attempt to steal passwords. However, it does not protect against MITM: the MITM can simply pass on the first and second credentials to the real server. Multi-factor authentication also does not protect against malware that can piggyback on an authenticated connection.

Key Point

Due to the ability of malware and other attacks to traverse an authenticated connection, we view the multi-factor authentication approach, at least by itself, as a weak response to the current network transaction insecurity climate.

As Bruce Schneier, chief security technology officer of telecommunications giant BT and renowned security expert, stated, "...it won't work for remote authentication over the Internet... in the end there will be a negligible drop in the amount of fraud and identity theft."[18]

Schneier's comments were perhaps prescient; in March 2011, RSA's SecurID system was compromised by hackers who allegedly used phishing attacks on RSA employees to install Adobe Flash malware that opened up RSA's servers to the attackers. The attackers were allegedly able to steal the SecurID seeds and use them to breach the networks of RSA SecurID users, including Lockheed Martin Corporation. One has to wonder, if these reports are true, why a security company like RSA would leave the keys to the kingdom on servers accessible to the Internet. It's not like hacking through firewalls, UTMs, and DMZs is not a daily occurrence across the Internet world.

6.1.4.4 Network Access Control

Network Access Control (NAC) attempts to protect a network from access by clients that do not fulfill a configurable profile. For example, a NAC server may reject or quarantine within a logically separate network any client that does not have the proper operating system patches applied and anti-malware software installed. The most basic form of NAC on Ethernet networks is the use of 802.1X authentication described in Chapter 5.

NAC is a useful enterprise tool because it enables system administrators to enforce a health policy on clients connecting to the corporate network. NAC is most useful when network clients are a somewhat captive and controlled population. NAC is not appropriate for open networks (i.e., the Internet) in which the server has no influence over client hardware and software. But the biggest problem with NAC is the inherent difficulty in assessing health. Similar to anti-malware technologies, NAC is reactive. Once a NAC client passes the pre-programmed test of its state, the client's undetected malware and social-engineering risks remain.

6.1.4.5 Secondary Device Verification

Mobile devices are increasingly taking part in critical network transactions. For example, near-field communication (NFC) contactless terminals built into a phone can be used as the network transaction human interface endpoint. In other cases, the mobile device can be used as a mechanism for two-factor authentication: instead of a separate key fob, the bank can send the user the identification number, called a mobile transaction number (mTAN) within an SMS message. In yet other cases, the mobile device is used as an out-of-band transaction verification: when a user performs a network transaction using a PC, the service provider can send an SMS message to obtain a final confirmation from the user before proceeding.

Key Point

Out-of-band transaction verification is a compelling approach because it enables both client and server to obtain a higher level of confidence in the authenticity of each other's identity as well as the actual information being transacted.

This theory is based on the assumption that attackers will have a harder time creating a coordinated attack of both the endpoint and out-of-band device.

Criticisms of secondary device verification include the requirement to use a separate mobile device itself, the potential lack of cellular coverage, and the additional cost of SMS messaging. Another problem is the security of the mobile device itself. As we discuss later in the chapter, with vulnerabilities in Google Android iPhone, BlackBerry, and Windows Mobile operating environments, there is little doubt that sophisticated attackers can and will devise methods to subvert mobile phone operating systems if they were to become widely used for network transactions. The server, too, is a potential weakness, in that a rogue or otherwise corrupted server can follow the protocol of sending an out-of-band verification that will incorrectly lead the user to believe that the transaction is safe.

6.1.4.6 Zone Trusted Information Channel

IBM's Zurich research laboratories created the Zone Trusted Information Channel (ZTIC), a USB device that provides its own minimal display and network security protocols (SSL/TLS).[19] ZTIC provides an authenticated and encrypted channel between client and server and a means for clients to verify transaction details from the server. However, instead of the aforementioned SMS message, ZTIC uses its own simple display and on-board buttons. ZTIC's out-of-band verification is, in theory, superior to an SMS message due to the more highly assured information channel upon which the verification message is transmitted.

ZTIC has not been certified to a high-assurance level, and therefore we must assume until proven otherwise that the software, albeit far simpler than a typical endpoint device, cannot be expected to withstand determined and sophisticated attackers.

Nonetheless, ZTIC suffers from the same drawback as the mobile phone approach: the user must learn to use and manage a separate device. Furthermore, it is unknown whether the costs to produce, distribute, and maintain this device would be acceptable to a majority of system designers. ZTIC plugs into the endpoint via USB and therefore is unsuitable for classes of embedded devices that lack a USB port.

Another criticism of ZTIC is that, while it holds a private key used for secure communications, the device itself does not have an interface for user authentication. A lost or stolen device is therefore more susceptible to malicious use. The device is enabled by user authentication on the PC in which it is plugged. But insecurity of the PC is the reason for using a ZTIC in the first place!

6.1.4.7 Virtualized Web Client

Virtualization enables an improved version of the secure browser concept discussed earlier in which a dedicated virtual web client appliance runs in one virtual machine (or in a native embedded microkernel process when using a microkernel Type 1 hypervisor), while the system's primary environment runs in another virtual machine. Critical to this approach is the use of high-assurance separation between the sandboxed browser and the primary user environment. Figure 6.1 shows a sample implementation in which the embedded system's primary environment is Linux, and the virtual web appliance consists of a ported Firefox browser. One can easily envision numerous other applications of this architecture, including the use of multiple virtual appliances and user environments. For example, some forms of anti-malware filtering software could be deployed as a virtual appliance, thereby preventing guest-infested malware or guest operating system vulnerabilities from adversely affecting the anti-malware agent.

6.1.4.8 Summary: Modern Approaches to Transaction Security

Although the convenience of the Internet for financial and other high-value transactions is compelling, predominant approaches to security are unable to provide a sufficient level of assurance against loss. Some enterprise and government organizations have already unplugged their enterprise Internet connections for fear of attack. The U.S. military banned USB thumb drives for the same reason.[20]

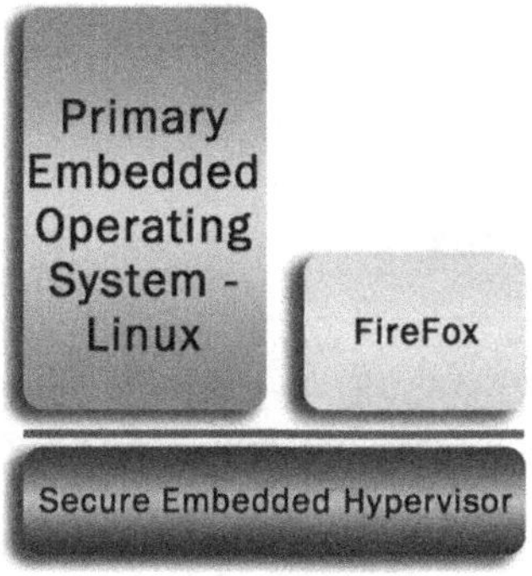

Figure 6.1:
Web browser virtual appliance.

In addition to a lack of high-assurance pedigree and certification, most approaches to transaction security are not sufficiently natural and easy to use for the general population. Clearly, a high-assurance approach that still allows the maximum reuse of expected legacy environments is needed.

6.1.5 Trustworthy Embedded Transaction Architecture

Let's take this sandboxing approach a step further. Virtualization can be used to host the system's guest Linux environment while employing two critical security subsystems that run in a separate virtual machine or within a native microkernel process (if applicable). The first subsystem, called the communications server, provides a trustworthy communications connection between the embedded system and the remote server. The second subsystem, called the transaction verifier, provides a trusted verification interface for the client (see Figure 6.2).

6.1.5.1 Communications Proxy

When the web client instantiates an HTTPS connection to a server, web traffic is filtered and redirected to a communications proxy that establishes a second cryptographically secured connection with the server. The key idea here is that this solution creates a protected connection, logically out-of-band from the Linux system, but still in-band with respect to the embedded system. Thus, a separate physical device is not needed. Because the second set of encryption keys, server certificates, and protocol software runs within native trusted processes, these critical data cannot be stolen or corrupted by the guest operating system, regardless of malware infiltration. Furthermore, the native security subsystem is able to take advantage of a Trusted Platform Module (TPM) or equivalent hardware co-processor capabilities, if present, for hardened storage of keys and for platform attestation.

The secure proxy defeats MITM attacks as well as malware attacks that would attempt to commandeer the cryptographic keys used for secure communications. The secure connection

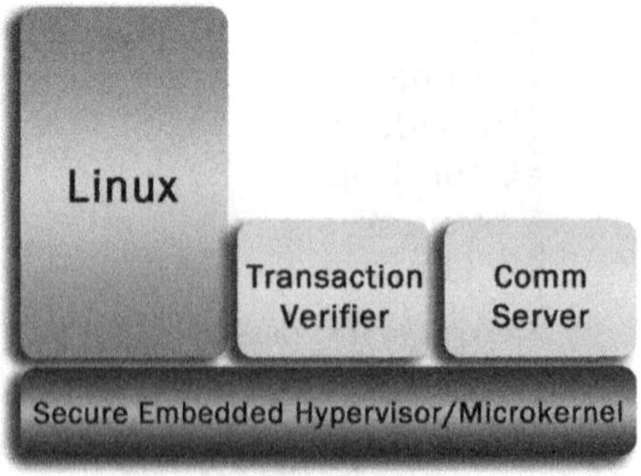

Figure 6.2:
Secure embedded transaction architecture.

also protects against pharming attacks because the client and server are able to achieve a high assurance in their mutual authentication protocols. Applying the same approach within the server endpoint can further augment overall transaction security.

6.1.5.2 Transaction Verifier

When a transaction is about to be executed, the server sends a verification request to the client. This request is also sent on the secure connection and received by the trusted hypervisor/operating system and thus is impervious to security attacks that afflict general-purpose operating software within common client endpoint systems. Note that, depending on the situation, the server application may need to be adapted to generate this final verification cycle. However, the existence of the SMS approach proves that service providers are willing to make such minor modifications. In fact, the same format message used in SMS verifications can be used for the secure connection.

Within an autonomous embedded system, the transaction verifier might execute pre-programmed rule checking to verify a transaction response and send back its confirmation to the server. Within client endpoints operated by a human user, verification often requires human confirmation. The verifier can take advantage of the computer's display to provide a verification dialog box that can't be spoofed or overwritten by the guest operating system. The transaction verifier is a client of a trusted graphical display manager hosted by the trusted hypervisor/operating system.

When a human is in the loop, a dedicated LED or audio device could be used to announce an incoming transaction response, but this may require endpoint hardware modifications. Another approach is to use a small area of the display screen, called the *verification sprite*, which is reserved for the transaction verifier. The guest operating system does not have direct access to the physical display device. Rather, the guest's accesses to the display are virtualized. The trusted graphical display manager will update the physical display with the guest's virtualized graphics content, with the exception of the non-overlapping verification sprite whose contents are controlled by the transaction verifier. When a transaction verification request comes in from the server, the sprite alters to notify the user that the request is pending. For example, the sprite can change colors.

The user can now click on the sprite (or use a special keyboard sequence) to invoke a secure confirmation dialog that can use the full power of the screen. The physical act of clicking the sprite or using a reserved keyboard sequence provides the user with a trusted path to the verification screen, a sample implementation of which is shown in Figure 6.3.

There are a number of important advantages to this approach that are lacking in existing transaction security solutions. The most important benefit is the ability to reuse the same embedded hardware platform and not force the user to adopt an external foreign device. Some embedded systems are fully contained as well as space and cost constrained, making it impractical to add additional hardware.

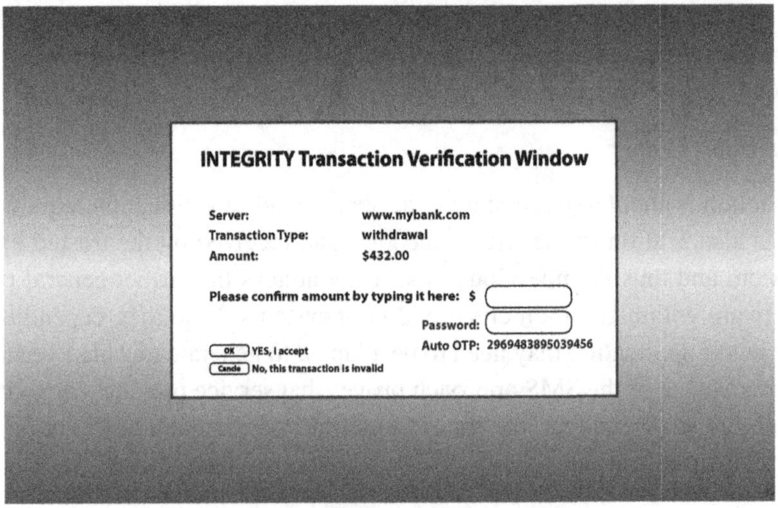

Figure 6.3:
Trusted transaction verification dialog.

When a human is in the loop, reuse of the graphical interface promotes a natural usability. The trusted display manager has the full definition of the endpoint device's graphics in which to craft its verification dialog. By taking maximum advantage of the display, the user is not only guaranteed to receive and notice verification requests from properly configured servers, but the user experience also is guaranteed to be far better than can be achieved in a small form-factor attached device.

End-to-end network transaction security will be improved by applying the same high-assurance methodology on the server side. This can be done either by using a network transaction appliance (containing the trusted hypervisor/operating system and communications server) in the server's network or by adopting a similar virtualization architecture used in the client endpoint. As seen in Figure 6.4, out-of-band secure communication subsystems on both ends of the transaction provide a connection that both client and server can trust.

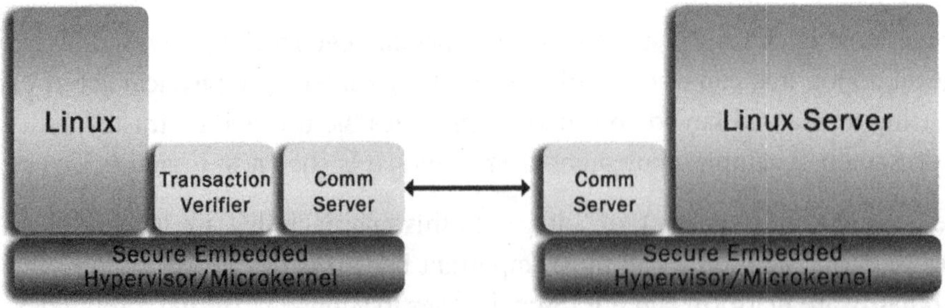

Figure 6.4:
Applying secure hypervisor/microkernel architecture to both endpoints.

Using a trusted operating system for the server-side security protects against MITM and malware by fully sandboxing the end-to-end secure communications components.

6.1.5.3 Cryptosystem and PKI Policy Enforcement

> **Key Point**
>
> End-to-end control of the transaction connection also can protect against PKI and cryptosystem weaknesses by enforcing the use of strong public key, bulk encryption, and Message Authentication Code (MAC) algorithms and key lengths instead of relying on the native web server and its default TLS implementation.

For example, the communications server might disallow the use of the weak MD5 hash algorithm and require the use of elliptic curve cryptography (ECC) with 384-bit or larger key lengths for the public key cryptosystem. Furthermore, because public key cryptography is used to negotiate and derive the symmetric key used for bulk encryption, the policy can ensure that public key length provides a strength that is commensurate with the underlying symmetric cryptosystems.

A trusted hypervisor/operating system-based infrastructure can even be used as its own certificate authority. Pharming attacks are unable to thwart transaction assurance because legitimate clients and servers always know when they are communicating with valid peers on the other end.

TLS is designed to be flexible, handling backward compatibility and dealing with the realities of a complex web world with many types of servers and clients. However, this flexibility leads to vulnerabilities, such as the weakness in TLS, discussed in Chapter 5, that enables an MITM to downgrade the cryptographic strength used in the protocol. Control of both ends of the connection via the secure communications server enables policy to be enforced at the TLS level. This policy can drastically improve overall system security. For example, the policy can require that the client authenticate itself to the server (client authentication is optional in the TLS standard).

Modern society has become increasingly dependent on the security of network transactions over wide area networks. Network transactions are used for commerce and banking, and to remotely control critical infrastructure. Embedded systems that employ insecure endpoint operating systems and applications are unable to provide the requisite security for high-value transactions. As evidenced by the growth of trends such as cloud computing and Software as a Service (SaaS), the security impact of successful cyber attacks is likely to grow. The good news is that high-assurance methodology and certified high-robustness technologies can break through the insecurity status quo and do so without sacrificing the world's legacy operating systems and applications. The only thing we need is the determination to think outside the box and apply these new techniques creatively and effectively.

6.2 Automotive Security

In Chapter 1, we discussed a troubling 2010 university research project in which a team of researchers successfully hacked critical safety-related systems (engine, breaks) over the low-security in-car network and how modern automobiles are now regularly being connected to wide area networks using wireless technologies. The combination of these situations results in viable attack vectors against national infrastructure by sophisticated threat agents. As we also discussed in Chapter 1, compounding the challenge for embedded systems developers is the incredible electronic control unit (ECU) complexity explosion in modern vehicles, with upwards of 200 microprocessors found in some high-end models.

In addition to embedded system count and complexity, multiple networks of varying type, including Controller Area Network (CAN), FlexRay, Local Interconnect Network (LIN), and Media Oriented Systems Transport (MOST), connect these ECUs. The car OEM integrates networked ECU components and software from dozens of Tier-1 and Tier-2 suppliers. While the OEM often defines requirements for these ECUs, it does not rigorously control their actual contents or development process.

It should come as no surprise that this situation has become untenable. OEMs are suffering from the longest pole in the tent syndrome: a single ECU, delivered late or with serious reliability problems, may be all that is needed to cause shipping delays or customer-visible failures that lead to recalls and poor reputation. Add to this the new challenge of security: a single vulnerability in a critical component, such as the gateway to safety-critical networks and functions, can allow in remote attackers.

While we mentioned a few key areas for improving security in this environment, we revisit automotive security in more detail here.

6.2.1 Vehicular Security Threats and Mitigations

The realm of security threats to cars can be coarsely classified in three domains: local-physical, remote, and internal-electronic. Combinations of these threat domains will often be required to inflict damage.

6.2.1.1 Local-Physical

Examples of local-physical threat would be someone physically tapping into the drivetrain's CAN network and disrupting communications or damaging an ECU via power surge or excessive heat application. Such an invasive attack can quite easily disable critical car functions. However, a local attacker, such as a disgruntled mechanic or maligned spouse, can harm only one car and is therefore unlikely to get the attention of security teams. Furthermore, a car's massively complex, distributed electronic system is simply impractical to protect from physical attack. So we generally punt on this class of threats.

There is, however, one exception, and it is an important one. Somewhere within one or more ECUs, private cryptographic keys are stored for use in creating protected communication channels and to provide local data protection services. Communications may include car to service center or other OEM infrastructure, car to multimedia provider, car to car, car to power grid (in electric vehicles), car to smartphone, or even car to bank. Figure 6.5 shows some examples of long-range radio connections in next-generation vehicles.

Data-at-rest protection may be required for automotive algorithms, multimedia content, and cryptographic material.

Private keys must be kept in storage that can withstand sophisticated physical attacks, both invasive and non-invasive, because the loss of even a single private key may enable an attacker to establish connections into remote infrastructures where widespread damage and property loss can ensue. OEMs must be able to achieve high assurance of key protection across the

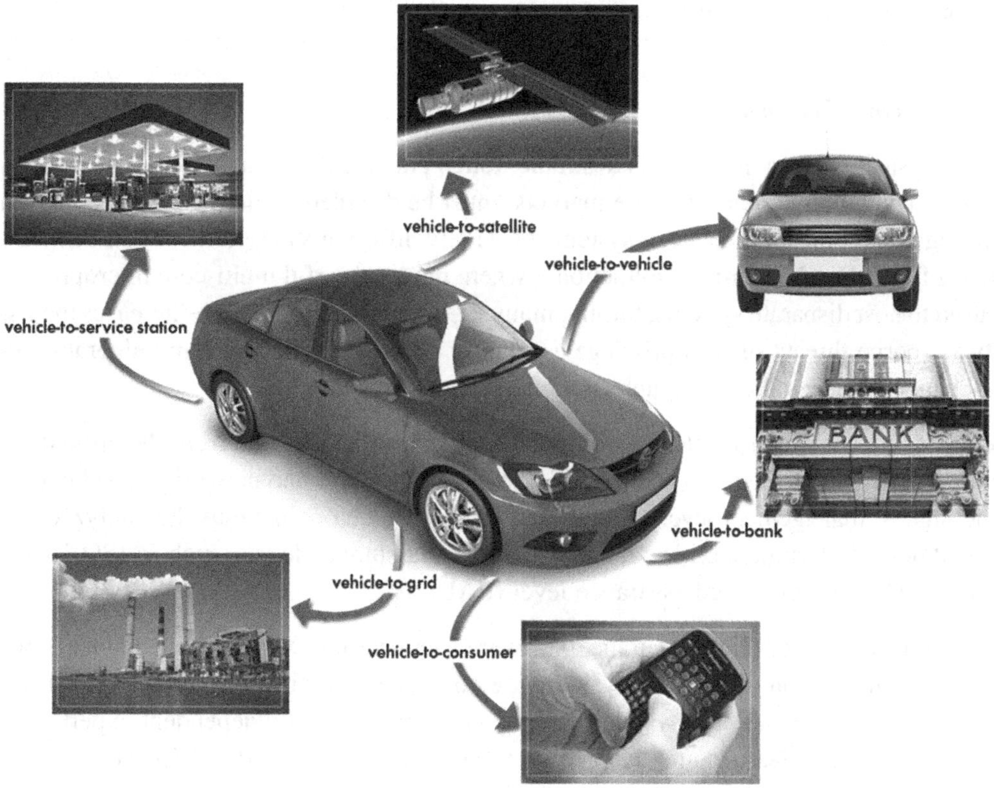

Figure 6.5:
Examples of next-generation extra-vehicular networks communications.

entire life cycle, from creation to embedment into ECUs, delivery and integration within the car, and in the field. Embedded systems cryptographic experts can help OEMs and their suppliers with guidance and oversight in this area.

6.2.1.2 Remote

Remote threats are the classics: a hacker tries to probe the car's long-range radio interfaces for vulnerabilities in network security protocols, web services, and applications to find a way into the internal electronics complex. Unlike high-end data centers, the car is unlikely to be outfitted with a full complement of IDS, IPS, firewalls, and UTMs. Regardless, recent intrusions at Sony, Citigroup, Amazon, Google, and RSA starkly demonstrate that these defense mechanisms are Swiss cheese against sophisticated attackers. Clearly, the car's critical systems must be strongly isolated from ECUs and networks not critical for safe operation.

> **Key Point**
>
> Whenever possible, complete physical isolation of critical automotive ECUs from network-connected vehicular infrastructure should be imposed.

6.2.1.3 Internal-Electronic

While physical network isolation is desirable, touch points will inevitably exist. For example, the car's navigation system, in some markets, must be disabled while the car is in motion, implying communication between systems of widely differing safety criticality. Furthermore, a strong future trend toward consolidation—where more powerful multi-core microprocessors are used to host disparate systems, turning many ECUs into virtual ECUs—increases the risk of software-borne threats such as privilege escalation due to operating system vulnerabilities, side-channel attacks on cryptography, and denials of service.

Therefore, the car's internal electronics architecture must be designed from the ground up for security. Touch points between critical and non-critical systems and networks must be justified at the highest management levels, and these electronic touch points must be analyzed and certified devoid of vulnerabilities at the very highest assurance levels, such as ISO 15408 (Common Criteria) evaluated assurance level (EAL) 6+.

PHASE—Principles of High-Assurance Software Engineering, described in Chapter 3, which espouses minimization of complexity, software component architecture, principle of least privilege, secure software and systems development process, and independent expert security validation—must be learned and adopted by OEMs and promulgated to ECU suppliers.

Car manufacturers and tier-1s may not have been thinking a lot about security when they designed the cars hitting roads today, but clearly that must change. Manufacturers must work closely with embedded security specialists early in the design and architecture of in-car electronics and networks and must raise the bar on security-driven engineering and software assurance.

6.3 Secure Android

At recent Embedded Systems Conferences, Android operating-system-related courses have met with startling popularity. Embedded systems designers are considering the use of Android for all forms of human-machine interfaces (HMIs) in practically all major embedded systems-related industries: automotive head units, medical device graphical interfaces, and home smart energy management panels, just to name a few. Android brings to the embedded systems community the power of open source Linux augmented with the graphical interfaces and app store infrastructure of one of the world's most popular mobile operating systems. In addition, the rapidly emerging market for Android Mobile Device Management (MDM) solutions provides embedded systems developers with the promise of a world-class remote device management infrastructure that can seamlessly tie in to traditional back-end IT systems. MDM functions include remote monitoring and auditing, firmware update, application configuration management and control, data-at-rest encryption, VPN services, remote wipe (e.g., when an embedded device is believed to be compromised), and more. It is no wonder embedded systems developers are intrigued, wondering how Android can make their next designs far more functional by taking advantage of the powerful open source Android movement.

A significant portion of expert instruction on embedded Android centers regarding how to achieve common critical embedded system requirements for real-time response and safety and/or security-critical operation using a mobile operating system is not designed for this. Since this book is about embedded security, let's focus on the security challenge.

6.3.1 Android Security Retrospective

As part of Android's original introduction in 2008, Google touted improved security in its smartphones. Google's website touted the platform's security: "A central design point of the Android security architecture is that no application, by default, has permission to perform any operations that would adversely impact other applications, the operating system, or the user."[21] Days after release of the first Android phone, the G1, a well-publicized, severe vulnerability was found in the phone's web browser. But the G1's security woes didn't end there. In November, hackers discovered a way to install arbitrary programs on the phone, prompting this lament from Google: "We tried really hard to secure Android. This is definitely a big bug. The reason why we consider it a large security issue is because root access on the device breaks our application sandbox."[22] In fact, the Android bug would silently and invisibly interpret every word typed as a command and execute the command with superuser privileges.[23] A ZDnet writer called this vulnerability the "Worst. Bug. Ever."

In late 2010, security researchers uploaded to the Android market a spoofed Angry Birds game application that invisibly downloads other apps without the user's approval or knowledge. The extra downloads were malicious, stealing the phone's location information and contacts and sending illicit text messages. This research demonstrated arguably the most deleterious example of least privilege principle failure: a completely untrusted application downloaded from the Internet has the privilege to bring down arbitrary software and access sensitive data on the device. As part of their work, the researchers reported numerous other weaknesses in Android, including a faulty use of SSL, lack of app authentication, an easy method of breaking out of the Android Dalvik virtual machine sandbox via native code, and the focus of the attack, a weak permissions architecture.[24]

Next, we revisit our favorite website, the U.S. CERT National Vulnerability Database. A search on Android turns up numerous vulnerabilities of varying severity. Here is a sampling of the worst offenders:

CVE-2011-0680: Allows ***remote attackers*** to read SMS messages intended for other recipients
CVE-2010-1807: Allows ***remote attackers*** to execute arbitrary code
CVE-2009-2999, -2656: Allows ***remote attackers*** to cause a denial of service (application restart and network disconnection)
CVE-2009-1754: Allows ***remote attackers*** to access application data
CVE-2009-0985, -0986: Buffer overflows allow ***remote attackers*** to execute arbitrary code

We point out these particular vulnerabilities because they fall into the most serious severity category of remote exploitability.

These vulnerabilities are specific to the Android stack that runs on top of Linux. Android is, of course, also susceptible to Linux kernel vulnerabilities. In Chapter 1, we presented a case study of Linux complexity, referring to a 2008 paper written by a few of the leading Linux kernel developers. Those authors have since published an update of their Linux kernel development statistical overview,[25] and the numbers are truly staggering. With more than 20,000 lines of code modified *per day* at times during the development cycle, 6,000 unique authors, and the rapid growth in its overall code base, it should come as no surprise that dozens of Linux kernel vulnerabilities are reported each year and that a steady stream of undiscovered vulnerabilities are latent in every Linux distribution deployed to the field. While a significant portion of the growth and churn in the Linux kernel code base is due to adding support for new microprocessors and peripherals, the core kernel itself, such as networking and file system support, also undergoes rapid change. CVE-2009-1185 documents a flaw in the Linux netlink socket implementation and is but one example of a Linux vulnerability that has been allegedly used to compromise Android devices.

6.3.2 Android Device Rooting

Android *rooting* (also known as *jailbreaking*) is the process of replacing the manufacturer-installed kernel (Linux) and/or its critical file system partitions. Once a device is rooted, the hacker can change Android's behavior to suit her particular desires. The term *rooting* originates from the UNIX concept of root privilege, needed to modify protected functions. The goals of Android hackers range from the hobbyist's desire to overclock a CPU for better performance (at the expense of battery life) and install custom applications to more malicious pursuits such as illegally obtaining carrier network services and installing key loggers and SMS snoopers. The collection of new and replaced files installed by the hacker is referred to as a custom *ROM*, another imperfect reference to the concept of firmware that is often deployed in read-only memory.

Hackers often use Android vulnerabilities to root Android phones.

Key Point

The rate of Android vulnerability discovery is such that practically every Android-based consumer device has been rooted within a short period of time, sometimes within a day or two of release.

In addition to software vulnerabilities, secure boot problems are another major source of Android rooting attacks. Some Android device makers, such as Barnes & Noble with its Nook Color, have permitted (if not encouraged) rooting to facilitate a wider developer community and device sales. In this case, rooting is usually accomplished with a form of side loading/booting using an SD card or USB to host or install the custom ROM. The manufacturer-installed boot loader does not cryptographically authenticate the Android firmware, paving the way for ROM execution.

Some device makers have gone to pains to prevent rooting for various reasons. Obviously, many embedded developers using Android will want to lock down the Android OS completely to prevent unauthorized medication and malicious tampering. One of the most high-profile secure boot failures in this realm is the Amazon Kindle. The presumed aim of locking down the Kindle is to force users to access Amazon content and require use of the Kindle e-reader software. The Amazon secure boot approach attempted to authenticate critical system files at startup using digital signature checks. Hackers used vulnerabilities in Linux to circumvent these checks and run malicious boot code, rooting the device. Information about the successful intrusion can be found on the public Internet. It is interesting to note that Amazon abandoned its anti-rooting posture with the Kindle Fire tablet.

Yes, we paint a grim picture of Android security. However, the picture is based on simple facts that ought not be surprising: Android was never designed to provide high assurance of security functions.

6.3.3 Mobile Phone Data Protection: A Case Study of Defense-in-Depth

Android's tremendous popularity juxtaposed with its lack of strong security has sparked a rigorous scramble by software vendors, device OEMs, systems integrators, and government security evaluators to find ways to retrofit improved system security to Android-based devices.

> **Key Point**
>
> One approach to raising the level of assurance in data protection within an Android-based device is to employ multiple encryption layers.

An Android smartphone, for example, can use a layer four (OSI model) SSL VPN client to establish a protected data communication session. An IPsec VPN application, running at layer three, can be used to create a second, independent connection between the smartphone and the remote endpoint (see Figure 6.6). This secondary connection uses independent public keys to represent the static identities of the endpoints. The data in transit is doubly encrypted by these two concurrent connections. This layered security approach is an example of *defense-in-depth*.

> **Key Point**
>
> The concept of defense-in-depth originates in the military: multiple layers of defense, such as a combination of mines and barbed wire, rather than just mines or barbed wire alone, to increase the probability of a successful defense as well as potentially to slow the progress of an attacker.

Defense-in-depth has been successfully applied in war since ancient times, and the concept is alive and well in the information security world.

Let's consider a few of the threats against an SSL data protection application. An attacker can directly attack the application, perhaps exploiting a flaw in the SSL software stack, to disable encryption entirely or steal the encryption keys residing in RAM during operation. An attacker

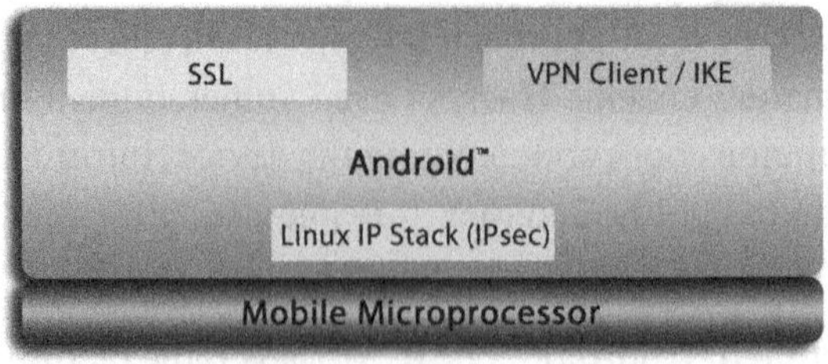

Figure 6.6:
SSL/IPsec within android's native sandboxes.

can try to steal the static public SSL keys stored on disk. If these keys are compromised, the attacker can impersonate the associated identity to gain access to the remote client over a malicious SSL session. Malware elsewhere in the Android system can use side-channel attacks to break the SSL encryption and recover its keys.

Layered SSL/IPsec data protection is a sensible application of defense-in-depth to counter these threats. If an attacker is able to break the SSL encryption, the IPsec layer will continue to protect the data. An attacker may be able to steal the SSL keys but not the IPsec keys. The attacker may be able to install malware into the SSL application but not the IPsec application. The SSL application may exhibit side-channel weaknesses to which the IPsec application is immune. To succeed, the attacker must break both the SSL and IPsec encryption layers.

Clearly, this layered approach depends on the independence of the layers. The runtime environment must provide strong isolation of the SSL and IPsec application layers, and the runtime environment itself must not provide an attack surface through which to break that isolation. Much of the research and product development aimed at Android security has focused, in one form or another, on providing sandboxes for data isolation and the protected execution of critical functions. Those sandboxes would be used to realize the layered encryption approach. Let's now compare and contrast the various approaches at Android sandboxing. Embedded developers considering the adoption of Android in their next-generation designs can use this comparison to make sensible security choices.

6.3.4 Android Sandboxing Approaches

6.3.4.1 Separate Hardware

One potential sandboxing approach is to have multiple microprocessors dedicated to the differing tasks. While Android smartphone OEMs are unlikely to add additional hardware cost to their designs, embedded systems developers may have more options depending on many factors, including form factor flexibility. For example, a PCI-capable design may be able to host an IPsec VPN card that wraps the second layer encryption around the main processor's Android SSL. In some cases, however, the additional hardware size, weight, power, and cost will be prohibitive for this approach.

6.3.4.2 Multi-Boot

The multi-boot concept has been attempted on a handful of laptops and netbooks over the years. In a dual-boot laptop scenario, a secondary operating system, typically a scaled-down Linux, can be launched in lieu of the main platform operating system. The scaled-down system is typically used only for web browsing, and the primary goal is to enable the user to begin browsing within a handful of seconds from cold boot. The secondary operating system resides in separate storage and is never running at the same time as the primary platform operating system. In some cases, the lightweight environment executes on a secondary microprocessor (e.g., an ARM SoC independent of the netbook's main Intel processor). On an Android mobile device, the primary Android can be hosted on internal NAND flash, and a secondary Android could be hosted on an inserted microSD card (see Figure 6.7).

The secondary operating system has good isolation from a safety perspective; however, the inconvenience of rebooting and inability to seamlessly switch between environments has severely limited adoption. The multi-boot option is also impractical for the layered encryption use case that requires concurrent execution of the sandboxes.

6.3.4.3 Webtop

The webtop concept provides a limited browsing environment (the webtop) independent from the primary operating system environment. However, instead of a dual boot, the webtop runs as a set of applications on top of the primary operating system. In the case of the Motorola Atrix Android smartphone, released in 2011, the webtop sandbox is an independent file system partition that contains a limited Ubuntu Linux-based personality (see Figure 6.8). The primary Android partition is located on the same internal NAND flash device within the phone. The Atrix webtop is intended to provide a desktop-like environment for users that dock the phone on a separately purchased keyboard/video/mouse (KVM) apparatus.

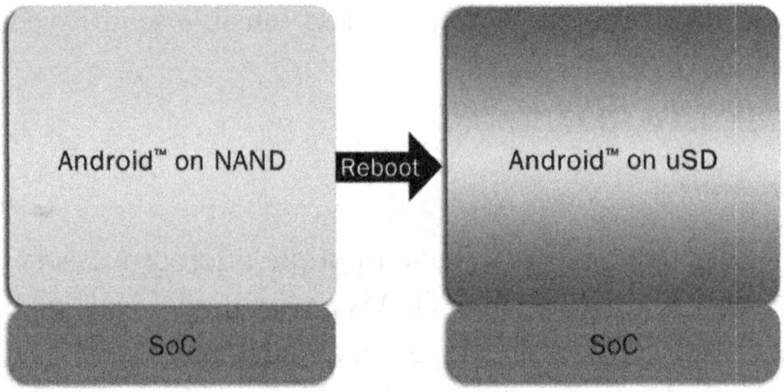

Figure 6.7:
Dual-boot android.

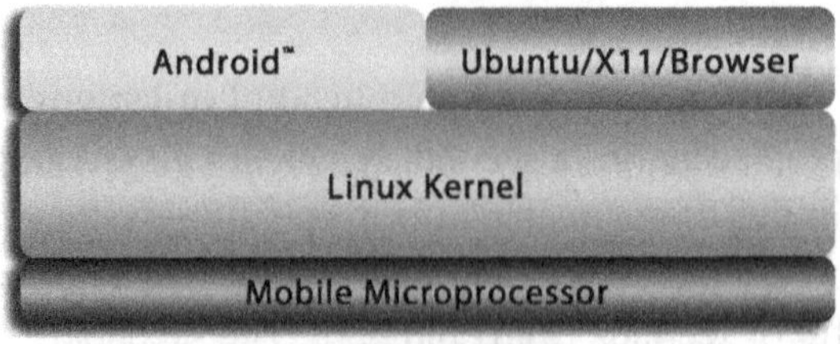

Figure 6.8:
Android webtop environment.

While webtop was most likely not intended as a security capability, one mapping of this approach to the layered encryption use case is to execute IPsec from the primary Android environment and an SSL-based web session from the webtop sandbox. The problem with this approach is that the isolation of the webtop's SSL from the Android IPsec depends on the entire Linux kernel, including its TCP/IP stack.

6.3.4.4 Mobile Device Management Encrypted Containers

The growing popularity of Android mobile devices and desire to use them in the workplace has spawned dozens of Mobile Device Management (MDM) products and companies.

> **Key Point**
>
> The two main purposes of Mobile Device Management are to provide mobile data protection and IT management services.

Manageability includes application configuration (ensuring that all employees have an approved set of pre-loaded software), auditing, document management, and remote wipe (disabling the handset when an employee leaves the company). Data protection covers both data at rest and data in transit (e.g., VPN to the corporate network).

Android MDM solutions often use application-level encryption. For example, an enterprise e-mail client may implement its own encryption protocol for the connection between mobile device and enterprise e-mail server and its own encryption of the e-mail folders resident on the phone.

Some MDM solutions use Android profiles to divide the Android system into two sets of applications: one for the user's personal environment and one for the enterprise-managed environment (see Figure 6.9). When the enterprise profile is invoked, the MDM product may automatically turn on encryption for data associated with that profile.

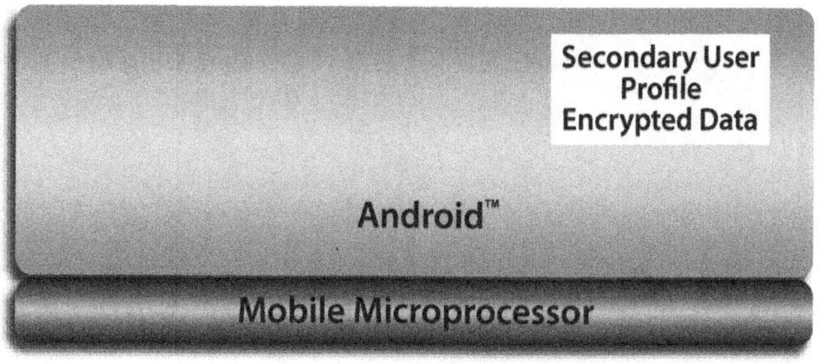

Figure 6.9:
MDM containers.

Clearly, this approach can be used to implement the layered encryption use case: the MDM application can create an SSL connection on top of the underlying Android's IPsec connection.

However, once again, the security of both layers relies on the underlying Android operating system.

6.3.4.5 Remoting

> **Key Point**
>
> One approach to enterprise data protection in Android is not to allow any of the enterprise data on the mobile device itself. Rather, the only way to access enterprise information is using a remote desktop and/or application virtualization.

When the device is not connected to the enterprise (e.g., offline operation), enterprise applications and services are unavailable.

Remoting precludes the requirement for local data protection; however, our use case for layered data-in-motion protection remains. The remoting application (see Figure 6.10) provides SSL encryption while the underlying Android runs IPsec.

However, once again, the security of both layers relies on the underlying Android operating system.

6.3.4.6 Type-2 Hypervisor

Type-2 hypervisors, introduced in Chapter 2, are similar to webtops and MDM containers in that the secondary environment runs as an application on top of the primary operating system. However, instead of hosting only a browser, the secondary persona is a full-fledged guest operating system running within a virtual machine created by the hypervisor application (see Figure 6.11). The hypervisor uses the primary operating system to handle I/O and other resource management functions.

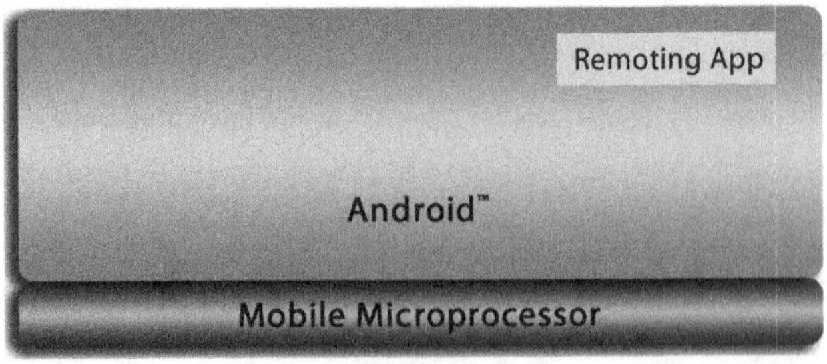

Figure 6.10:
Remoting.

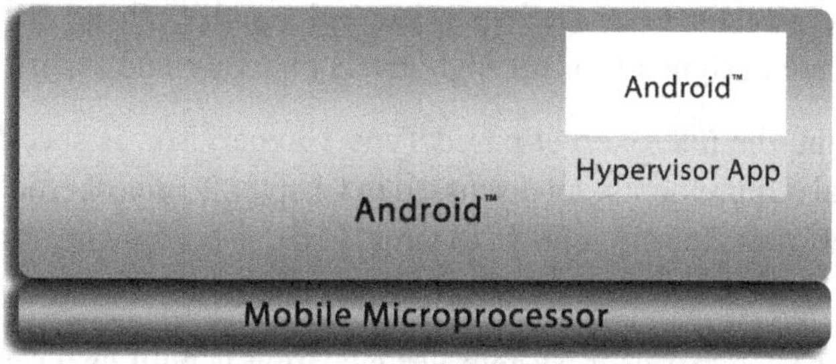

Figure 6.11:
Type-2 hypervisor.

Type-2 mobile hypervisor products are used to provide an enterprise management persona on top of the primary Android environment. The virtualized Android can use an SSL connection to the enterprise while the underlying Android's IPsec is also used to wrap the communication between endpoints.

However, once again, the Type-2 model fails to provide strong isolation. Faults or security vulnerabilities in the primary general-purpose operating system will impact the critical functions running in the virtual machine. Furthermore, Type-2 hypervisor applications deployed in the enterprise space have themselves been found to contain vulnerabilities that break the sandbox.

6.3.4.7 Sandboxes Built on Sand

Constant readers, if you have diligently been absorbing the concepts of systems software security (Chapter 2), secure software development (Chapter 3), and the remainder of this book, then we hope you observe as obvious the common weakness among all the common sandboxing approaches previously described. Multiple Android applications, MDM containers, remoting applications, webtops, and Type-2 hypervisors all attempt to retrofit security to the Android environment itself. As we demonstrated in detail earlier in this chapter, the Android/Linux system, while providing rich multimedia functionality of which embedded designs can take good advantage, is riddled with security vulnerabilities that simply cannot be fixed. High-assurance security must be designed from the beginning.

However, while high assurance cannot be retrofitted to Android itself, it can be retrofitted at a system level. Let's look at how.

6.3.4.8 Type-1 Hypervisor

Type-1 hypervisors, introduced in Chapter 2, also provide functional completeness and concurrent execution of a secondary enterprise persona. However, because the hypervisor runs on the bare metal, persona isolation cannot be violated by weaknesses in the persona operating system. Thus, a Type-1 hypervisor represents a promising approach from both a functionality and security perspective. However, the hypervisor vulnerability threat still exists, and not all Type-1 hypervisors are designed to meet high levels of safety and security.

One particular variant, the microkernel-based Type-1 hypervisor, is specifically designed to meet security-critical requirements. As discussed in Chapter 2, microkernels provide a superior architecture for safety and security relative to large, general-purpose operating systems such as Linux and Android.

In a microkernel Type-1 hypervisor, system virtualization is built as a service on the microkernel. Thus, in addition to isolated virtual machines, the microkernel provides an open standard interface for lightweight critical applications, which cannot be entrusted to a general-purpose guest. For example, SSL can be hosted as a microkernel application, providing the highest possible level of assurance for this encryption layer. IPsec packets originating from Android are doubly encrypted with the high-assurance SSL layer service before transmission over the wireless interface (see Figure 6.12).

The real-time microkernel is an excellent choice for embedded systems since the microkernel can host any embedded, real-time applications not appropriate for the Android/Linux environment.

The microkernel Type-1 hypervisor typically uses the microprocessor MMU to isolate the memory spaces of the primary Android environment and the native SSL encryption application. However, device drivers in Android may use DMA that can violate the memory partitioning by bypassing the MMU entirely. Potential approaches to guarding this attack vector include running the hypervisor in TrustZone on an applicable ARM-based microprocessor, using an IOMMU, or mediating all DMA bus masters. TrustZone and IOMMUs are discussed further in Chapter 2.

The isolation properties of some secure microkernels can even protect against sophisticated covert and side-channel software-borne attacks.

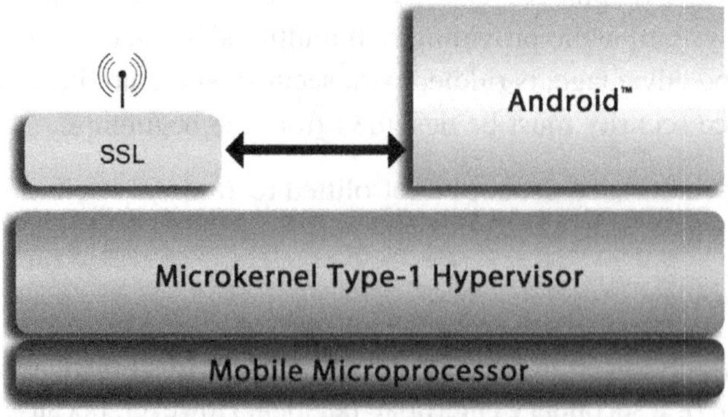

Figure 6.12:
Microkernel Type-1 hypervisor approach to layered data-in-motion encryption.

6.3.4.9 Physical Security

> **Key Point**
>
> Once we have an approach that prevents software attacks from breaking the sandbox between protection layers, the embedded device may benefit from taking defense-in-depth a step further to protect the layered encryption system from physical attacks using an attached secure element.

A lost or stolen mobile device in the hands of a sophisticated attacker, for example, is susceptible to memory snooping, power analysis, and other invasive and non-invasive physical attacks. While physical protection of the entire embedded system may not be practical, targeted physical protections can make a huge difference in overall system security.

As discussed in Chapter 4, a secure element can be used to provide physical protection of critical parameters, including private keys. The native microkernel SSL service's static public key can be housed in a secure element. For example, a smart card can be incorporated into a microSD device (that can simultaneously be used for additional secure storage) and attached to a smartphone or embedded system (see Figure 6.13). Of course, implementations will vary depending on the types and sophistication of physical protections available. Embedded systems designers can examine the FIPS 140 certification level and product documentation to compare and contrast offerings.

> **Key Point**
>
> Layered encryption as a defense-in-depth strategy is a sensible approach to increasing the assurance of Android-based data protection services; however, what is not sensible is to run both layers within the Android environment itself.

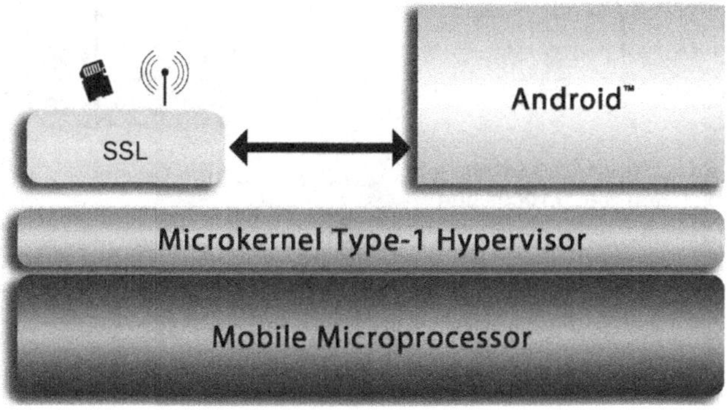

Figure 6.13:
Adding physical security protection via attached smart card to the microkernel Type-1 hypervisor.

There is simply too much vulnerability to prevent both layers from being simultaneously subverted. Embedded systems designers considering Android must also carefully sandbox critical security functions outside the Android system. Modern embedded microprocessors and system software solutions provide the requisite features to get the best of both worlds: the power of Android's multimedia and applications deployment infrastructure alongside, but securely separated from, critical system security functions.

6.4 Next-Generation Software-Defined Radio

6.4.1 Red-Black Separation

Software-defined radios (SDRs) that process classified information are typically architected with a standard red-black separation in which a red side is responsible for sensitive information processing and cryptographic functions, and the black-side processor is responsible for communication stacks and drivers. The red and black sides are hosted on separate hardware components. In fact, the red side is usually composed of both a general-purpose processor and a separate cryptographic processor.

On egress, classified information originated in the red side is encrypted and sent over some interconnect to the black side for transmission. On ingress, information received by black-side drivers is passed across the interconnect for decryption and any other red-side processing, such as guards and authentication. This general architecture is shown in Figure 6.14.

Note that the logical red-side crypto component may include an attached special-purpose hardware device, such as an FPGA, for cryptographic algorithm execution, key storage, and physical redundancy. Any software running within the cryptographic boundary falls under the

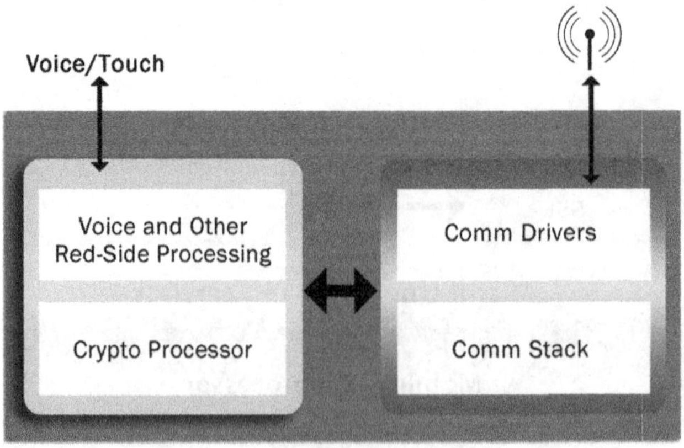

Figure 6.14:
Traditional communications device red-black architecture.

intense scrutiny of the NSA during Type-1 certification. This naturally imposes a strict assurance demand and complexity limit upon the red-side software.

6.4.2 Software-Defined Radio Architecture

Now let's look at the red-black architecture in more detail, using the example of a software-defined radio (SDR). More specifically, let's consider devices that employ the Software Communications Architecture (SCA), an open framework used to develop SDRs. The SCA provides standardized APIs for managing waveforms and other radio-relevant resources. The SCA lives on top of an SCA-compliant operating system. In an SDR, the SCA may be used in both the red and black sides. The black side will typically include components required to manage radio communications, including the waveforms themselves. Therefore, a real-time operating system (RTOS) is often essential.

The red side may use less of the SCA's functionality. At a minimum, SCA-standardized message passing and component management functions would be included. Of critical note, however, is the role of the red side in managing plaintext. Human-machine interface (HMI) components, such as keyboard drivers, voice codecs, and touch-screen management software, will be included in the red-side processing. For example, a handheld radio operator may speak voice data that must be collected by the red side for cryptographic processing before it can be sent across to the black side for transmission.

As with any Type-1 communications device, minimizing the certification-relevant software content in the red side is a strategic goal of SDR developers as well as the certifying agency. Because of its real-time and security-critical requirements, a trusted RTOS is a natural fit. A sample architecture showing the major red- and black-side SDR components is shown in Figure 6.15.

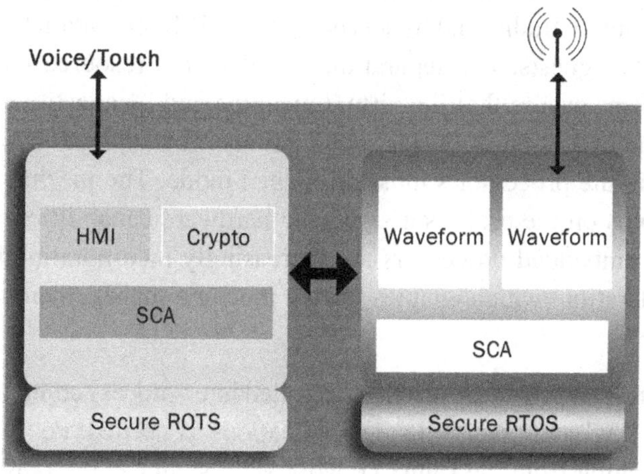

Figure 6.15:
Traditional software-defined radio architecture.

6.4.3 Enter Linux

While an analog voice radio has a relatively simple HMI, multi-purpose digital voice/data radios may incorporate a far more sophisticated HMI. Linux and its variants are obvious choices for implementation of the HMI due to the availability of popular HMI frameworks widely deployed in the consumer electronics industry. As we discussed in the preceding Android case study and in Chapter 1, Linux, a monolithic operating system, does not meet high-assurance certification requirements and is not appropriate for low-latency, hard real-time tasks. To incorporate Linux into the red-side HMI of secure communications devices without bringing its entire multi-million-line source code base into the cryptographic boundary, an obvious choice is to incorporate a second applications processor dedicated to Linux. Adding a second processor, however, increases footprint, power, production cost, and system complexity. This is especially problematic in resource-constrained devices, such as battery-powered radios.

When they have been incorporated using secondary red-side processors, general-purpose operating systems are typically used only for unclassified communications. Classified communications require custom HMI interfaces to achieve high assurance of the integrity and availability of sensitive data and commands. The Holy Grail, of course, is to incorporate Linux into the original red processor without sacrificing real-time performance; jeopardizing security; increasing certification overhead; or growing system footprint, weight, power, and cost. Furthermore, we want to use Linux for classified as well as unclassified interfaces, enabling the user to fully realize the rich productivity benefits of Linux.

This goal is made possible using a microkernel-based Type-1 hypervisor, where the microkernel is an MILS separation kernel. Chapter 2 describes MILS and MILS separation kernels. The separation kernel can host Linux in a securely compartmentalized virtual machine. However, unlike traditional hypervisors, the MILS separation kernel can host native applications as well as guests. The separation kernel's strict resource scheduling and protection mechanisms ensure that the virtual machine and its constituent applications are unable to impact the execution of critical applications. The separation kernel is the only software that runs in the processor's most privileged mode. The mechanism of system virtualization depends on the processor's specific hardware capabilities. As discussed in Chapter 2, modern embedded processors are increasingly incorporating hardware virtualization acceleration, which enables virtual machine management to be as simple and efficient as possible.

The separation kernel can provide a strictly controlled inter-process communication (IPC) path between the Linux HMI and other red-side applications as needed. For example, Internet Protocol data originating from the Linux network subsystem can be transmitted over the IPC pipe to the crypto subsystem for encryption and transmission. Application of the MILS virtualization concept to the SDR example is shown in Figure 6.16.

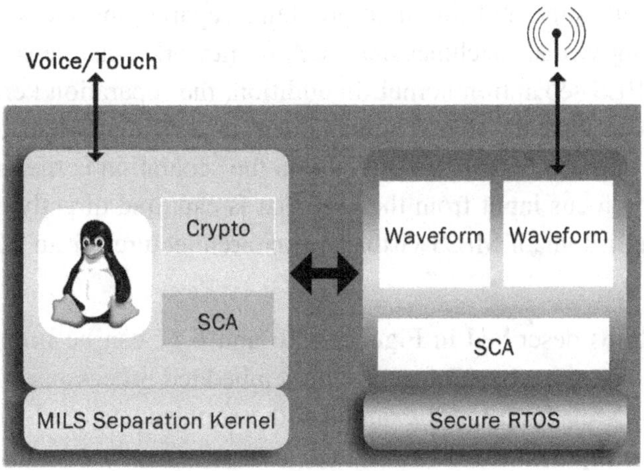

Figure 6.16:
MILS virtualization of linux in SDR.

The difficulty of using Linux for managing classified communications, however, remains. For example, let's consider the use of a Linux touch-screen widget to enter the sensitive command "zeroize all cryptographic keys now." Mission success may depend on the command information passing correctly through Linux to the cryptographic subsystem. Yet the low assurance of Linux makes it impossible to make the necessary integrity and availability guarantees. Again, MILS virtualization provides a solution. In the MILS virtualization architecture, the physical graphics device is controlled exclusively by a trusted application running on the MILS separation kernel; Linux has access only to virtualized devices. Thus, when a command is entered through Linux and across the virtualized interface, a trusted application can check the command. The command is committed only if the user is satisfied that Linux has faithfully transmitted his intent. This verification stage provides an MILS-enforced trusted path from user to high-assurance components; Linux is completely out of the loop.

6.4.4 Multi-Domain Radio

Some software-defined radios must manage information at varying classification levels. Default policy requires that information at different levels be kept physically isolated. Typically, this is accomplished in one of two ways. One method is to require a cold restart, in which writable hardware resources are zeroized so as to safely switch the device between security levels. This approach is unfriendly to users, and the reboot could delay communications and impact mission effectiveness. The second approach is to incorporate a discrete red-side processor for each security level and only require a secure switch of the human interface devices (e.g., display, keyboard). Once again, the extra processing elements will increase the footprint, production cost, and system complexity of the device. MILS virtualization also solves the multi-domain problem. Separate instances of any required subsystems, including virtual machines and SCA frameworks, can be strictly isolated and scheduled by the MILS separation kernel. In addition, the separation kernel's ability to host native applications can solve the shared human interface device problem: a trusted multi-level secure (MLS) HMI application can run directly on the separation kernel, multiplexing multi-domain I/O based on focus input from the user that is captured directly by the MLS component. The multi-domain MILS virtualization architecture for an SDR is shown in Figure 6.17.

While the architectures described in Figures 6.16 and 6.17 can be implemented using a single-core processor, the advent of multi-core embedded processors can make them more practical. As can be seen from these examples, the sophisticated red-side processing includes numerous components, many of which can execute concurrently on multiple cores, under supervision of a multi-core-aware separation kernel. This extra horsepower enables good performance for virtual machines while ensuring real-time behavior for critical applications.

The rich functionality of the latest multimedia software packages, such as Linux, can be incorporated into even the most demanding real-time, secure military communications devices without increasing hardware footprint, cost, or certification burden. The key innovation is MILS virtualization, which exploits the resource management capabilities, native applications environment, and assurance pedigree of a trusted separation kernel and modern hypervisor techniques to effectively consolidate general-purpose and critical subsystems.

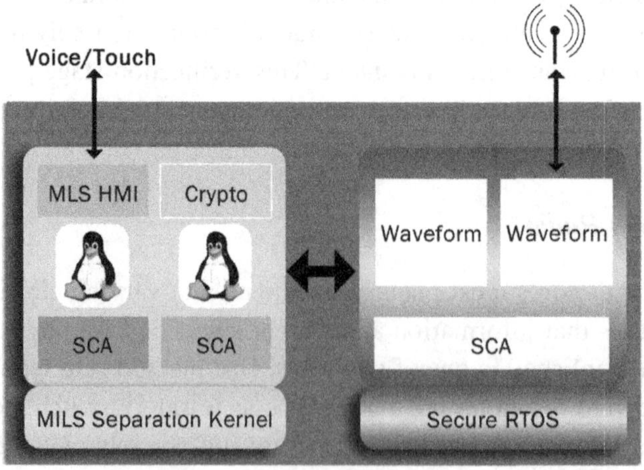

Figure 6.17:
Multi-domain SDR with multi-instance linux virtual machines and high-assurance MLS components.

6.5 Key Points

1. Much of the world's commerce and critical infrastructure control depends on the security of network transactions.
2. For as little as $25,000, a sophisticated hacker will break into most any network, even the most strenuously protected enterprise data centers.
3. Transaction security is defined by the ability of both client and server to have trust in the authenticity, integrity, and confidentiality of the messages that are sent and received by both parties throughout the transaction.
4. Man-in-the-middle attacks are possible not only on unencrypted links, but also with encrypted links, such as an SSL-protected HTTP connection, by exploiting weaknesses in the cryptographic infrastructure or underlying operating systems.
5. In general, social engineering attacks are increasingly prevalent and effective, due in part to the growing sophistication of hackers who are willing to mine the Internet for personal data that can be used to launch attacks and the increasing use of social networking sites like Facebook that draw more personal data into the public domain and make that data more readily accessible.
6. Similar to man-in-the-middle attacks, DNS-based pharming attacks are particularly dangerous because neither the client nor the server must necessarily be subverted for the attack to succeed.
7. Without a dramatic improvement in secure design and assurance, the dual trend of increasing software defects and attacker sophistication converges to a total loss of confidence in shared networks such as the Internet as a medium for commerce or other critical network-based computing.
8. The problem with anti-malware software is that it is fundamentally a reactive technology; as new vulnerabilities and attack methods are discovered, the anti-malware protocols quickly become obsolete.
9. An encouraging approach to transaction security involves sandboxing the embedded web client from the rest of the computing environment; the theory is that a sandboxed browser is less likely to be infected with malware and therefore more likely to be trusted for secure transactions.
10. Out-of-band transaction verification is a compelling approach because it enables both client and server to obtain a higher level of confidence in the authenticity of each other's identity as well as the actual information being transacted.
11. End-to-end control of the transaction connection also can protect against PKI and cryptosystem weaknesses by enforcing the use of strong public key, bulk encryption, and Message Authentication Code (MAC) algorithms and key lengths instead of relying on the native web server and its default TLS implementation.
12. Whenever possible, complete physical isolation of critical automotive ECUs from network-connected vehicular infrastructure should be imposed.

13. The rate of Android vulnerability discovery is such that practically every Android-based consumer device has been rooted within a short period of time, sometimes within a day or two of release.
14. One approach to raising the level of assurance in data protection within an Android-based device is to employ multiple encryption layers.
15. The concept of defense-in-depth originates in the military: multiple layers of defense, such as a combination of mines and barbed wire, rather than just mines or barbed wire alone, to increase the probability of a successful defense as well as potentially to slow the progress of an attacker.
16. The two main purposes of Mobile Device Management are to provide mobile data protection and IT management services.
17. One approach to enterprise data protection in Android is not to allow any of the enterprise data on the mobile device itself. Rather, the only way to access enterprise information is using a remote desktop and/or application virtualization.
18. Once we have an approach that prevents software attacks from breaking the sandbox between protection layers, the embedded device may benefit from taking defense-in-depth a step further to protect the layered encryption system from physical attacks using an attached secure element.
19. Layered encryption as a defense-in-depth strategy is a sensible approach to increasing the assurance of Android-based data protection services; however, what is not sensible is to run both layers within the Android environment itself.

6.6 Bibliography and Notes

1. What is Endpoint Security? http://www.endpointsecurity.org/Documents/What_is_endpointsecurity.pdf.
2. Epstein K, Elgin B. *Network Security Breaches Plague NASA*. http://www.businessweek.com/print/magazine/content/08_48/b4110072404167.htm; November 20, 2008.
3. Bohn K, Todd B. *Obama, McCain Campaigns' Computers Hacked for Policy Data*. http://www.cnn.com/2008/TECH/11/06/campaign.computers.hacked; November 6, 2008.
4. Greenemeier L. *The TJX Effect*. http://www.informationweek.com/news/security/cybercrime/showArticle.jhtml?articleID=201400171; August 11, 2007.
5. Zetter K. *Miley Cyrus Hacker Raided by FBI*. http://blog.wired.com/27bstroke6/2008/10/miley-cyrus-hac.html; October 20, 2008.
6. Kirk J. *'Evil Twin' Wi-Fi Access Points Proliferate*. http://www.networkworld.com/news/2007/042507-infosec-evil-twin-wi-fi-access.html; April 25, 2007.
7. Sotirov A, et al. *MD5 Considered Harmful Today*. http://www.win.tue.nl/hashclash/rogue-ca; December 30, 2008.
8. McMillan R. *CheckFree Warns 5 Million Customers After Hack*. http://csoonline.com/article/474365/CheckFree_Warns_Million_Customers_After_Hack; January 7, 2009.
9. Krebs B. *Citibank Phish Spoofs 2-Factor Authentication*. http://blog.washingtonpost.com/securityfix/2006/07/citibank_phish_spoofs_2factor_1.html; July 10, 2006.
10. Kristensen T. *Symantec Beats the Competition....* http://secunia.com/blog/29; October 13, 2008.
11. Skoudis Ed, Carpenter M. Seven Desktop Security Suites Reviewed. *Information Security Magazine*; November 2007; http://searchsecurity.techtarget.com/magazineContent/Product-review-Seven-integrated-endpoint-security-products.

12. National Vulnerability Database, search on vendors Symantec. http://web.nvd.nist.gov/view/vuln/search?execution=e2s1
13. *McAfee Antivirus Vulnerability Published.* http://news.zdnet.co.uk/security/0,1000000189,39191831,00.htm; March 18, 2005.
14. http://www.ironkey.com
15. Claburn T. *Another Google Chrome Security Flaw Identified.* http://www.informationweek.com/news/internet/google/showArticle.jhtml?articleID=210500290; September 5, 2008.
16. King ST, et al. *Secure Web Browsing with the OP Web Browser.* http://www.cs.uiuc.edu/homes/kingst/Research_files/grier08.pdf; 2008.
17. National Vulnerability Database, search for VMware and/or Xen. http://web.nvd.nist.gov/view/vuln/search?execution=e2s1
18. Schneier B. *Two-Factor Authentication: Too Little, Too Late.* http://www.schneier.com/essay-083.html; April 2005.
19. Weigold T, et al. *The Zurich Trusted Information Channel—An Efficient Defence against Man-in-the-Middle and Malicious Software Attacks;* TRUST 2008, LNCS 4968; 2008. 75−91.
20. Shachtman N. *Under Worm Assault, Military Bans Disks, USB Drives.* http://blog.wired.com/defense/2008/11/army-bans-usb-d.html; November 19, 2008.
21. Security and Permissions in Android. http://code.google.com/android/devel/security.html.
22. Shankland S. *Google Details 'Reboot' Bug, Android Security Fixes.* http://news.cnet.com/8301-1009_3-10093573-83.html; November 11, 2008.
23. Burnette Ed. *Worst. Bug. Ever.* http://blogs.zdnet.com/Burnette/?p=680; November 7, 2008.
24. Oberheide J, Lanier Z. *TEAM JOCH vs. Android.* Washington, DC: Presentation made at ShmooCon 2011 Conference. http://n0where.org/talks/shmoo11-teamjoch_v4.pdf; January 28-30, 2011.
25. Kroah-Hartman G, et al. *Linux Kernel Development: How Fast It Is Going, Who Is Doing It, What They Are Doing, and Who Is Sponsoring It: An August 2009 Update.* http://www.linuxfoundation.org/sites/main/files/publications/whowriteslinux.pdf; August 2009.

Index

T